AF807434

TEACHING BIOLOGY IN SCHOOLS

An indispensable tool for biology teacher educators, researchers, graduate students, and practising teachers, this book presents up-to-date research, addresses common misconceptions, and discusses the pedagogical content knowledge necessary for effective teaching of key topics in biology. Chapters cover core subjects such as molecular biology, genetics, ecology, and biotechnology, and tackle broader issues that cut across topics, such as learning environments, worldviews, and the nature of scientific inquiry and explanation. Written by leading experts on their respective topics from a range of countries across the world, this international book transcends national curricula and highlights global issues, problems, and trends in biology literacy.

Kostas Kampourakis is a researcher in science education and a lecturer at the University of Geneva, Switzerland.

Michael J. Reiss is Professor of Science Education at UCL Institute of Education, University College London, UK.

Teaching and Learning in Science Series

Norman G. Lederman, Series Editor

Responsive Teaching in Science and Mathematics
Edited by Amy D. Robertson, Rachel Scherr, David Hammer

Reconceptualizing STEM Education
The Central Role of Practices
Edited by Richard A. Duschl, Amber S. Bismack

Supporting K-12 English Language Learners in Science
Putting Research into Teaching Practice
Edited by Cory Buxton, Martha Allexsaht-Snider

Reframing Science Teaching and Learning
Students and Educators Co-developing Science Practices
In and Out of School
Edited by David Stroupe

Representations of Nature of Science in School Science Textbooks
A Global Perspective
Edited by Christine V. McDonald, Fouad Abd-El-Khalick

Teaching Biology in Schools
Global Research, Issues, and Trends
Edited by Kostas Kampourakis, Michael Reiss

For more information about this series, please visit: www.routledge.com/Teaching-and-Learning-in-Science-Series/book-series/LEATLSS

TEACHING BIOLOGY IN SCHOOLS

Global Research, Issues, and Trends

Edited by
Kostas Kampourakis
and Michael J. Reiss

Routledge
Taylor & Francis Group
NEW YORK AND LONDON

First published 2018
by Routledge
711 Third Avenue, New York, NY 10017

and by Routledge
2 Park Square, Milton Park, Abingdon, Oxon, OX14 4RN

Routledge is an imprint of the Taylor & Francis Group, an informa business

© 2018 Taylor & Francis

The right of Kostas Kampourakis and Michael J. Reiss to be identified as the authors of the editorial material, and of the authors for their individual chapters, has been asserted in accordance with sections 77 and 78 of the Copyright, Designs and Patents Act 1988.

All rights reserved. No part of this book may be reprinted or reproduced or utilised in any form or by any electronic, mechanical, or other means, now known or hereafter invented, including photocopying and recording, or in any information storage or retrieval system, without permission in writing from the publishers.

Trademark notice: Product or corporate names may be trademarks or registered trademarks, and are used only for identification and explanation without intent to infringe.

Library of Congress Cataloging-in-Publication Data
Names: Kampourakis, Kostas, editor.
Title: Teaching biology in schools : global research, issues, and trends / edited by Kostas Kampourakis and Michael J. Reiss.
Description: New York : Routledge, 2018. | Series: Teaching and learning in science series | Includes bibliographical references and index.
Identifiers: LCCN 2018002876 | ISBN 9781138087941 (hbk) | ISBN 9781138087989 (pbk) | ISBN 9781315110158 (ebk)
Subjects: LCSH: Biology—Study and teaching (Secondary)
Classification: LCC QH315 .T36 2018 | DDC 570.71—dc23
LC record available at https://lccn.loc.gov/2018002876

ISBN: 978-1-138-08794-1 (hbk)
ISBN: 978-1-138-08798-9 (pbk)
ISBN: 978-1-315-11015-8 (ebk)

Typeset in Bembo
by Apex CoVantage, LLC

To Vasso Zogza for introducing me to Biology Didactics (KK)
To Gunnhildur Óskarsdóttir for many years of friendship (MJR)

CONTENTS

FIGURES

TABLES

PREFACE

The discipline of biology is growing at an unprecedented rate as more is understood of life at every level—from molecular biology to interactions at the global scale. Such developments in human knowledge about biology place great demands on teachers who, when they were at school, may well have been taught certain things that would now be regarded as being misleading or even wrong.

This book is therefore intended to help teachers of biology, whether they are long-experienced or just beginning their teaching career. The majority of chapters deal with specific areas of the biology curriculum, looking at the misconceptions that students often have, at the rationale for teaching in the area and at the principal concepts that students should learn if they are to manifest sound understanding. In addition, as described later in more detail, we have a number of chapters, rarely seen in books about science teaching, that are intended to help readers deal with important aspects of pedagogy that apply regardless of the specific content area one is focusing on.

An Introduction discusses the notion of biology didactics that contributors to this book share, describes the aims of the book and provides an outline of its contents. A brief account is given of issues in biology didactics. The chapter also explains why this book will be useful to science teachers and educators and how it can be used both as a resource for teaching and as a major sourcebook in teacher preparation courses.

Chapter 2 focuses on the foundational role that intuitive reasoning, such as essentialist and teleological reasoning, and intuitive language forms, such as generics (e.g. 'all birds have feathers'), play in the development of students' understanding of biology. Intuitive reasoning and language are frequently castigated, yet they can play a valuable role in the acquisition of more scientifically valid knowledge.

Chapter 3 highlights children's potential to engage in meaningful learning about the biological world when at kindergarten or primary school. It draws on research in developmental psychology about the emergence of biological knowledge and on biology education research about children's own ideas and the design of learning environments that may contribute to their elaboration.

Arguably the most significant changes in biology over the past century have been in molecular biology. Schools and higher education institutions are thus tasked with finding ways to help students develop deep and meaningful understandings of the myriad of molecular mechanisms that drive life. Chapter 4 discuss what is important to learn in this domain. It presents a progression for deepening students' ideas about molecular biology over the course of schooling and examines the role of technology, instruction and curriculum in fostering student learning.

Chapter 5 presents a consideration of cell biology, a key unifying concept in biology. A recommendation is made for how cell biology instruction might be scaffolded, and there is an extensive review of students' understandings and conceptions related to cell biology, as well as an outline of typical laboratory experiences to support learning about the cell. Specific recommendations are made about designing instruction plans and teaching cell biology.

Chapter 6 examines how human biology might be taught, considering the human body as a complex system. Complex systems are often challenging subjects, because they are characterised by many elusive and non-intuitive elements (emergence, multilevel organisation, interconnections, heterogeneous components and invisible dynamic processes) that can be difficult for teachers to explain and for students to perceive. This chapter offers several principles for overcoming this challenge.

Health and wellbeing is a broad and diverse topic. Teachers need to be well versed in the many aspects that this topic covers, from the effects of communicable and non-communicable diseases and the need to understand and maintain a healthy lifestyle by having a balanced diet and taking regular exercise, to the effects of drugs on physiological functioning. Chapter 7 explores how health and disease are defined, some of the barriers to teaching the subject in schools and some misconceptions held by students. It outlines some examples of successful interventions that support students' knowledge attitudes, and behaviours relating to health.

Chapter 8 looks at reproduction and sex education. School biology often pays little or no attention to reproduction in taxa other than humans and a small number of 'typical' flowering plants; it is argued that there is still much value in a traditional introduction to life cycles and reproduction in a large range of taxa. The chapter goes on to consider what makes for an appropriate sex education course, considering such issues as comprehensive sex education versus abstinence education and how school sex education can address questions to do with sexual orientation.

Biology education often focuses on adult forms when describing structure, function, physiology, adaptations, heredity and more. However, the features of all adult body parts are the outcomes of complex developmental processes, and they also evolve through complex evolutionary processes. In Chapter 9, it is therefore argued that the teaching and learning of development should occupy a central place in biology education, although this is not often the case.

In Chapter 10, on genetics, a central question that is addressed is the relationship between classical (Mendelian) genetics and molecular genetics. A second important issue is that school genetics consists of both knowledge about genetics and knowledge of doing genetics. It is therefore important to examine whether genetics is taught so as to enable scientific literacy or academic knowledge.

As is widely acknowledged, many school students find plant biology uninteresting. In Chapter 11, a case for teaching plant science in the secondary curriculum is presented using the 'big ideas' approach to school biology. In so doing, an assemblage of teaching approaches, in which student-centred learning complements practical experiments, is suggested for one strand of plant science, namely the process of photosynthesis.

The concept of energy is central for learning biology at school. Photosynthesis and respiration are the two fundamental, complementary biological processes that describe the fixation, processing and release of energy, interwoven with the synthesis and decomposition of organic matter. These processes are highly complex and abstract and relate to different system levels, which makes them very difficult to learn. Chapter 12 explores difficulties in learning, proposes conceptualisations of these concepts and provides some recommendations for teaching energy, photosynthesis and respiration, drawing on findings from empirical science education research.

The teaching of ecology poses important challenges to educators due to the nature of ecological relationships and processes. Deficiencies in teaching approaches that result in generating and transmitting misconceptions through ecology education are discussed in Chapter 13, and an educational framework for teaching ecology is proposed. The proposed framework combines outdoor activities, computer simulations and laboratory experiments in a series of learning sequences to help students grasp the complexity of ecosystems' functions and human-nature interactions.

Evolution is one of the most challenging biology topics to teach, learn and assess. Chapter 14 focuses on conceptual issues relating to the teaching and learning of evolutionary change, and is grounded in research on student thinking. Teleology, typological thinking, intentionality, use-inheritance, evolutionary 'pressures' and other intuitive ideas are widely documented barriers to evolutionary understanding. Robust explanatory models have been shown to be the exception in novice learners. Cross-case comparisons, highlighting surface feature dissimilarity while also illustrating causal unity, can foster robust understanding of evolutionary change.

The field of microbiology has led to significant advances in health and in industrial, agricultural and environmental processes. The rapidly expanding knowledge base in this area has resulted in some concern about how microbiology should best be taught in school science. For example, there is a perceived concern about safety when students grow and handle microorganisms and the controversial nature of some human activity related to microorganisms (e.g., germ warfare) is sometimes considered too difficult to manage in the classroom. Yet, the complex nature of this topic and the fascinating intersection with human beings and society provides an opportunity to engage students with school science. Chapter 15 discusses what good teaching about microbiology might look like.

Chapter 16 focuses on teaching biotechnology in the context of socio-scientific issues. This involves fostering students' understanding as to what biotechnology is and what techniques and processes are used, the values that are involved and what benefits and risks are associated with specific biotechnological techniques and applications, as well as what multiple (often competing) political, social, legal, economic and ethical claims are made for and against biotechnology.

Chapter 17 examines how authentic texts can be used to promote disciplinary literacy in biology. It advocates the use of Adapted Primary Literature articles. Such articles are designed to enable the use of scientific research articles in high school. They represent authentically the structure, claims, content and uncertainties that are found in the original research articles, and may therefore serve as models of scientific reasoning and communication.

Chapter 18 deals with the nature of scientific knowledge and the nature of scientific inquiry in biology teaching. It focuses on empirically based approaches to facilitating students' understandings of these, and argues that scientific literacy requires that an individual understands subject matter, the nature of scientific knowledge and the nature of scientific inquiry.

Chapter 19 describes the nature and structure of biological explanations. It then focuses on causal explanations in biology, and distinguishes between developmental and evolutionary explanations for the origin of biological phenomena and between causes of effects and the processes through which these effects are brought about.

Chapter 20 discusses research findings about argumentation in biology education. It asks how the practice of argumentation relates to biological knowledge and addresses questions related to why students should engage in argumentation in biology and to how teachers can support this engagement.

Finally, Chapter 21 examines the worth of the concept of 'worldviews' for understanding why some students choose to reject certain elements of what they are taught in biology. One notable topic where an approach based on worldviews has proved useful is evolutionary biology, which students from fundamentalist religious backgrounds often reject. Other biological topics that may benefit from a worldviews approach include anthropogenic climate change, vaccination advocacy and the use of animals for human purposes.

ACKNOWLEDGEMENTS

Editing a book is in several ways quite similar to but in other ways very different from writing a book. Perhaps the most important difference between the two procedures is the difficulties in coordinating the contents of the book, in managing the contributors and, above all, in ensuring high quality contributions. However, we faced no such difficulties, and actually no difficulties at all, while editing this book. For this, we are indebted to all the contributors to this book for their diligent work. As ever, we are also very grateful for the patience and support of our families while working on this book. Last, but not least, we are indebted to each other for a productive collaboration, which has also been very enjoyable.

1

INTRODUCTION

Biology Didactics

Michael J. Reiss and Kostas Kampourakis

Introduction

Policy documents in various countries all over the world often suggest that scientific literacy should be a major aim of education at schools. This is especially important for biology, as there are many socio-scientific questions that students will encounter during their lives, and which may demand relevant decisions, that require a good understanding of biology if they are to be answered. Such questions may be related to environmental issues, genetic testing, infectious diseases and more. Biological research in the twenty-first century has an impact on various aspects of human life, with important implications for how we understand health, disease and identity. Therefore, literacy in biology is a core component of scientific literacy. Future citizens who are literate about science should definitely have an appropriate understanding of biology for medical, environmental and other purposes.

However, research in teaching and learning of biology, as well as in its public understanding, reveals serious shortcomings. It seems that—generally speaking—it is possible for many citizens to leave school holding inaccurate conceptions and having serious misunderstandings of biological phenomena (Tekkaya, 2002; Coley & Tanner, 2015; Stern & Kampourakis, 2017). There are various reasons for this. Biological research advances at a fast pace and it is often difficult for school biology to keep up with it. Teachers also often lack the necessary pedagogical content knowledge (PCK). This could, in principle, be acquired through teacher preparation courses, but too often this is not an option for teachers. In part, this is because there is so much to cover in teacher preparation courses that, even for those who benefit from them, there is only a limited amount of time to spend on how to teach biology well.

The Rationale and Scope of This Book

Given the above issues with the teaching of biology in schools, there is a need for a book that can guide both in-service biology teachers and those about to join the profession by presenting and discussing the relevant research. This book is intended to serve this function. It can be used as a course book in biology initial teacher education courses, and it can also be used by individual teachers as an introduction to research in teaching and learning of biology. The book therefore covers the most important topics in biology education, both those that are usually discussed in teacher preparation courses (e.g. evolution, ecology, genetics, human biology, health and disease) and others that are equally important but are not often considered (e.g. cognitive biases, explanations, sex education, energy, worldviews, nature of science, development). Another way of categorising the chapters is that some are concerned with specific biological topics while others are concerned with broader issues that cut across the various topics.

The contributors to this book are leading experts on their respective topics, and they are actively involved in research in teaching and learning biology. Coming from a variety of countries across the world, they have contributed to a truly international book that transcends national curricula and recommendations and highlights global issues, problems and trends in teaching and learning biology. No other book on biology didactics from an international perspective currently exists, and this book aims at filling this gap. This book is therefore intended for pre-service and in-service biology teachers, who could use it either on their own or in teacher training courses. It should also be useful to biology educators involved in teacher preparations programs. The book should also be of interest more broadly to researchers in science education and anyone interested in the teaching and learning of biology.

Biology Didactics

The title of this book is *Teaching Biology in Schools*. In this chapter, though, we have used the terms 'biology didactics'. The two of us are long-standing members of the European Researchers in Didactics of Biology professional organisation (ERIDOB). In the early years of this organisation, almost every conference had a discussion, at some point, about the meaning of the term 'didactics' and how this related to (and differed from) the Anglophone term 'teaching' (with its Greek-derived antecedent 'pedagogy').

By now, these debates look somewhat dated. The key issue is that high quality didactics/teaching requires consideration of the following two matters:

- The knowledge, skills and/or dispositions that the teacher wishes the student(s) to attain;
- The present cognitive and psychomotor states and abilities of the student(s) and their motivation.

To exemplify, suppose that a teacher is interested in teaching a class of 7–8-year-old students about food webs in the context of woodland ecology. The principal intention of the lesson is for students to gain knowledge. Ideally, for each student (and writing this emphasises the compromises and simplifications that every school teacher has to make) the teacher would assess their prior knowledge (cognitive state), their ability (which would affect both the rate at which the teacher could realistically decide to proceed with the teaching and the mode(s) of learning—e.g. through words, through drawings and/or through other representations) and their motivation to learn (e.g. if motivation is low, the teacher might decide to show the class a video of a woodland showing some of the feeding relationships or might decide to visit a woodland with a class to look for evidence of feeding relationships). Comparably, a lesson for secondary school biology students on the use of light microscopes to view cell structure might have more of an emphasis on the psychomotor skills necessary to operate a compound light microscope, while a lesson on pollution might have as one of its aims that students develop the disposition neither to contribute to excessive pollution themselves nor to acquiesce in the polluting activities of others.

In this book, the contributors mainly concentrate on biological knowledge rather than skills and dispositions. There are books on biology practical work—though that is to understand the term 'skills' somewhat narrowly; a broader conceptualisation of skills would see them as including what are sometimes referred to as 'soft' skills, e.g. group work and communication. It is also open to debate as to whether reasoning (both scientifically and ethically) is considered a skill or not. In terms of dispositions, very little has been written about the role of biology education in their promotion, though there is a growing literature (some of which has existed for some time) on the role of education in the promotion of the virtues and good behaviour (e.g. Straughan, 1988; Bebeau et al., 1999; Kristjánsson, 2015). Finally, we need to keep in mind the purpose of our teaching (cf. Kincheloe & Steinberg, 2004, who encourage researchers to ask the research questions that will make a difference to students' lives).

The Chapters

As outlined above, the chapters in this book fall into two categories: those that each deal with a particular topic; and those that address cross-cutting issues.

The particular topics that are dealt with are molecular biology (Chapter 4), cell biology (Chapter 5), human biology (Chapter 6), health and disease (Chapter 7), reproduction and sex education (Chapter 8), development (Chapter 9), genetics (Chapter 10), plant biology (Chapter 11), energy, photosynthesis and respiration (Chapter 12), ecology (Chapter 13), evolution (Chapter 14), microbiology (Chapter 15) and biotechnology (Chapter 16). In these chapters, many authors have adopted a common structure in which, after an introduction, there are sections on students' misconceptions about the topic, difficulties in teaching about the topic, issues to do with conceptualising the topic, recommendations for teaching about the topic and, finally, conclusions.

These section headings are self-explanatory, though it is worth making a few points about student misconceptions. There are those who dislike the term 'misconceptions', feeling that it is somewhat patronising to learners, and so prefer such terms as 'alternative conceptions', 'alternative frameworks' or 'pre-scientific understanding'. We have no objection to such terms; indeed, we have considerable sympathy with the thinking behind their adoption. Nevertheless, there are many instances where students do have what can straightforwardly be described as erroneous knowledge—which is why we are comfortable with the term misconceptions. The student who thinks that energy circulates in ecosystems as matter does, the student who thinks that conception in humans is most likely at the time of menstruation and the student who thinks that venous blood is blue are all mistaken.

This being noted, we agree that students' preconceptions, that is, their initial conceptions formed through experience, are important for the educational process and must be carefully considered before designing teaching. Some of these preconceptions are wrong (and these are the ones that we call misconceptions; in this sense, preconceptions form a broader category than misconceptions), but others are not and actually often help students correctly make sense of the world, for instance by facilitating them to observe patterns. Conceptual change research supports the conclusion that for effective learning, students' misconceptions must be explicitly addressed, and their other preconceptions must be appropriately restructured. But in this book, we have paid more attention to misconceptions because we think that these are the ones that teachers should primarily be aware of and try to address.

The aforementioned central chapters on biological topics are surrounded by others that tackle issues that cut across biology topics:

- Chapter 2 focuses on the foundational role that intuitive reasoning, such as essentialist and teleological reasoning, and intuitive language forms, such as generics (e.g. 'all birds have feathers'), play in the development of students' understanding of biology. Intuitive reasoning and language are frequently castigated yet, as the authors of this chapter point out, they can play a valuable role in the acquisition of more scientifically valid knowledge.
- Chapter 3 highlights children's potential to engage in meaningful learning about the biological world when at kindergarten or primary school. It draws on developmental psychology research about the emergence of biological knowledge and on biology education research about children's own ideas and the design of learning environments that may contribute to their elaboration.
- Chapter 17 examines how authentic texts can be used to promote disciplinary literacy in biology. It advocates the use of Adapted Primary Literature articles. Such articles are specifically designed to enable the use of scientific research

articles in high school. They represent authentically the structure, claims, content and uncertainty that are found in the original research articles, and may therefore serve as models of scientific reasoning and communication.

- Chapter 18 deals with the nature of scientific knowledge and the nature of scientific inquiry in biology teaching. It focuses on empirically based approaches to facilitating students' understandings of these and argues that scientific literacy requires that an individual understands subject matter, the nature of scientific knowledge and the nature of scientific inquiry.
- Chapter 19 describes the nature and structure of biological explanations. It then focuses on causal explanations in biology, and distinguishes between developmental and evolutionary explanations for the origin of biological phenomena and between causes of effects and the processes through which these effects are brought about.
- Chapter 20 discusses research findings about argumentation in biology education. It asks how the practice of argumentation relates to biological knowledge and addresses questions related to why students should engage in argumentation in biology and to how teachers can support this engagement.
- Chapter 21 examines the worth of the concept of 'worldviews' for understanding why some students choose to reject certain elements of what they are taught in biology. The classic instance where an approach based on worldviews has proved useful is evolutionary biology, which students from fundamentalist religious backgrounds typically reject. Other biological topics that may benefit from a worldviews approach include anthropogenic climate change, vaccination advocacy and the use of animals for human ends.

What You Will Not Find in This Book

Finally, it may be profitable to say something about what is not included in this book. This is not, we emphasise, because we are not interested in such questions or because we consider them unimportant. Every book—unless it is to be so long that it will be difficult to get published, let alone read—needs to delimit its boundaries.

The Purpose of Practical Work and Its Specifics

Science education, including biology education, differs from almost all other subjects taught in school in that it involves practical lessons that are, generally speaking, undertaken in specifically designed and, in many cases, purpose-built laboratories (White, 1988). By 'practical lesson' is meant any lesson in which the learners are involved in manipulating and/or observing real (as opposed to virtual) objects and materials (Abrahams & Reiss, 2017). Although some of the chapters in this book do consider both the purpose of practical work in the teaching of particular biological topics and certain specifics as to which practicals might be undertaken, we do not provide a systematic coverage of the role of practical work in biology didactics. To do so would include covering the functions of practical work (e.g. in promoting learning or engaging students), its nature (e.g. directed versus open-ended) and specifics as to how it can be undertaken for different biology topics and different ages of students.

Whether Biology Should be Taught Through Contexts

Much of school science has the reputation of being difficult, dull, out-of-touch with students' aspirations and irrelevant to society as a whole (Osborne et al., 2003). Specifications have traditionally been constructed from a scientist's viewpoint, with the concepts being developed in ways that are seen to be sensible by a scientist. So, for example, a biology course for 16–19-year-olds might start with biochemistry and cell biology and go onto such topics as physiology, genetics, ecology and evolution. But many students see things differently and want teachers to show them *why* the various biological concepts are important (Reiss, 2006). One possibility is to make the context—or storyline—the driving force.

A number of reasons can be given as to why teaching in context makes sense. For a start, it can motivate certain students (though other students may prefer to 'cut to the chase' and get onto what they see as the 'real' biology). A second reason is that the history of science suggests that much fundamental work in biology is context-driven. Today, governments and private companies (for example in the pharmaceutical industry) spend huge amounts of money on applied research. Often such applied research leads to fundamental scientific advances. It is a mistake to think that the path from fundamental science to technology is simple and unidirectional. Teaching through contexts may often be a more realistic way of modelling the way that some scientific knowledge is produced.

Teaching Ethical Issues in Biology

Few would deny that today's biology raises a wide range of ethical issues. One only has to think of *in vitro* fertilisation, therapeutic cloning, genetic engineering, global climate change, conservation, animal experimentation and end-of-life issues to appreciate that. But what precisely might be the aims of teaching ethics to biology students? Based on Davis (1999), at least four can be suggested (Reiss, 1999).

First, such teaching might heighten the ethical *sensitivity* of participants. For example, students who have never thought about whether certain breeds of dogs should have their tails removed ('docked') or whether a foetus that is not capable of feeling anything can be said to have rights might be encouraged to think about such issues. Such thinking can result in students becoming more aware and thus more sensitive. Secondly, such teaching might increase the ethical *knowledge* of students. For example, appropriate teaching about the issue of rights might help students to distinguish between legal and moral rights, to understand something of the connections between rights and duties and to be able to identify fallacies in arguments for or against the notion of animal rights. Thirdly, such teaching might improve the ethical *judgement* of students. Finally, and perhaps most ambitiously, such teaching of ethics might make students *better people* in the sense of making them more virtuous or otherwise more likely to implement normatively right choices.

The Links That Biology Teaching Can Make With Other Subjects

There are a number of subjects on which biology draws without making comparable contributions—mathematics and chemistry being prime instances. Then there are subjects with which biology overlaps—geography being an example (for instance, in the teaching of ecology). Finally, there are subjects to which biology contributes. Psychology and sociology are prime examples—though neither subject is much taught to younger students. E. O. Wilson (1975) famously argued that biology would eventually take over both subjects as the biological bases of human behavior became better understood, though this is not a view with which we agree.

Conclusion

Biology is a discipline that is making rapid advances. Teaching it can be very satisfying, but such teaching makes deep demands on teachers. The intention of this book is that the chapters it contains will make a substantial contribution to biology didactics and, thus, to student learning and engagement with the subject.

References

Abrahams, I. & Reiss, M. J. (Eds) (2017). *Enhancing Learning With Effective Practical Science 11–16*. London: Bloomsbury.

Bebeau, M. J., Rest, J. R. & Narvaez, D. (1999). Beyond the promise: a perspective on research in moral education. *Educational Researcher*, 28(4), 18–26.

Coley, J. D. & Tanner, K. (2015). Relations between intuitive biological thinking and biological misconceptions in biology majors and nonmajors. *CBE-Life Sciences Education*, 14(1), p.ar8. doi:10.1187/cbe.14–06–0094.

Davis, M. (1999). *Ethics and the University*. London: Routledge.

Kincheloe, J. & Steinberg, S. R. (Eds) (2004). *19 Urban Questions: Teaching in the City*. New York: Peter Lang.

Kristjánsson, K. (2015). *Aristotelian Character Education*. Abingdon: Routledge.

Osborne, J., Simon, S. & Collins, S. (2003). Attitudes towards science: a review of the literature and its implications. *International Journal of Science Education*, 25, 1049–1079.

Reiss, M. J. (1999). Teaching ethics in science. *Studies in Science Education*, 34, 115–140.

Reiss, M.J. (2006). Teacher education and the new biology. *Teaching Education*, 17, 121–131.
Stern, F. & Kampourakis, K. (2017). Teaching for genetics literacy in the post-genomic era. *Studies in Science Education*, 53(2), 193–225.
Straughan, R. (1988). *Can We Teach Children to be Good? Basic Issues in Moral, Personal and Social Education*, 2nd edn. Milton Keynes: Open University Press.
Tekkaya, C. (2002). Misconceptions as barrier to understanding biology. *Hacettepe Üniversitesi Eğitim Fakültesi Dergisi*, 23, 259–266.
White, R.T. (1988). *Learning Science*. Oxford: Blackwell.
Wilson, E. O. (1975). *Sociobiology: The New Synthesis*. Cambridge, MA: Belknap Press.

2

COGNITIVE BIASES OR COGNITIVE BRIDGES?

Intuitive Reasoning in Biology

Evelyn Margaret Evans and Karl S. Rosengren

Introduction

"One can take almost any advance, either in evolutionary biology or systematics, and show that it did not depend as much on discoveries *as on the introduction of improved concepts*" (Mayr, 1982, p. 24, lines 11–13 [*our italics*]). Ernst Mayr's precept is the focus of this chapter. Articulating the developmental and instructional trajectory of those concepts may be an arduous task, but one that is crucial if we are to improve students' understanding of biology. In his historical treatise on the growth of biological thought, Mayr (1982) was one of a few scholars (see Gruber & Barrett, 1974) to explicitly integrate studies of the historical background of biological discoveries with philosophical and psychological analyses of the thinking processes of the major players, especially of Darwin and his contemporaries. Though he would hardly consider himself a psychologist, Mayr's (1982) description of the role of essentialist and teleological concepts in early-stage scientists' biological thought provided psychologists with the basis for a cognitive-developmental analysis of the transition from the folk- or intuitive biologies of children, students, and lay adults to the scientific thinking of contemporary biologists (e.g., Evans, 2001; Gelman, 2003; Kelemen, 2012; Shtulman & Schulz, 2008).

While Mayr's focus was on the emergence of ideas about ultimate evolutionary-causation, the essentialist and teleological reasoning patterns he described play a key role in many of the misconceptions found in contemporary students' understanding of biology. It should be noted, however, that there are drawbacks to the use of early-stage scientists' concepts as proxies for contemporary students' reasoning, one being the debate about the accuracy of the historical depictions (see Wilkins, 2013). Nonetheless, an exchange of ideas between philosophers of science and researchers is potentially productive, yielding a more accurate understanding of conceptual change in both early-stage scientists and students (Kampourakis, 2015). Our goal in this chapter is to go beyond the idea that misconceptions are necessarily impediments to conceptual change, even though they are widespread and persistent. In some cases, we shall argue, they may provide a link between students' folk intuitions and a scientific understanding of biological processes. However, the connection between intuitive reasoning and scientific understanding may need to be made explicit for students to benefit from this link.

First, we describe and integrate two theoretical accounts, the dual process theory of cognition and the cognitive constraints perspective on reasoning. This integration potentially provides a useful way of constructing a bridge between intuitive and scientific reasoning. Cognitive constraints refer to biases that either limit or facilitate thinking across a variety of domains of knowledge. Intuitive theories are commonsense or folk theories used to reason about a specific domain and, like scientific theories, they are organized around an underlying causal model, though they differ in their coherence and lack of public scrutiny. Vosniadou and Brewer (1992) propose that initial intuitive theories are influenced more strongly by constraints and for this reason may be relatively uniform across individuals and cultures, while the *synthetic* theories that emerge from an interaction between initial intuitions and general experience may be more idiosyncratic. After describing our theoretical perspective, we focus on some key ideas from the extensive research on essentialist and teleological reasoning, specifically emphasizing how folk intuitions play out in students' understanding of biological topics. We provide some examples of when and how these forms of cognitive biases might function as impediments and when they might act as bridges to more advanced reasoning. Finally, we address the issue of what future research is needed.

Theoretical Perspectives: Dual Process Theory, Cognitive Constraints

A haphazard collection of intuitive ideas is unlikely to provide the basis for a coherent curriculum in biology or any other subject (see diSessa et al., 2004). Thus, a challenge for researchers and educators is to articulate how intuitive ideas might relate to each other and to scientific reasoning more generally. Given that there is no single approach that provides the necessary theoretical framework for establishing a coherent model, we summarize and integrate two candidates: dual process theory and (more briefly) cognitive constraints. There are several variants of the dual process framework, but in this chapter we focus on the approach taken by Evans, Stanovich, and their colleagues (e.g., Evans, J. & Stanovich, 2013). They take the perspective that there are essentially "two minds in one brain" (Evans, J., 2010, pp. 20–21); one is an evolutionarily old rapid processer of belief-based associative information, the intuitive brain—the System I mode of thinking—and the other is a more reflective serial information processer—System II thinking (Evans, J. & Stanovich, 2013). These two systems of thought were separately proposed by

Tversky and Kahneman (1981) and popularized by Kahneman (2011) in his book, *Thinking Fast and Slow*. For the purposes of this chapter, however, we will refer to these two ways of thinking as intuitive and reflective or scientific (depending on the content area). A key issue is the nature of the relationship between these two systems.

One obvious prediction from dual process theory is that the young learner or the novice student is more likely to utilize intuitive reasoning and the older or more experienced student will utilize the more reflective or scientific mode (Evans, 2018). However, lay adults and even scientists default to an intuitive reasoning mode under some conditions (e.g., time constraints; see Shtulman & Valcarcel, 2012), indicating that the intuitive mode is not supplanted with increasing expertise (see J. Evans, 2010, p. 48). Moreover, research suggests that it is commonplace for students and adults to switch between these modes and even to incorporate both forms in the same response, often without appearing to notice any inherent contradictions (Evans, 2001; Evans et al., 2011; Legare et al., 2012; Nehm et al., 2010). Overall, intuitive reasoning is more likely to be used in situations where an immediate action plan is required (Kahneman, 2011, p. 20), whereas reflective reasoning is used when the reasoner is exercising effortful and mindful control. Moreover, skills that were initially effortful, such as driving, may become routinized and intuitive, no longer making demands on the reflective mind (J. Evans, 2010, p. 56). Taken together, this research suggests that the two reasoning modes interact, with individuals shifting between them based on experience and contextual factors, including the nature of the specific task being performed.

We argue that in order to map out the ways in which learning occurs, an interactive approach is the one most likely to be fruitful. The historical and contemporary research to date suggests that learning in any domain is unlikely to be a linear progression from intuitive to scientific reasoning. In some cases, intuitive reasoning may jumpstart the process of learning a new domain (Lakoff & Nunez, 2000); in other cases, some intuitions may have to be consciously supplanted (e.g., Kelemen, 2012); and in yet other cases, there is likely to be a constant interchange between the two modes until the learner has a sound grasp of the new domain and it becomes unconsciously and intuitively implemented whenever the problem is encountered, just like driving (Evans, 2018). This latter idea is similar to that of Siegler's (1996) overlapping wave model of strategy use. He has shown that when students acquire a new mathematical strategy it does not replace an existing one, but is added to their overall repertoire of strategies. Even as students acquire high-powered strategies to solve problems, they still fall back on less sophisticated ones. A key idea of our interactive perspective is that intuitive biological reasoning is not replaced by scientific reasoning, but continues to exist in modified or synthetic forms, even in the minds of scientists. Figure 2.1 shows a graphical representation of this idea based on Siegler's model. This line of theorizing suggests that a simplistic approach to the relationship between intuitive and scientific reasoning would be counterproductive.

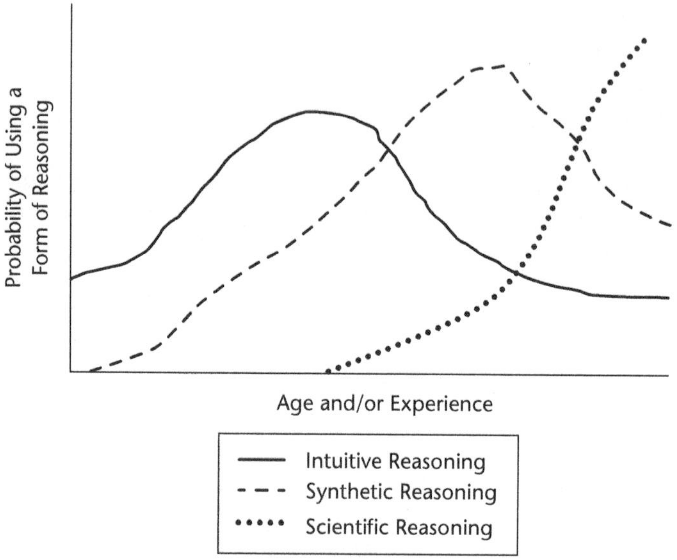

FIGURE 2.1 Relationship between intuitive, synthetic, and scientific reasoning. Young children and individuals with less experience (on the left of the X-axis) exhibit primarily intuitive reasoning. Scientists, on the far right of the X-axis, exhibit a mixture of reasoning patterns, depending on the context

Relatedly, researchers working in cognitive development and education have proposed that cognitive constraints shape the way that young children, students, adults, and even scientists reason about material they learn in school and confront in the wider world (Arenson & Coley, 2017; Emmons & Kelemen, 2015; Evans, 2001; Shtulman & Valcarcel, 2012). These constraints both limit certain kinds of reasoning and facilitate others. One can think of them as operating like a channel that funnels water rapidly away from a city, thus both constraining the flow of water away from the city and facilitating rapid flow in the channel. Two key cognitive biases constrain biological reasoning, essentialism and teleology, both of which can be thought of as intuitive reasoning modes within the dual process framework. We have proposed a developmental learning trajectory (Evans, 2013, 2018; Evans et al., 2012) that would allow us to map the likely sequence of changes that occur as learners of different ages and experiences encounter new material in the biological domain. In the rest of this chapter, we will apply this approach to essentialist and teleological reasoning.

Essentialist Reasoning

Essentialism: Helpful or Harmful?

Markers of essentialist reasoning are evident as early as the preschool years (Gelman, 2003) and are still part of the conceptual repertoire of the non-scientist adult (and scientists, as well). This suggests that essentialism in some form is functional and adaptive and may well bootstrap aspects of biological reasoning. Further, an essentialist bias may be adaptive in that it helps individuals rapidly categorize things in the world, separating dangerous animals from non-dangerous ones, or edible plants from inedible ones. This speedy classification system could be highly beneficial for survival, helping a forager to obtain resources efficiently and to avoid harm. If essentialism is adaptive, then it is unlikely to be easily jettisoned, nor would it be wise to do so.

Lay essentialist thinking carves the biological world into stable and distinct kinds (or species) with immutable boundaries, as dictated by nature (e.g., Gelman, 2003). From this perspective, variability within a species is seen as superficial, as species are characterized by an undefined "essence," the unobservable cause of group membership. Typically, lay adults conceive of genes as placeholders for essences, but they still apply lay essentialist thinking: a genetic essentialism (Heine et al., 2017). While collectors of biological specimens or breeders may realize that no two individuals in a species are alike and that these characteristics are heritable (Mayr, 1982, p. 46), we have found that neither school-age children nor typical adults recognize the significance of variability within a species; they are more likely to focus on similarities (French, Menendez, Herrmann, Evans, & Rosengren, in press; Herrmann et al., 2013). This underestimation of within-species variability is seen as key constraint, hindering students' understanding of natural selection and common descent (Evans et al., 2010; Emmons & Kelemen, 2015; Shtulman & Schulz, 2008).

To understand how essentialism might be beneficial for learning, we consider in more depth what it means to "essentialize." Generally, researchers in psychology have considered essentialist reasoning as an overall cognitive constraint (Medin & Ortony, 1989). This perspective differs from the historic view of biological essentialism with its assumption that there *actually* exists an underlying essence, which makes an organism what it is (e.g., a cat essence causes an organism to be a cat). The premise of psychological essentialism is that people *act as if they believe* natural categories have essences (Medin & Ortony, 1989).

There Is No Single Essence

Several researchers (Gelman & Rhodes, 2012; Heine et al., 2017; Wilkins, 2013) argue that essentialism is multifaceted. Gelman and Rhodes (2012), for example, propose five subcomponents or cognitive constraints. The first is *boundary intensification*, which leads people to behave as if species membership is all or none, and treat members of different species as more dissimilar than they are. A second component is *immutability*, the notion that species membership is stable over time and across natural transformations, such as growth and metamorphosis. The remaining subcomponents include *innate potential* (i.e., category membership is determined at birth), *non-obvious properties* (i.e., species have an unobservable essence that determines observable characteristics), and *inductive potential* (i.e., knowledge of species membership enables inferences about members of the species). French and colleagues (French, Menendez, Herrmann, Evans, & Rosengren, in press; Herrmann et al., 2013) have also suggested *featural stability*, which leads people to assume that features of a species remain stable over time, both across the lifespan and over generations.

Components of Essentialism Have Different Development Pathways

For the most part, researchers have assumed that the subcomponents of psychological essentialism play a consistent role in individuals' reasoning about the biological world. However, French et al. (in press) have proposed that these components may emerge at different times over development. Specifically, they view *featural stability* as an early-emerging component that causes young children to assume that offspring and adults of a species will look highly similar, and to reject metamorphosis as a natural change. When reasoning about unfamiliar insects, even adults assumed that juveniles would grow bigger but not undergo dramatic changes, such as metamorphosis, though, in reality, almost all insects undergo metamorphosis. This result suggests that a bias to assume featural stability emerges early and is a default pattern in adults. Based on these results, we argue that featural stability is likely to be a System 1 form of reasoning. Further, *boundary intensification* is also likely an early emerging System 1 form of intuitive essentialist reasoning that helps elicit rapid category judgments. Featural stability and boundary intensification may work in a complementary fashion to lead individuals to expect little variation within a species.

Other components of essentialist reasoning, most notably *immutability* and *innate potential*, appear to be later developing components that require a certain level of knowledge about biological change. We argue this because young children (~3 years) generally reject metamorphosis as a natural type of change (Rosengren et al., 1991) and also respond as if an animal shifts category membership when superficial external properties are changed (DeVries, 1969; Keil, 1989, p. 178). Slightly older children respond that species membership is maintained over these changes. These findings indicate that as children gain knowledge about what types of biological change are possible, the essentialist bias of innate potential leads them to reason that species identity is determined at birth and is stable over time and transformations. As students gain knowledge it appears to trigger a more reflective System 2 form of reasoning about identity that is an important bridge to scientific understanding.

Generics, Language, Visualization, and Essentialism: Shifting Boundaries

A key assumption of our integrated perspective is that language serves to both enhance and inhibit essentialist reasoning, with different subcomponents triggered by specific language forms (Gelman & Rhodes, 2012). Generics (e.g., "all birds have feathers" or "all wolves hunt in packs"; see Gelman et al., 2010) may focus an individual's attention on similarities within a species, implicitly contrasting them with other species. In this way, generics may increase boundary intensification, consistent with System 1 reasoning.

In contrast, shifting the level at which an animal is labeled may highlight individual differences or cross-category similarity. For example, labels that highlight an individual member of a species (e.g., this is a dog named Rover) versus a species (e.g., butterflies drink nectar from flowers) versus a class (e.g., mammals have hair) focus a student's attention on different groupings of animals, fostering taxonomic reasoning (see Coley & Muratore, 2012). That is, with its focus on a particular level of classification, specific language forms may lead to different inductive potentials—from a single individual, to a species, to a class of animals. In this way, language may shift an individual from System 1 intuitive reasoning about within-species homogeneity and strict category boundaries, to System 2 reasoning, with a focus on cross-category similarity, blurring, for example, the sharp boundaries between species (e.g., cats versus dogs) that are members of the same class (e.g., mammals).

By combining visualization and language forms museum exhibits and scientific media can potentially decrease boundary intensification by highlighting, for example, the ubiquity of DNA, blurring the boundaries between all living kinds. In an experimental study, pairs of adolescent museum visitors explored the relationships between species at a digital touch table that displayed 70,000 species in an interactive tree of life (Horn et al., 2016). At the shared ancestral node of any two species, such as a dandelion and a human, a glowing double helix, the iconic visualization of DNA, signaled the presence of a common ancestor. In comparison with a control group, the adolescent explorers were more likely to endorse the idea that all living kinds had an ancestor in common and to exhibit a better understanding of tree-of-life concepts. Analyses of the adolescents' explorations revealed that the quality of their touch patterns and of their conversation separately predicted improved understanding. In particular, the number of times the youth activated the DNA icon at the shared nodes of diverse species, the frequency of their tree-of-life language (e.g., "related"), and their interest level (the "wow" factor) positively predicted the study outcomes, regardless of youth age, parent educational level, or prior knowledge. This powerful combination of visualization and language forms appeared to elicit a boundary shift in adolescents' essentialist reasoning: from a focus on species—System 1 intuitive processing—to a focus on all living kinds—the "expanded" essentialism of a System 2 reasoner. An expanded essentialism may bridge the gap between genetic essentialism and the theory of common descent (Evans, 2018; Horn et al., 2016).

Teleological Reasoning

Teleology: Helpful or Harmful?

The detection of recurring patterns in nature is a basic survival strategy for prey and predator alike. Noticing the relationship between the characteristics of particular species, their habitat, and their food sources provides information that aids survival. Explaining those patterns, though, is a peculiarity of humans. For the early naturalists, the intricate wings of a butterfly or the beak of a bird were the hallmarks of intelligent design, as these features seemed perfectly adapted to a particular behavior, flying or eating seeds (Evans, 2013). While the early naturalists, many of whom were clergy, may have resorted to theistic teleological explanations ("God created it"), the lay person is more likely to use a related but more intuitive reasoning stance, teleological realism, a form of lay-teleology that links to a commonsense or folk psychology (Sehon, 2005). Two distinct forms of teleological explanation are evident historically and in contemporary lay populations: a theistic teleology and a natural teleology in early-stage scientists (Lennox & Kampourakis, 2013); creationism and teleological realism in lay-teleological reasoning (Evans, 2013). While the historic forms do not map directly onto the lay-teleological forms, they do share key characteristics. Both theistic teleology and contemporary creationism acknowledge the role of a purposeful creator of the natural world. Natural teleology and teleological realism, in contrast, explain the adaptive features of organisms in non-theistic terms; they reference the needs of the organisms, not God. Our focus is on intuitive explanations, especially teleological realism, as these have the potential to bridge intuitive and scientific explanations,

Creationist Teleology Versus Teleological Realism

A teleological realist's explanation for a feature such as giraffe's long neck is that the long neck exists because it is needed in order for the giraffe to eat the leaves of tall trees; the long neck improves the survivability of giraffes (Evans, 2013). A full explanation might link to a commonsense psychology, in particular the psychological states that help the giraffe achieve this outcome (the giraffe "wants" a long neck, because. . .). This description highlights a key difference between creationism and teleological realism. Creationism emphasizes the goals and purposes of an intentional designer, whereas for the teleological realist, features of living kinds serve the needs of the organisms themselves: the long neck benefits the giraffe (Evans, 2013; Kampourakis & Zogza, 2007; Keil, 1994; Lennox & Kampourakis, 2013). For adolescents and adults (but not 7–9-year-olds), creationism is closely tied to religious worldviews (Evans, 2001, 2013); teleological-realist explanations, in contrast, are widespread, from preschoolers (Kelemen, 2012) to biology teachers (Nehm et al., 2010) to adult visitors in natural history museums (Evans et al., 2010). For example, when asked to explain the different beak sizes of the Galapagos Island finches, 72% of adult museum visitors responded with teleological-realist or synthetic explanations such as:

> Well in order to survive, their body parts had to adjust to certain things, similar to the way giraffes' necks probably grew long as they reached for the

> plants at the top of the trees, so the beak grew longer in order to deal with the tougher seeds.
>
> *(Evans et al., 2010, p. 342)*

While teleological realism is unrelated to either acceptance of common descent or the religious views of the students, it does predict an understanding of natural selection; creationist beliefs show the opposite pattern, negatively predicting students' acceptance of common descent, but unrelated to natural selection understanding (Barnes et al., 2017).

From Teleological Realism to Natural Selection

Modern science focuses on mechanistic explanations. The scientific mechanism that explains adaptive systems is natural selection, with no purpose implied. In the Galapagos Islands, for example, those few finches who already had larger beaks were more likely to thrive when large seeds were abundant; further, with the extra nourishment they were more likely than the smaller-beaked cohorts to reproduce; thus, over time, the population of finches would have larger beaks (Grant & Grant, 2014). How does a student shift from a purpose-based teleological-realist explanation, with no mechanism implied, to a purposeless natural selection? A series of studies has demonstrated the likely stepping stones that make this shift possible.

A complete teleological realist explanation encompasses two components: a reference to the survival needs of the organism (finches need beaks to eat seeds, otherwise they die) and a reference to the psychological states thought necessary to bring about this outcome, such as the desire or intention of the finch to develop the beak in order to eat the seeds. Several research studies have demonstrated the existence of a *restricted* teleology in lay-adults' reasoning, in which the survival needs of the organism are addressed, without any reference to psychological states. Moreover, there is a clear developmental and experiential trajectory. In studies where child, student, and adult participants are asked to endorse intuitive and scientific explanations for biological adaptations, the older and more knowledgeable the participant, the more likely they are to reject psychological explanations ("they *want* to change") and accept a *restricted* teleological realist explanation, exclusively ("they *need* to change"). Younger children endorse both. Furthermore, the shift to a restricted teleology is associated with a better grasp of natural selection (Horn et al., 2016; Legare et al., 2013). This restricted teleology, however, only solves part of the problem. Study participants are still likely to argue that the feature benefits all members of the population. The next step is to eschew a strict essentialism by acknowledging the variability in the population and that only those members of the population with the feature (e.g., bigger beaks) will likely survive—differential survival. Further stepping stones include the understanding that features are differentially inherited and that, over time, the population will change because of selection processes; natural selection is the mechanism that explains adaptation (Evans et al., 2010; Evans, 2013; Evans et al., 2012).

While teleological realism is undoubtedly a System 1 intuitive reasoning pattern, an argument can be made that the need- or purpose-based reasoning of a *restricted* or natural teleology when applied to adaptive systems is more likely to fall under the scope of System 2, a reflective reasoning pattern. The developmental and experiential shift from teleological realism to the need-based reasoning of a restricted teleology involves the rejection of a commonsense or folk psychology. Similar to the components of essentialism, described earlier, language use can signal this shift. By drawing attention to students' use of mental state terms, students can be made metacognitively aware that intentions and desires are not viable 'biological' causes in the sense that they cannot explain the emergence of adaptive systems. Need-based reasoning, on the other hand, signals an awareness of the kind of adaptive system an organism needs to survive, without recourse to a misconstrued psychological mechanism. As just described, this provides a potential anchor, leveraging the scientific explanation of natural selection (Evans, 2018) and, arguably, the basis for scientific inquiry (Lennox & Kampourakis, 2013).

Future Research

Learning trajectories that map the learning that occurs in a particular classroom over a period of time are reasonably well established in the field of science education but, in general, they do not take a developmental approach (Evans et al., 2012). With development, children might become more adept at integrating intuitive and scientific reasoning. In particular, their increasing ability to reason metacognitively would enable them to monitor their intuitions and recognize when they might hinder or help (Evans, 2018). From this perspective, synthetic patterns (or misconceptions) are not necessarily evidence of muddled thinking; instead, they may signal increasing competence—as markers of learning-in-progress. If the research community could develop a series of guides in which they identify core intuitions that demonstrate increasing understanding of key concepts, then students could be encouraged to self-monitor their learning.

Educators in general might consider ways of bridging the gap between intuitive theories and scientific ones. Teachers might begin lessons having students articulate their intuitive theories, highlighting differences among students and between their intuitions and established scientific theory; students could map out changes in their understanding over the semester. Teachers might also become familiar with common cognitive constraints that influence everyday reasoning in different domains and help students recognize bias in their own and others' reasoning. For example, rejection of a strict essentialism and recognition of the functional significance of variability within species appears to signal a key shift from intuitive to scientific reasoning in evolutionary biology. This understanding could be scaffolded by encouraging students to report on the variability among the birds in their neighborhood or in domestic species, or to criticize representations of species in books or media that portray a single exemplar. By supporting students' meta-cognitive reasoning, giving them opportunities to reflect on their own thinking, science teachers could provide students with a lifelong approach that would enhance their understanding of the natural world.

References

Arenson, M. & Coley, J. D. (2017). Anthropocentric by default? attribution of familiar and novel properties to living things. *Cognitive Science*, 42, 253–285.

Barnes, M. E., Evans, E. M., Hazel, A., Brownell, E. R. & Nesse, R. (2017). Teleological reasoning, not acceptance of evolution, impacts students' ability to learn natural selection. *Evolution: Education and Outreach*, 10, 2–12.

Coley, J. D. & Muratore, T. M. (2012). Trees, fish, and other fictions: folk biological thought and its implications for understanding evolutionary biology. In K. S. Rosengren, S. K. Brem, E. M. Evans & G. M. Sinatra (Eds.), *Evolution Challenges: Integrating Research and Practice in Teaching and Learning About Evolution*. New York: Oxford University Press, 22–47.

DeVries, R. (1969). Constancy of generic identity in the years three to six. *Monographs of the Society for Research in Child Development*, 34(3), iii–67.

diSessa, A. A., Gillespie, N. M. & Esterly, J. B. (2004). Coherence versus fragmentation in the development of the concept of force. *Cognitive Science*, 28, 843–900.

Emmons, N. A. & Kelemen, D. A. (2015). Young children's acceptance of within-species variation: implications for essentialism and teaching evolution. *Journal of Experimental Child Psychology*, 139, 148–160.

Evans, E. M. (2001). Cognitive and contextual factors in the emergence of diverse belief systems: creation versus evolution. *Cognitive Psychology*, 42, 217–266.

Evans, E. M. (2013). Conceptual change and evolutionary biology: taking a developmental perspective. In S. Vosniadou (Ed.), *International Handbook of Research on Conceptual Change*, 2nd edn. New York: Routledge, 220–239.

Evans, E. M. (2018). Bridging the gap: from intuitive to scientific reasoning - the case of evolution. In K. Rutten, S. Blancke & R. Soetaert (Eds.), *Perspectives on Science and Culture*. West Lafayette, IN: Purdue University Press, 131–148.

Evans, E. M., Legare, C. & Rosengren, K. S. (2011). Engaging multiple epistemologies. In R. Taylor & M. Ferrari (Eds.), *Epistemology and Science Education: Understanding the Evolution vs. Intelligent Design Controversy*. London: Routledge, 111–139.

Evans, E. M., Rosengren, K. S., Lane, J. D. & Price, K. S. (2012). Encountering counterintuitive ideas: constructing a developmental learning progression for evolution understanding. In K. S. Rosengren, S. Brem, E. M. Evans & G. Sinatra (Eds.), *Evolution Challenges: Integrating Research and Practice in Teaching and Learning About Evolution*. Oxford: Oxford University Press, 174–199.

Evans, E. M., Spiegel, A., Gram, W., Frazier, B. F., Tare, M., Thompson, S. & Diamond, J. (2010). A conceptual guide to natural history museum visitors' understanding of evolution. *Journal of Research in Science Teaching*, 47, 326–353.

Evans, J. St. B. T. (2010). *Thinking Twice: Two Minds in One Brain*. Oxford, England: Oxford University Press.

Evans, J. St. B. T. & Stanovich, K. E. (2013). Dual-process theories of higher cognition: advancing the debate. *Perspectives on Psychological Science*, 8(3), 223–241.

French, J., Menendez, D., Herrmann, P., Evans, E. M. & Rosengren, K. S. (in press). Cognitive constraints influence the understanding of lifecycle change. *Journal of Experimental Child Psychology.*

Gelman, S. A. (2003). *The Essential Child: Origins of Essentialism in Everyday Thought.* Cambridge: Oxford University Press.

Gelman, S. A. & Rhodes, M. (2012). "Two thousand years of stasis": how psychological essentialism impedes evolutionary understanding. In K. S. Rosengren, S. K. Brem, E. M. Evans & G. M. Sinatra (Eds.), *Evolution Challenges: Integration Research and Practice in Teaching and Learning About Evolution.* Cambridge: Oxford University Press, 3–21.

Gelman, S. A., Ware, E. A. & Kleinberg, F. (2010). Effects of generic language on category content and structure. *Cognitive Psychology*, 61(3), 273–301.

Grant, P. B. & Grant, B. R. (2014). *40 Years of Evolution: Darwin's Finches on Daphne Major Island.* Princeton, NJ: Princeton University Press.

Gruber, H. E. & Barrett, P. H. (1974). *Darwin on Man: A Psychological Study of Scientific Creativity.* New York: EP Dutton.

Heine, S. J., Dar-Nimrod, I., Cheung, B. Y. & Proulx, T. (2017). Essentially biased: why people are fatalistic about genes. *Advances in Experimental Social Psychology*, 55, 137–192.

Herrmann, P. A., French, J. A., DeHart, G. B. & Rosengren, K. S. (2013). Essentialist reasoning and knowledge effects on biological reasoning in young children. *Merrill-Palmer Quarterly*, 59(2), 198–220.

Horn, M. S., Phillips, B., Evans, E. M., Block, F., Diamond, J. & Shen, C. (2016). Visualizing biological data in museums: visitor learning at an interactive tree of life exhibit. *Journal of Research in Science Teaching*, 53, 895–918.

Kahneman, D. (2011). *Thinking, Fast and Slow.* London: Macmillan.

Kampourakis, K. (2015). The need for interdisciplinary dialog in evolution education: a comment on the responses by Ware & Gelman and Shtulman. *Cognitive Science*, 39(4), 846–848.

Kampourakis, K. & Zogza, V. (2007). Students' preconceptions about evolution: how accurate is the characterization as "Lamarckian" when considering the history of evolutionary thought? *Science & Education*, 16, 393–422.

Keil, F. C. (1989). *Concepts, Kinds, and Cognitive Development.* The MIT Press series in learning, development, and conceptual change. Cambridge, MA: The MIT Press, 234–254.

Keil, F. C. (1994). The birth and nurturance of concepts by domains: the origins of concepts of living things. In L. A. Hirschfeld & S. A. Gelman (Eds.), *Mapping the Mind: Domain Specificity in Cognition and Culture.* New York: Cambridge University Press, 234-254.

Kelemen, D. (2012). Teleological minds: how natural intuitions about agency and purpose influence learning about evolution. In K. S. Rosengren, S. K. Brem, E. M. Evans & G. M. Sinatra (Eds.), *Evolution Challenges: Integration Research and Practice in Teaching and Learning About Evolution.* Cambridge: Oxford University Press, 66–92.

Lakoff, G. & Nunez, R. (2000). *Where Mathematics Comes From: How the Embodied Mind Brings Mathematics Into Being.* New York: Basic Books.

Legare, C. H., Evans, E. M., Rosengren, K. S. & Harris, P. L. (2012). The coexistence of natural and supernatural explanations across cultures and development. *Child Development*, 83, 779–793.

Legare, C. H., Lane, J. & Evans, E. M. (2013). Anthropomorphizing science: how does it affect the development of evolutionary concepts? *Merrill-Palmer Quarterly*, 29(2), 168–197. (Special Issue on Evolution.)

Lennox, J. G. & Kampourakis, K. (2013). Biological teleology: the need for history. In K. Kampourakis (Ed.), *The Philosophy of Biology: A Companion for Educators.* Dordrecht: Springer, Vol. 1, 421–454.

Mayr, E. (1982). *The Growth of Biological Thought: Diversity, Evolution, and Inheritance*. Cambridge, MA: Harvard University Press.

Medin, D. L. & Ortony, A. (1989). Psychological essentialism. In S. Vosniadou, A. Ortony, (Eds.), *Similarity and Analogical Reasoning*. New York, NY: Cambridge University Press, 179–195.

Nehm, R. H., Kim, S. Y. & Sheppard, K. (2010). Academic preparation in biology and advocacy for teaching evolution: biology versus non-biology teachers. *Science Education*, 93, 1122–1146.

Rosengren, K. S., Gelman, S. A., Kalish, C. & McCormick, M. (1991). As time goes by: children's early understanding of biological growth. *Child Development*, 62, 1302–1320.

Sehon, S. (2005). *Teleological Realism: Mind, Agency, and Explanation*. Cambridge, MA: MIT Press.

Shtulman, A. & Schulz, L. (2008). The relation between essentialist beliefs and evolutionary reasoning. *Cognitive Science*, 32, 1049–1062.

Shtulman, A. & Valcarcel, J. (2012). Scientific knowledge suppresses but does not supplant earlier intuitions. *Cognition*, 124(2), 209–215.

Siegler, R. S. (1996). *Emerging Minds: The Process of Children's Thinking*. Oxford: Oxford University Press.

Tversky, A. & Kahneman, D. (1981). The framing of decisions and the psychology of choice. *Science*, 211, 453–458.

Vosniadou, S. & Brewer, W. F. (1992). Mental models of the earth: a study of conceptual change in childhood. *Cognitive Psychology*, 24, 535–585.

Wilkins, J. S. (2013). Essentialism in biology. In K. Kampourakis (Ed.), *The Philosophy of Biology: A Companion for Educators*. Dordrecht: Springer, Vol. 1, 395–419.

3

BIOLOGY AND YOUNG CHILDREN

Marida Ergazaki

Introduction

Young children are extremely curious about the biological world. When entering kindergarten, they already have a 'naïve biological theory'; its core is made of key ontological distinctions and reasoning devices (Inagaki & Hatano, 2002), whereas its body includes children's ideas about several topics with regard to living things.

According to Inagaki and Hatano (2002), young children seem to differentiate between non-living things and animals or plants by appealing to animals' capacity to move and grow, or plants' capacity to grow and recover from damage by re-growing. Moreover, they seem to draw upon animals' need for food and plants' need for water in order to differentiate between non-living things and 'plants/animals' as a whole. Finally, they seem to realize that a psychological function like our wish or intention cannot affect biological functions like pain, hunger, or the rate our stomach or heart works. On the other hand, as suggested by Inagaki and Hatano, young children have reasoning devices that allow them to (a) predict the properties or behaviors of living things by drawing upon their knowledge about humans ('personifying reasoning'); (b) explain living things' body functions by appealing to the purpose of remaining alive and to the 'vital power' required for this purpose ('teleological-vitalistic reasoning'); and (c) attribute a common, internal, unchanged essence to members of a biological kind ('essentialist reasoning').

Shifting focus from the core of young children's 'naïve biology' to its body, it should be noted that this is quite rich. According to the findings of biology education research or developmental psychology research on the emergence of biological knowledge, young children seem to have their own ideas about several biological topics, for instance, growth and life cycles (Rosengren et al., 1991; Shepardson, 2002), habitat, food relationships, and decomposition (Ergazaki &

Andriotou, 2010; Leach et al., 1996), the body's internal structure (Prokop et al., 2006; Reiss & Tunnicliffe, 2001), germs and illness (Ergazaki et al., 2010; Solomon & Cassimatis, 1999), reproduction and inheritance (Zogza & Christopioulou, 2005; Ergazaki et al., 2014; Solomon et al., 1996), and the origin of species and evolution (Evans, 2000; Legare et al., 2013).

Of course, as time goes by, 'naïve biology' needs to be transformed to 'school biology'. Nevertheless, the fact that children possess a 'naïve biology' so early indicates their interest in understanding the biological world and their potential to do so. The latter is also highlighted by biology education research concerned with designing and testing learning environments that take into account children's preconceptions and challenge them through active learning. Figure 3.1 gives just a very brief example of how children's representations of biological entities may change after taking part in such learning environments (from Ergazaki et al., 2010).

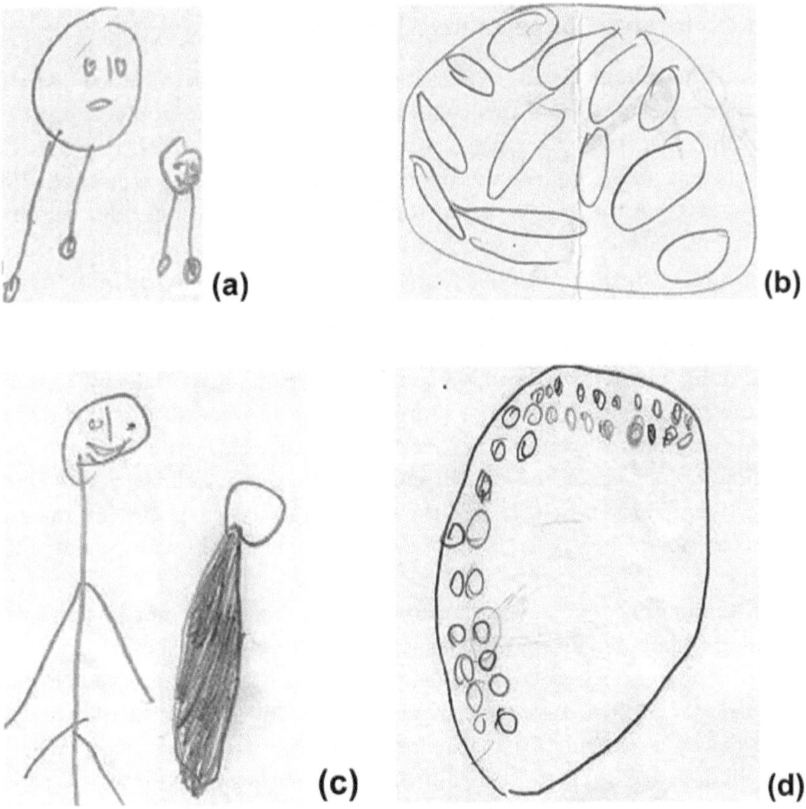

FIGURE 3.1 An example of young children's potential to understand the biological world when appropriately supported: germ representations may change from (a) or (c) to (b) or (d), respectively

As shown, human-like (a) or bug-like (c) representations of germs can change to more realistic ones as seen through a microscope ((b) or (d), respectively).

In order to give a much better idea of young children's ability to engage in meaningful biological learning, I will use the topic of genetic inheritance as a concrete example. Why is that? If young children's potential as biology learners can become evident in the context of such a demanding topic, then the argument *for* their potential gets even stronger. Genetics is a very challenging topic, and if young children can actually cope with it even at a very basic level, it seems important to exploit this potential and start building a solid base for the proper introduction of the relevant concepts later in secondary school. So, in the following sections I will discuss young children's conceptions about inheritance, difficulties in teaching inheritance to them, concepts that are involved, and teaching recommendations that research has to offer. It should be noted that throughout the chapter the term 'inheritance' is used to refer to genetic inheritance in particular.

Young Children's Conceptions About Inheritance

There is a quite long debate in developmental psychology regarding whether young children have a naïve 'theory of kinship' or not; some studies suggest that they do (Springer, 1992; Springer & Keil, 1991; Hirschfeld, 1994; Terwogt et al., 2003), whereas others suggest the opposite (Carey, 1995; Solomon et al., 1996; Solomon, 2002). Although the latter studies may provide a better description of children's misconceptions, it is worth discussing the former, too.

According to Springer (1992), young children believe that the body traits of the parents are more likely to be found in their offspring rather than in their friends or other unrelated people. Moreover, according to Springer and Keil (1991), young children prefer to explain the offspring's body traits by appealing not to the mother's wish, but to a more mechanistic scenario that involves the transfer of tiny pieces from the mother's body to that of the foetus body during pregnancy. Finally, according to Hirschfeld (1994) and Gelman and Wellman (1991), young children believe that the offspring's race or species are the same as their parents' and cannot change because of the social environment where they grow.

Such findings, no matter how promising they seem, may not be enough to give young children credit for possessing a 'naïve theory of kinship'. Carey (1995) suggested that to do so, researchers need to be conclusive about whether young children can offer birth-driven arguments for the body traits of the offspring and nurture-driven arguments for the behavioral ones; for instance, about whether young children can attribute the offspring's eye colour to the biological parents and the fact that they gave *birth* to them, but attribute the offspring's love for animals to the adoptive parents and the fact that they *nurture* them. According to Solomon et al. (1996), this is not the case for children under the age of 7 years old. Engaged in adoption tasks, their young informants were not able to link the offspring's body traits with those of the biological parents and the offspring's behavioral traits with those of the adoptive parents by appealing to birth and nurture respectively.

The idea that young children do not show 'trait type'-dependent reasoning about the parent-offspring resemblance was further supported by the findings of other researchers such as Williams and Smith (2006) and Ergazaki et al. (2014). Nevertheless, this does not necessarily seem problematic from the point of view of biology didactics. Firstly, the nature of the behavioral traits is more often than not puzzling even to scientists; an early introduction of inheritance that includes such traits may not be a very good idea after all. Secondly, the fact that *birth*-driven arguments seem to be very appealing to young children for explaining, although in a promiscuous manner, the offspring's traits, reveals that they *do* recognize a *biological* dimension in family resemblance. Such recognition, although not accurate enough, might be helpful in the process of constructing a more solid understanding of genetic inheritance (Ergazaki et al., 2014).

Preschoolers have other misconceptions about the origin of the offspring's traits as well. They may explain it by drawing upon the offspring's wish or the trait's nature (good/bad) (Ergazaki et al., 2014). So, they may claim, even correctly, that the offspring would have their parents' brown eyes or their parents' strong heart, but the justifications for these claims may be that the offspring wish to look like their parents or that strong hearts are good to have. Ideas like the offspring's wish or the trait's nature, even if leading to correct claims about the offspring's traits, render the claims irrelevant to inheritance and need to be appropriately addressed in teaching.

Preschoolers' ability to recognize the combined contribution of both biological parents to the offspring's body traits seems important from the point of view of biology didactics. According to Terwogt et al. (2003), 4–5-year-olds show a 'mother bias' for predicting the offspring's phenotype: they often choose the 'mother-like' phenotype for the offspring instead of the 'father-like' or the 'combined' one. Moreover, their reasoning about the offspring's body traits seems vulnerable to 'social', 'psychological', 'gender', and 'superiority' biases, too. As Terwogt suggested, even children who consistently choose the 'combined' phenotype for the offspring *do* change their minds when provided with extra, irrelevant information. For instance, knowing that the father has more friends or a more desirable phenotype than the mother can shift children from the 'combined' phenotype prediction to the 'father-like' one. Obviously, such biases need to be addressed if a more meaningful understanding of family resemblance is to be built by young children.

As already mentioned, young children appear ready for a wish-free, mother-based model that explains the offspring's body traits through the transfer of tiny pieces from the mother's body to that of the foetus during pregnancy (Springer & Keil, 1991). But how might they respond to a more advanced, wish-free model that explains the offspring's body traits by including both parents and their 'genes'-containing 'egg' and 'sperm'? According to Ergazaki et al. (2014), preschoolers find this 'pre-biological' model much more appealing than that of 'pieces-transfer' when asked to choose between the two in a forced-choice task. More interestingly, they tend to use the 'pre-biological' model quite frequently when, after having completed the forced-choice task, they are asked to articulate a freer explanation for the parent-offspring physical resemblance. This seems to indicate young children's potential to reach a much better understanding of inheritance.

Trying to fill in the research gap regarding children's understanding about inheritance from middle childhood to adolescence, Venville et al. (2005) found that 9–12-year-old students, although having a 'theory of kinship', actually lack a 'theory of genetics'. In other words, they do differentiate between the origin of the offspring's body traits and behavioral traits, but they have significant difficulties in explaining the underlying mechanisms. According to Venville et al. (2005), children in upper primary school may already be familiar with the words 'genes' or 'DNA', but they are not familiar with the relevant concepts: they don't know what genes or DNA really are, where they may be found, or how they function. An interesting misconception is that genes and traits are the same thing, or that genes are found only in some parts of the body. Moreover, children do not seem to understand that they can distinguish between living and non-living things by appealing to genes or DNA: they may believe that cars or computers have DNA, whereas plants don't.

Donovan and Venville (2014) re-traced these ideas along with others: 'genes and DNA are two different things', 'genes are responsible for family resemblance whereas DNA for everyone's uniqueness', 'DNA is mainly found in blood, skin or hair', and 'its role is to solve crimes'. Moreover, they suggested that misconceptions like these may be linked to the entertainment mass media that 10–12-year-olds use. Knowing more about what DNA can do outside the body than what it naturally does in it needs to be addressed if at least a rough theory of genetics is to be shaped in primary school.

Difficulties in Teaching Inheritance to Young Children

Inheritance is one of the most difficult topics to teach and learn throughout education. Significant difficulties may arise from the abstract and complex nature of genetic phenomena; in other words, from their invisibility and association with different levels of biological organization (Knippels, 2002). When it comes to early childhood education in particular, abstractness and complexity are even more problematic. Young children do better with concrete thinking, but in order to learn how the parent-offspring resemblance they experience in their everyday lives arises, they need to think about 'things' they never saw and cannot see anyway. In fact, they need to explain something they perceive with their senses, by drawing on ideas that their senses cannot provide access to. Children cannot perform empirical tests of their beliefs about family resemblance, unlike what happens in other topics like e.g. plant growth or decomposition. Nevertheless, the educational activities still need to promote active learning in order to allow children challenge their intuitive explanations about family resemblance themselves.

But how easy can it be to convince young children about the inadequacy of their explanations when they seem absolutely plausible to them? In fact, this may be a major difficulty, especially if empirical tests are not an option because of the topic's nature. And, of course, things may get even harder when children's explanations are grounded in their everyday experiences. Why would they challenge their 'mother-bias' regarding family resemblance if it is in line with their experiences of a new baby's arrival? Young children are familiar with mums carrying babies in their bellies, but they do not seem to really know how babies get in there in the first place. Even when they recognize that the father has something to do with it, most of the times they consider his supposed contribution as irrelevant to the process of reproduction.

As nicely summarized by Volbert (2000), several developmental psychology studies in the period from the 1970s until the 1990s showed that 4–7-year-olds have almost no accurate knowledge on how babies are conceived. Quite similar conclusions have been reached more recently, as well. According to Zogza and Christopoulou (2005), preschoolers usually do not recognize their dad's contribution to the creation of a baby. In fact, God, doctors, or mum and her great appetite for food are very often thought to be responsible for the baby arising in mum's belly. Nevertheless, preschoolers may sometimes think that dad *does* contribute to mum's pregnancy, although not in exclusively biological ways: dad puts the baby into mum's belly, or just marries mum and, later on, helps with the housework, takes her to the hospital, helps doctors to get the baby out, and pays the hospital bills. Finally, unlike what earlier studies suggest (Volbert, 2000), preschoolers may sometimes know that dad's contribution concerns providing his 'sperm', which creates a baby along with mum's 'egg' (Zogza & Christopoulou, 2005; Ergazaki et al., 2015).

Moreover, according to Bernstein and Cowan (1975), the concept of reproduction is associated with the concept of 'identity'; in order to think about the origin of babies, young children need to understand that organisms preserve their identity despite changes of appearance. Becoming a typical baby while starting as a tiny, non-invisible 'egg and sperm'-creature is a quite huge change even at the adults' eyes, and this may be a source of difficulties that teachers need to be aware of when teaching inheritance to young children.

Finally, although important for teaching family resemblance to young children (see also the next section), human reproduction is a sensitive, if not taboo, topic. Some teachers may have reservations about it and about how it should be introduced in preschool or lower primary school. In fact, the didactic transformation of the topic for young children calls for extra caution in order to be informative enough, while at the same time taking into account emotional issues and refraining from reproducing outdated stereotypes.

Conceptualizing Inheritance as a Topic for Young Children

Inheritance may be conceptualized as the physical resemblance between parents and offspring, due to the transmission of genes from the former to the latter through the process of reproduction. But what do 'physical resemblance', 'genes', 'gene-transmission' or 'reproduction' actually mean to young children?

The concept of 'physical resemblance' is accessible through senses; offspring's body traits can be observed and compared to those of their parents rather easily. In fact, young children are aware of physical resemblance within families; what they need is to articulate a better explanation of it. Here comes the idea of the *biological* relationship between the offspring and *both* their parents, which can be highlighted through very simplified versions of the concepts of 'reproduction', 'genes', and 'gene-transmission'. According to these, parents can create babies with *material* entities that come from their own bodies; it is mum's 'egg' and dad's 'sperm' that give rise to their baby, not their wish or somebody else like God, for instance. Moreover, mum's 'egg' and dad's 'sperm' contain tiny 'things', mum's 'genes' and dad's 'genes', which pass to the 'egg and sperm'-formed offspring. These 'tiny things' make mum and dad members of a particular species (e.g. 'human-genes' make them humans, while 'bear-genes' make them bears) and also have certain body traits (e.g. 'long nose-genes' make them have long noses); and as they pass on to their offspring, mum's genes and dad's genes become their offspring's genes and make them look like their parents, too.

The teaching objective that derives from this brief conceptual analysis concerns facilitating young children to develop 'pre-biological' reasoning about the parent-offspring resemblance by acquiring a rudimentary idea of genes and gene-transmission during reproduction. *Material* entities originating from the bodies of both parents (mum's 'egg' and dad's 'sperm') are supposed to explain how babies are created, whereas other *material* entities ('genes') within mum's egg and dad's sperm are supposed to explain why babies look like their parents. It should be absolutely clear that the idea of 'genes' can serve just as a 'placeholder', in other words, as a 'conceptual peg' (Solomon & Johnson, 2000) that may help young children keep a better explanation about family resemblance 'in place'.

Next, I present teaching interventions that have been designed within developmental psychology or biology didactics in order to explore the introduction of inheritance to young children. One of these, in line with the previous conceptual analysis of the topic, uses both the concepts of 'reproduction' and 'genes' (Ergazaki et al., 2015), while others use either the former (Schroeder et al., 2007) or the latter concept (Solomon & Johnson, 2000), and others none (Springer, 1995; Williams & Affleck, 1999).

Recommendations for Teaching Inheritance to Young Children

Williams and Affleck (1999) designed a teaching intervention for 4- and 7-year-olds, based on Springer's (1995) 'essential facts' about intra-uterine life: (a) babies grow inside their mum's belly before their birth; (b) when there, babies are not affected by the outside world; and (c) the close contact between babies and their mums during pregnancy makes possible the transmission of traits from mums to babies. Although Springer suggested that these ideas could help children improve their reasoning about inheritance, William and Affleck did not find a significant improvement when they tested their teaching intervention.

Solomon and Johnson (2000) designed a very short intervention for 5–6-year-olds, drawing also upon Springer (1995). In fact, they designed a 20-minute session to introduce Springer's 'essential facts' along with a very simplified idea of 'genes' that might help children 'hold in place'/organize these facts. What they tried to do was to familiarize them with the idea that the 'rabbit-genes', for instance, are transmitted from the 'rabbit-mum' to her child during pregnancy, and eventually make the child a little rabbit too. Of course, this idea is not scientifically accurate, but still it is more elaborated than the one of traits' transmission during pregnancy. Children who participated in the intervention were more likely than those who did not to (a) appeal to birth in order to justify parent-offspring resemblance regarding body traits, (b) appeal to learning in order to justify parent-offspring resemblance regarding behavioral traits, and (c) recognize that parent-offspring resemblance regarding body traits has to do with inborn parental traits and not acquired ones.

Solomon and Johnson's (2000) intervention was the first one to introduce a rudimentary idea of 'genes' to young children. Nevertheless, apart from being short, it was restricted to the mother's genes only, leaving the father's contribution out of focus. On the contrary, Schroeder et al. (2007) attempted to introduce 4–5-year-olds to the idea that *both* parents contribute to the offspring's creation and physical appearance through dad's 'sperm' and mum's 'egg', but they left out of focus the idea of 'genes' as developmentally inappropriate. Testing the intervention of fifteen 20-minute sessions they designed, Schroeder et al. (2007) found that inheritance can be discussed with preschoolers quite effectively, if put in the context of reproduction and not necessarily with 'genes' as a 'placeholder'.

But how might young children respond to a teaching intervention that draws upon a rudimentary idea of both parents' genes and their transmission through the process of reproduction? Ergazaki et al. (2015) addressed this question, trying to help young children develop a more advanced, 'pre-biological' reasoning about the resemblance between the offspring and their parents. Testing their intervention with small groups of 5–5.5-year-olds, they found that it worked quite well. The ten activities of the intervention, which can be performed in three 60- to 90-minute sessions in kindergarten or lower primary school, take into account children's young age and the importance of engaging them in active learning (for an overview, see Table 3.1).

Focusing on primary school, Venville and Donovan (2008) suggested that not only 5th graders, but 2nd graders as well may be introduced to the physical relationship between DNA, genes, alleles, and chromosomes, if they are given the opportunity to actively explore concrete representations of these concepts. So, in their case studies, they tested a one-session intervention that included an introduction, an activity based on a wool-model of DNA, and a guided discussion.

TABLE 3.1 An overview of a teaching intervention about inheritance for young children (Ergazaki et al., 2015)

Activities per session			*Target ideas*
1.1	'Make the animal families'	Guided card-sorting & discussion	'X-gives-birth-to-X'
1.2	'How the rabbit-couple had their baby-rabbit'	Interactive card-based story	How babies are born: mum's egg & dad's sperm give rise to a baby
1.3	'Why the rabbit-couple can only have baby-rabbits and not . . . puppies?'	Brainstorming	Children lack a good explanation about why X gives birth to X
1.4	'The rabbit-couple can only have baby-rabbits because of the rabbit-genes in their egg/sperm'	Interactive card-based story	Why X gives birth to X: 'species-genes' within parents' egg and sperm
1.5	'Draw the species-genes in mum's egg & dad's sperm that will give rise to their baby'	Guided drawing & discussion	Why X gives birth to X: 'species-genes' within parents' egg and sperm
1.6	'Can the rabbit-couple have a baby-tiger, frog or mouse if they wish for it very much?'	Role-play & discussion	Why X can't give birth to Y: parental 'species-genes' vs parental wish
2.1	'How I came to the world with blue eyes, while my parents wished that I came with brown?'	Interactive card-based story	Parental 'trait-genes' & parental 'trait-genes' vs parental wish
2.2	'Guess the traits of the wishing-parents' child'	Floor-game with justified predictions	Parental 'trait-genes' vs parental wish
3.1	'Parents of species X adopt a child of species Y'	Interactive puppet-show	Parental 'species-genes' vs offspring's social interaction and wish
3.2	'Parents with trait A adopt a child with trait B'	Interactive card-based story	Parental 'trait-genes' vs offspring's social interaction and wish

- The introduction addressed the question of why babies look like their parents. Children could not actually respond, so DNA was discussed as a small, invisible substance, exclusively found in all living organisms. Children were encouraged to realize that they would need to use their imagination in order to learn about DNA by working with a wool-model.
- In the context of the model-based activity, children were provided with two woollen yarns. One of them was pre-assembled to represent a double strand of DNA ('chromosome') with different genes (sections of different length/color). Children were asked to assemble the second yarn as a second double strand of DNA that matches the first ('paired chromosome'). Then, they were encouraged to look at each other's 'wool DNA' ('pair of chromosomes'), see if their 'wool DNAs differed and try to explain why. When they realized that each gene in their 'wool DNA' has a different color-shade than the corresponding gene of their classmates' 'wool DNA', they were introduced to the idea that genes may have different forms ('alleles').
- The guided discussion focused on where a person's DNA comes from and how it controls their appearance. Brief explanations were introduced by the teacher whereas children were given time to ask their own questions. The idea of 'chromosomes' was briefly introduced.

The findings of Donovan's and Venville's case studies (2008) are encouraging. Nevertheless, in praxis, their 'wool-model' appeared to work differently for the 2nd and the 5th graders, as the researchers suggested in their article. For the older children, the model *did* serve as a tool for understanding the physical relationship between DNA, genes, and chromosomes, and also for placing inheritance in a broader theory about living things. For the younger children, though, the model served just as a tool for explaining how the familiar phenomenon of family resemblance may occur. So, it appears that the very essentials of inheritance may be enough for children at lower primary school. Helping them to build a solid, introductory understanding about why offspring resemble their parents and realize that this resemblance is unique to living things seems meaningful and feasible as well.

Finally, it is worth mentioning that Duncan et al. (2009) have designed a learning progression that spans several grades, from 5th grade to upper high school. Their suggestion was that early versions of the three different models (genetic, meiotic, and molecular) about genetic phenomena should be all introduced from the beginning and get gradually elaborated. The 'big ideas' that pop up when unpacking these models lie at the heart of Duncan et al.'s (2009) design. A detailed description of these ideas and how they are supposed to progress in time is provided in their article. We should notice, though, that all of them are associated with the central questions of how living things' appearance and function are affected by genes and how they differ.

Conclusions

Inheritance is one of the most demanding biological topics throughout school education. So, it seems to provide a rather illustrative 'test' for young children's learning potential. As already shown, the results of this 'test' are actually promising. Young children appear surprisingly ready to engage in meaningful learning about the biological world when appropriately supported, and those involved in early childhood biology education need to take this into account. Educational models that promote active learning may help educators invest in young children's potential, if they consider children's developmental level as a factor that should inform their instruction rather than limit it (Kuhn, 1997). The model of inquiry-based science education, for instance, can give them the chance to build on children's curiosity and early competencies to help them reach a better conceptual understanding and also become familiar with the scientific practices and even the nature of scientific knowledge, as Metz (2004) suggests. So, when it comes to biology and young children, it all seems to be a matter of informed choice and systematic effort. Choosing to be pessimistic is certainly *not* in line with the findings of contemporary research about what young children can achieve in their science lessons. As very aptly put by Keil (2011), "*science starts early*".

References

Bernstein, A. & Cowan, P. (1975). Children's concepts of how people get babies. *Child Development*, 46, 77–91.

Carey, S. (1995). On the origin of causal understanding. In D. Sperber, D. Premack & A. J. Premack (Eds.), *Causal Cognition: A Multidisciplinary Debate*. Oxford: Clarendon Press, 268–302.

Donovan, J. & Venville, G. (2014). Blood and bones: the influence of the mass media on Australian primary school children's understandings of genes and DNA. *Science & Education*, 23, 325–360.

Duncan, R. G., Rogat, A. D. & Yarden, A. (2009). A learning progression for deepening students' understandings of modern genetics across the 5th–10th grades. *Journal of Research in Science Teaching*, 46, 655–674.

Ergazaki, M., Alexaki, A., Papadopoulou, C. & Kalpakiori, M. (2014). Young children's reasoning about physical & behavioural family resemblance: is there a place for a precursor model of inheritance? *Science & Education*, 23(2), 303–323.

Ergazaki, M. & Andriotou, E. (2010). From "forest fires" and "hunting" to disturbing "habitats" and "food chains": do young children come up with any ecological interpretations of human interventions within a forest? *Research in Science Education*, 40(2), 187–201.

Ergazaki, M., Saltapida, K. & Zogza, V. (2010). From young children's ideas about germs to ideas shaping a learning environment. *Research in Science Education*, 40(5), 699–715.

Ergazaki, M., Valanidou, E., Kasimati, M. C. & Kalantzi, M. (2015). Introducing a precursor model of inheritance to young children. *International Journal of Science Education*, 37(18), 3118–3142.

Evans, E. M. (2000). The emergence of beliefs about the origin of species in school-age children. *Merrill Palmer Quarterly*, 46, 221–254.

Gelman, S. A. & Wellman, H. M. (1991). Insides and essences: early understandings of the non obvious. *Cognition*, 38, 213–244.

Hirschfeld, L. A. (1994). The child's representation of human groups. In D. Medin (Ed.), *The Psychology of Learning and Motivation: Advances in Research and Theory*. San Diego: Academic Press, Vol. 31, 133–183.

Inagaki, K. & Hatano, G. (2002). *Young Children's Thinking About the Biological World*. New York: Psychology Press.

Keil, F. C. (2011). Science starts early. *Science*, 331, 1022–1023.

Knippels, M. C. P. J. (2002). *Coping With the Abstract and Complex Nature of Genetics in Biology Education*. Utrecht, Netherlands: CD-Press.

Kuhn, D. (1997). Constraints of guideposts? developmental psychology and science education. *Review of Educational Research*, 67, 141–150.

Leach, J., Driver, R., Scott, P. & Wood-Robinson, C. (1996). Children's ideas about ecology 2: ideas found in children aged 5–16 about the interdependency of organisms. *International Journal of Science Education*, 18(2), 129–141.

Legare, C. H. & Lane, J. D. & Evans, E. M. (2013). Anthropomorphizing science: how does it affect the development of evolutionary concepts? *Merrill-Palmer Quarterly*, 59(2), 168–197.

Metz, K. (2004). Children's understanding of scientific inquiry: their conceptualization of uncertainty in investigations of their own design. *Cognition and Instruction*, 22(2), 219–290.

Prokop, P., Prokop, M., Tunnicliffe, S. D. & Diran, C. (2006). Children's ideas of animals' internal structures. *Journal of Biological Education*, 41, 1–6.

Reiss, M. J. & Tunnicliffe, S. D. (2001). Students' understandings of human organs and organ systems. *Research in Science Education*, 31, 383–399.

Rosengren, K., Gelman, S., Kalish, C. & McCormick, M. (1991). As time goes by: children's early understanding of growth in animals. *Child Development*, 62, 1302–1320.

Shepardson, D. P. (2002). Bugs, butterflies, and spiders: children's understandings about insects. *International Journal of Science Education*, 24(6), 627–643.

Solomon, G. E. A. (2002). Birth, kind and naive biology. *Developmental Science*, 5, 213–218.

Solomon, G. E. A. & Cassimatis, N. L. (1999). On facts and conceptual systems: young children's integration of their understanding of germs and contagion. *Developmental Psychology*, 35, 113–126.

Solomon, G. E. A. & Johnson, S. C. (2000). Conceptual change in the classroom: teaching young children to understand biological inheritance. *British Journal of Developmental Psychology*, 18(1), 81–96.

Solomon, G. E. A., Johnson, S. C., Zaitchic, D. & Carey, S. (1996). Like father, like son: young children's understanding of how and why offspring resemble their parents. *Child Development*, 67, 151–171.

Springer, K. (1992). Children's awareness of the biological implications of kinship. *Child Development*, 63, 950–959.

Springer, K. (1995). Acquiring a naive theory of kinship through inference. *Child Development*, 66, 547–558.

Springer, K. & Keil, F. C. (1991). Early differentiation of causal mechanisms appropriate to biological and non biological kinds. *Child Development*, 62, 767–781.

Terwogt, M. M., Stegge, H. & Rieffe, C. (2003). Children's understanding of inherited resemblance: the case of two parents. *International Journal of Behavioural Development*, 27(4), 366–374.

Venville, G. & Donovan, J. (2008). How pupils use a model for abstract concepts in genetics. *Journal of Biological Education*, 43(1), 6–14.

Venville, G., Gribble, S. J. & Donovan, J. (2005). An exploration of young children's understandings of genetics concepts from ontological and epistemological perspectives. *Science Education*, 89, 614–633.

Volbert, R. (2000). Sexual knowledge of preschool children. *Journal of Psychology and Human Sexuality*, 12, 5–26.

Williams, J. M. & Affleck, G. (1999). The effects of an age-appropriate intervention on youngchildren's understanding of inheritance. *Educational Psychology*, 19(3), 259–275.

Williams, J.M. & Smith, L.A. (2006). Social and experiential influences on the development of inheritance concepts. *International Journal of Behavioural Development*, 30(2), 148–157.

Zogza, V. & Christopoulou, G. (2005). Where do babies come from?': ideas of five years old children about reproduction and birth. In M. Ergazaki, J. Lewis & V. Zogza (Eds.), *Trends in Biology Education Research in the New Biology Era*. Patras: Patras University Press, 59–71.

4

MOLECULAR BIOLOGY

Ravit Golan Duncan and Dirk Jan Boerwinkel

Introduction: What Is Molecular Biology?

The term 'molecular biology' suggests a sub-discipline of biology, such as ecology or genetics. However, molecular biology can better be described as a perspective that biologists use to explain the phenomena they are studying. This is visible in sub-disciplines such as molecular cell biology, molecular genetics and molecular evolution.

Morange has suggested that molecular biology should not be considered as a sub-discipline of biology, but as a specific level of explanation (Morange, 2000). All biological sub-disciplines include molecular explanations of phenomena on the cell, organism or population levels. In biological research, these explanations have created deeper insights into these phenomena without making research on higher levels obsolete. Phenomena studied by molecular biologists therefore almost always refer to phenomena on cellular and higher levels of organization, of which the molecular level is the lowest level of explanation. An example is the interaction of proteins in a muscle fiber, explaining the phenomenon of muscle contraction on the cellular and organ level.

The same counts for molecular biology in biology education; here too, molecular biology can be described as the lowest level of explanation. The part molecular biology plays in biology education is therefore about asking and answering 'how questions' when studying phenomena at the cellular level (van Mil et al., 2016). This also means that molecular biology is biology and not chemistry. Although the molecular phenomena follow the laws of chemistry and physics, the questions that drive molecular biological research are biological ones. These questions ask for the functions of molecules on cellular and higher levels, and for the structures of molecules explaining mechanisms in the cell. Nevertheless, without the input of chemists and physicists, these explanations would not have been developed, because they depend on understanding the chemical interactions about biological molecules. Molecular biology is therefore a field where many disciplines meet (Huang, 2000).

Molecular biology provides unity in biological phenomena. One of the first steps in this process was that molecules found in organisms could be synthetized in the lab, and therefore did not have any special properties and were not a separate kind of chemistry. The universality of the building blocks of cells, of the genetic code and of many metabolic pathways made us aware of the extent to which all organisms are related. Even more importantly, many molecular explanations could fully explain the mechanisms causing the phenomena on higher levels. This meant the end of hypotheses on 'vital forces' and led to the materialistic stance that denies the existence of "any kind of stuff in the world other than the stuff described by physics and chemistry" (Dupré, 2009, p. 33). For these reasons, molecular biology has transformed biology into a more coherent scientific discipline.

In this chapter, we include traditional biochemistry contents in our description of molecular biology. The distinction between molecular biology and biochemistry is not clear. Courses in biochemistry tend to focus more on the chemistry of molecules in the cell, such as enzyme kinetics and redox potentials, whereas courses in molecular biology focus more on gene and protein function in the cell, but most textbooks on either biochemistry or molecular biology will cover both aspects.

Why Teach Molecular Biology?

Based on the curriculum emphases formulated by Roberts (1982), several types of arguments can be furnished to teach molecular biology, such as everyday engagement with science, societal decision-making and correct explanations.

First, many concepts and terms in molecular biology like DNA, proteins and enzymes are commonplace in everyday language. These terms are mentioned in movies and on the news; they can be found on the labels of consumable products and are discussed by health providers. We are supposed to understand information on processes such as enzymatic breakdown and diffusion, and information on dangerous or healthy substances acting at the cellular level.

Second, on a societal level, issues such as environmental pollution, malnutrition, drug abuse and development of new therapies or biofuels involve the mentioning of molecules, either of specific substances such as carbon dioxide and glucose or of groups of molecules indicated as stimulants, nutrients or antigens. As for everyday situations, making informed decisions on these issues requires a level of 'molecular literacy', because a deficient understanding of molecules might hinder comprehension of the issue. For example, the idea that molecules from organisms are 'more natural' and therefore healthier than synthetic substances might lead to wrong conclusions. A functional molecular literacy includes not only 'knowledge in science' but also 'knowledge about science': the ways molecular biological knowledge is generated and applied in societal contexts (Tibell & Rundgren, 2010).

Third, regarding the role of molecular biology education in the formulation of correct explanations, an understanding of molecular phenomena is fundamental to understanding key processes on higher levels, and is essential for all life science studies. As Duncan and Reiser (2007, p. 939) stated, "understanding genetic phenomena entails understanding how mechanisms and interactions at the molecular (genes, proteins) and microlevels (cells) bring about effects at the macrolevel (organism, population)". Furthermore, knowledge of molecular biology integrates knowledge of biology with knowledge of the other sciences.

What Should We Teach About Molecular Biology?

The fact that molecular biology concepts are discussed over different topics in biology education makes it difficult to decide which elements are essential in a molecular biology curriculum. Rather than learning the molecular details of dissimilation, protein synthesis and cellular transport, for a better understanding students should learn the mechanisms and entities that underlie these processes. Howitt et al. (2008) argued that molecular biology education should contribute to the development of big ideas in this domain and have developed a concept inventory for molecular biology that captures these ideas. In their proposition, the most basic concept is that of equilibrium, with which other concepts are linked, such as regulation, information and communication, energy and organization, compartmentalization, catalysis, self-assembly and molecular evolution. Based on what was mentioned earlier about what molecular biology is and why it should be taught, we offer a small set of core concepts for a molecular biology curriculum.

a. *Molecules are the only components of which organisms are built.*
   All organisms are made of molecules and only of molecules. The molecules of living organisms can be synthetized and are made of the same atoms as the universe. However, the molecular composition of organisms differs from the inorganic environment in the composition and size of the molecules, largely consisting of only a few elements (C, H, O, N). Most of the molecules are dissolved in an aqueous internal environment that is separated from the external environment and can therefore create specific conditions for reactions and electrochemical gradients.
b. *Cellular activities can in principle be explained by interactions of molecules, in particular proteins.*
   Students need some basic knowledge about how activities and interactions of biomolecules are grounded in physical and chemical changes. Fundamentally, they need to understand the constant movement and collision of molecules

in a cell. For some simple phenomena, such as diffusion, this already explains the respective phenomenon sufficiently. For other phenomena, more complex interactions have to be understood. Molecular properties such as being 'energy rich' or 'poisonous' exist only through the interaction with other molecules. Without a proton acceptor such as oxygen, glucose would not be 'energy rich', and without a receptor that can be blocked, a substance has no poisonous effect. Phospholipids form double layers only through their interaction with water molecules.

c. *Interactions between molecules in the cell depend on a limited set of characteristics.*
Molecules differ in size, polarity, charge, stability and binding properties, which explain their interactions and activities. Van Mil et al. (2016) provided a simplified account for the interactions and activities of biomolecules based on a combination of intelligible changes: colliding, binding and changing shape. Despite its limitations, this account makes clear that a limited set of basic changes can form the basis for very complex structures and functions in cells.

d. *Relating cellular structures and functions to the properties and interactions of their molecular constituents offers a deeper understanding of 'how it works'.*
Some examples of structure-function relations are:

- Properties of DNA afford key features like coding and accurate replication—functions needed for the information repository of the cell.
- Proteins' three-dimensional shapes and chemical properties (charge, hydrophobicity) afford and constrain the kinds of functions they can carry out.
- Size and charge influence permeability of molecules through membranes.
- Many interactions entail a 'lock-and-key' process (enzymatic reactions, signal transmitting, drug effects, hormonal influence, immune reactions).
- The polarity in phospholipids explains the formation of layers in an aqueous environment.

e. *There are several levels between cells and molecules.*
Subcellular structures such as chromosomes and ribosomes are known and can be made visible with high-powered microscopes. However, a cell has many other phenomena between individual molecules and the whole cell. Many molecular modules can be distinguished in a cell, regulating hormonal influence in the cell or chromosome separation during mitosis. These are temporary complexes and therefore often not visible, but they consist of closely and specifically cooperating molecular structures. Furthermore, in a cell different compartments with different conditions exist.

f. *Interactions between molecular and higher levels go both ways.*
The statement that molecular processes form the 'lowest level of explanation' does not mean that molecular explanations are the most fundamental explanations or that organisms are fully governed by molecular processes.

> Interactions between molecular processes and higher-level processes go both ways. Not only do molecular interactions explain the mechanisms of cellular processes, but processes on organismal and cellular levels influence at the molecular level which genes are expressed, which hormones are released or which cells will divide or die.

In the next section, we focus on protein interactions. These can explain a whole range of cellular phenomena, as there are different functions that proteins carry out in and out of the cell, such as serving as structural components, transport, catalysis, signaling, receiving signals, etc. Proteins are thus a central explanatory element of molecular biology explanations and illustrate all of the points mentioned above.

What Is Difficult to Learn About Molecular Biology?

There are several aspects of molecular biology phenomena that make them challenging to learn. Molecular biology phenomena occur at spatial scales that are very small (enzymes can be as small as 10 nm) and occur at very high speeds (100 or so reactions per second for some enzymes). These invisible and fast phenomena are intractable for students, making reasoning about them challenging. Compounding the intangible nature of molecular biology entities and processes, such as proteins and the many functions they carry out, is the lack of attention to these entities and processes in instruction (Duncan & Reiser, 2007; Marbach-Ad, 2001). Unlike the more extensive focus on genes and DNA, there is much less attention to proteins and their biological role in biology textbooks and instruction at the secondary level (Thörne & Gericke, 2014; van Mil et al., 2013). Thus, even though they are important entities in molecular biology, proteins remain relatively invisible to students throughout their schooling.

Aside from being invisible, molecular biology phenomena are also difficult to understand because they comprise several levels of organization that are ontologically distinct: physical and informational (Duncan & Reiser, 2007). Physical levels include cells, organelles, protein complexes, DNA molecules and other molecules. Interactions within and across these levels result in phenotypes at the level of the cell (and consequently tissue, organ and organism levels). The information level in this system consists of genes, which do reside in a physical molecule (DNA). However, their role in the system can be conceived as being ontologically distinct from the other physical entities as they provide instructions for making proteins. Thus, the information and physical levels intersect at the protein level (rather than the organelle or cell levels, which in contrast are not directly coded by the information level). Understanding how this informational level interacts and intersects with the physical levels is not trivial for learners. Students, especially in the secondary grades (middle and high school), tend to think that genes code directly for the actual phenotype we can observe (e.g. structures and functions at the organ and organism levels) and may even conflate genes and traits (Lewis & Kattmann, 2004).

Venville and Treagust (1998) showed that students progress through a trajectory for conceptual change related to the concept of gene from an initial view of genes as passive particles associated with traits to a view of genes as informational (containing information for 'whole' traits) to the most advanced view of genes as containing productive instructions for proteins. The use of the word productive here denotes the idea that the instructions lead to the formation of a physical product (protein). The shift from conceptualizing genes as all-inclusive instructions (blueprint) to viewing them as productive instructions for proteins is a critical one in this progression (Shea & Duncan, 2013). Such a shift motivates the need for proteins as the mediating mechanisms between the informational level of genes and the subsequent (physical) organization levels, e.g. the cell level and beyond. Students who conceive of the genetic information more broadly as a blueprint for everything about us do not see a need for any mediating mechanisms; they essentially circumvent the protein level by asserting that genes code directly for whatever organismal structure or function for which they are trying to account (Duncan & Reiser, 2007).

Reasoning about protein function, and cellular processes more generally, poses additional challenges in that these processes are often complex and emergent. The notion that complex structures like the clathrin cage shown in Figure 4.1 can self-assemble through random interactions in the cell seems counterintuitive and perplexing to students. However, through the interaction of specific adaptor proteins that facilitate the assembly of the clathrin monomers into a cage, under the right pH conditions and in the presence of the adaptor proteins these molecules can assemble into a cage in a test tube. Research on students' understanding of complex systems has shown that students struggle to conceive of complex structures and behaviors at the aggregate level as emerging from simple interactions at lower organization levels (Jacobson, 2001; Wilensky & Resnick, 1999).

High school and even college students do not readily conceive of how protein-based mechanisms can give rise to complex functions like sensing and transmitting signals, controlling gene expression, movement, etc. (Duncan & Reiser, 2007; van Mil et al., 2013). On the one hand, students struggle to envision how simple proteins can carry out complex functions; yet, on the other hand, they tend to imbue molecular and cellular structures with goals and intentional behaviors (Dreyfus & Jungwirth, 1990) or extrapolate functions from organs and organ systems onto molecular and cellular entities (Flores et al., 2003). These tendencies can engender anthropomorphic explanations of molecular processes—for example, assuming that molecules move directionally and intentionally (e.g. substrates move towards meeting their enzymes), rather than randomly (Stieff & Wilensky, 2003). Representing the movement of molecules is in some ways an issue of reasoning about

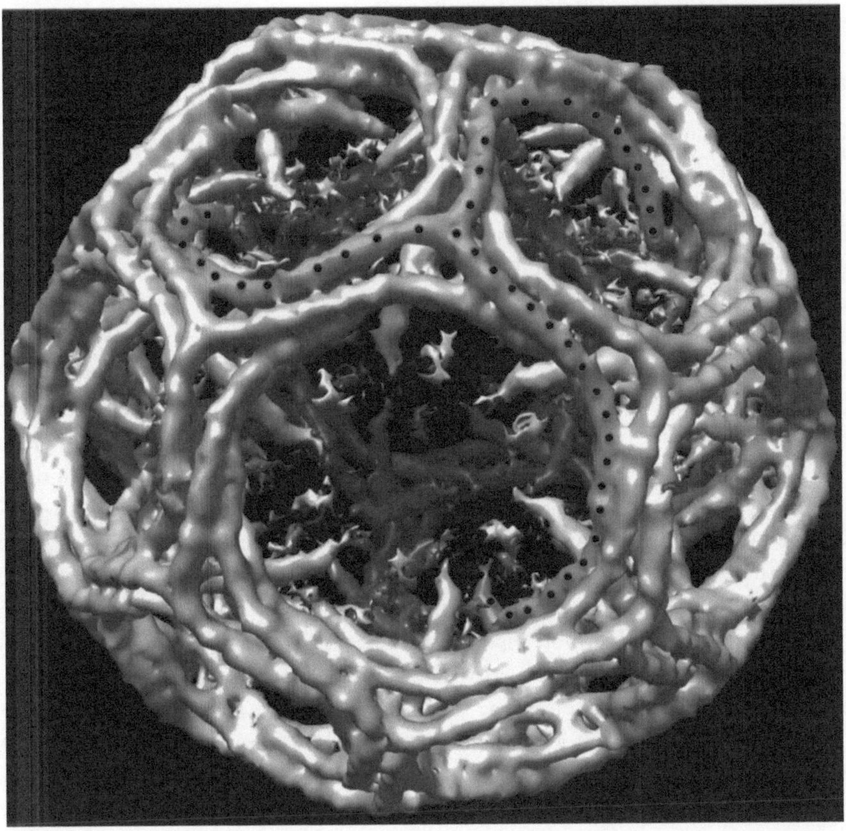

FIGURE 4.1 Clathrin cage composed of many clathrin proteins, with single clathrin protein, indicated with black dots on the image

Attribution—By Mazuraan (Own work) [CC BY-SA 4.0 (http://creativecommons.org/licenses/by-sa/4.0)], via Wikimedia Commons

chemical interactions. Similarly, students tend to assume that the properties of the whole are the same as the properties of the molecules—for example, that the molecules of a blue substance are blue themselves. Thus, students' understandings of chemistry also impact their ability to reason about molecular biology entities and processes.

In summary, reasoning about molecular biology phenomena presents several challenges to students given their ontologically distinct levels, unfamiliar entities and the emergent nature of their structures and processes. In the next section, we discuss some of the instructional implications that stem from these difficulties in terms of how and when we should teach molecular biology ideas.

Recommendations for Teaching Molecular Biology

How Should We Teach Molecular Biology?

One of the core implications is that students need to become a lot more familiar with the functions and interactions of proteins in biological processes. Students need to develop a toolkit of protein functions that they can use to reason generatively about molecular processes, both familiar and novel ones. Such a toolkit will allow students to provide plausible protein-based explanations of molecular processes. In addition, students need to come to view molecular interactions as occurring through random collisions that lead to binding of molecules and ultimately to conformation changes in these molecules that afford specific functions (van Mil et al., 2016). Combining the notion of a protein-function toolkit with a framework for understanding the nature of protein interactions (colliding, binding, changing shape) is likely to yield deeper and more powerful understandings of molecular biology processes. In particular, it should support students in generatively developing explanations of phenomena that are novel and unfamiliar to them. These explanations may not be accurate (in the sense of postulating the actual mechanism involved), but they should be plausible in that they postulate a biologically viable and sensible mechanism. Given the educational goals of producing scientifically literate citizens as well as future scientists, being able to explain and reason about novel phenomena is an important achievement for students.

We also advocate for focusing on proteins first before teaching about DNA and the central dogma. As we noted earlier, students who learn about DNA as a 'blueprint' do not see a need for a mediating mechanism and thus do not view proteins as having a central role in mediating the effects of genes. In fact, even learning the details of the central dogma and how genes are translated to proteins does not ensure the subsequent understanding that proteins are therefore involved in all genetic phenomena in one way or another (Marbach-Ad & Stavy, 2000). Attending to protein structure and function and the roles proteins (and protein complexes) play in and out of cells first prompts the need for instructions for making these entities. Learning about genes and DNA is thus better motivated after the role of proteins has been established.

Our emphasis on protein-based mechanisms and developing plausible explanations of molecular processes implies that instruction should engage students in figuring out molecular biology phenomena. Inquiry-based approaches to science instruction in which students engage in the exploration, investigation and explanation of phenomena are well suited for achieving the goals we posited. The choice of phenomena for investigation, however, needs to be well thought-out. We argue that in order to help students develop a toolkit of protein functions and understanding protein interactions, it is helpful to select phenomena in which proteins carry out functions with analogues in the macro-world. For example, protein functions such as channels, receptors, some structural proteins ('stretchy' muscle proteins) and transporters are simpler to reason about, as students are familiar with structures and functions in their everyday worlds that are analogous to protein functions (doors are similar to channels, transporters are similar to bicycles, muscle proteins are like rubber bands, etc.). While enzymes play a critical role in biology, and are often used as examples in instruction, they are not as accessible to students as there is no real analogue to speeding up reactions in their everyday worlds. Moreover, the structure of enzymes (unlike structures of channels or muscle proteins) does not suggest their function; one cannot see why a specific change in the structure will result in the enzyme not working beyond invoking the lock and key model. It seems to us that proteins other than enzymes may be a better entry point to understanding protein structure and function relationships and the variety of roles proteins play in cells. For example, as noted earlier students can readily reason about membrane channel proteins (viewing them as analogous to selective 'doors' in the membrane), and the topic of cell transport is usually one that is taught fairly early on in middle school biology. In terms of a channel protein context, a compelling phenomenon is lead poisoning, as lead enters the cell by mimicking calcium and is actively transported into the cell through the calcium channel. In fact, in order for students to develop a robust understanding of the myriad of functions proteins carry out, they should study a variety of proteins and related phenomena.

Students' investigations of molecular biology can be bolstered by viewing animations and using interactive simulations (Rotbain et al., 2006; Rotbain et al., 2008). Such simulations make visible the invisible and can illustrate molecular interactions in terms of the random nature of collisions and the importance of direct binding and conformational changes. However, understanding visualizations depends on domain knowledge and familiarity with symbolism (Kozma, 2003). In addition, students sometimes treat visualizations as realistic depictions of phenomena rather than abstract models (Harrison & Treagust, 1996). Therefore, it is important to attend to students' capacity to process and understand the visualizations used and support them in making sense of what they are seeing and manipulating.

Agent-based modeling environments (such as NetLogo below) can also help students understand emergent interactions. There are several useful software simulations and animations that can support student learning, including the Molecular Workbench simulations from Concord Consortium (http://mw.concord.org/modeler/); NetLogo simulations, particularly for related chemistry ideas (https://ccl.northwestern.edu/netlogo/); and the genetics science learning center, which has various activities and animation for multiple topics in molecular genetics and cell biology (http://learn.genetics.utah.edu/).

When Should We Teach Molecular Biology?

The last issue we wish to discuss is how we can build students' understanding of molecular biology over time and across grades using the learning progressions approach. Learning progressions (LPs) are theoretical models of learning over time. LPs in science education are organized around a few core disciplinary ideas and practices and describe the development of students' understandings as intermediate steps or levels between initial and final states (Corcoran et al., 2009; Smith, et al., 2006). Descriptions of these levels are extensively grounded in research on student learning in the domain, and progress along the levels is mediated by targeted instruction and curriculum. It is important to note that LPs by their very nature are hypothetical; they are conjectural models of learning over time that need to be empirically validated. Currently there are no LPs in molecular biology; however, there are several in genetics that deal with molecular genetics concepts relevant to molecular biology (Duncan et al., 2009; Elmesky, 2013; Roseman et al., 2006; Todd & Kenyon, 2016).

A core proposal of these progressions is that students should begin learning about proteins and their functions earlier in their schooling, rather than waiting to introduce these ideas only in high school. Research on this progression has shown that even middle school students (age 12–14) can reason about protein structure and function and develop protein-based explanations (Shea & Duncan, 2013). At this stage, students are expected to understand that genes contain instructions for making proteins (but not for how proteins themselves are synthesized, for which other proteins are required). Thus, students can begin to reason about the ways in which changing the instructions (mutations) might affect a protein's general structure and consequently its ability to function. At this stage, proteins are seen as little machines that can carry out functions and are part of structures in the cell. Their structure is only discussed in general terms as 'being stretchy', traversing the cell membrane, having an opening that allows specific molecules to pass through, etc. Teaching about the amino acid composition and other chemical properties of proteins is not really necessary at this level.

Building on these early understandings of proteins as little machines that can do the work of the cell, high school level molecular biology instruction can then delve into the central dogma and how the genetic code is actually 'read' by cellular machinery to produce proteins. At the high school level, students' ideas about the nature of specific genetic mutations (missense, nonsense, frameshift, etc.) can be deepened and connected to the actual composition and structure of proteins in terms of amino acid sequence and three-dimensional structure. In simple form, the idea that the regulation of gene expression is something that is controlled in the cell is also within the grasp of high school students. Phenomena like tanning in humans and ripening in fruit can be discussed as mechanisms in which an external signal can trigger, in a protein-mediated way, the expression of genes that result in the making of new types of proteins in the cell that confer new functions or structures (resulting in new phenotypes). We would argue that more complex concepts like metabolic pathways, signaling pathways and transcription regulation in development (e.g. homeobox genes) are best addressed in either advanced high school courses or in post-secondary classes.

We also wish to point out and emphasize that teaching students the details of processes such as transcription and translation in protein synthesis, stages of meiosis, structure of DNA, Krebs cycle, etc. does not guarantee an understanding of the bigger ideas (Duncan, 2007). Rather, such a focus tends to result in students 'missing the forest for the trees' because their fragmented understandings of the details of processes and structures do not cohere into robust and generative understandings of the underlying ideas. That is, students can describe the process of translation of RNA into proteins but still propose that genes can also directly code for other molecules (carbohydrates) and whole structures (organelles). If we want high school graduates who can reason productively in molecular biology, we are better off having them figure out the core ideas in this domain rather than teaching the "gory" details of structures and processes.

Learning progressions related to core understandings in chemistry such as the progression for matter and atomic molecular theory in the early grades (Smith et al., 2006) and at the secondary and post-secondary levels (Stevens et al., 2010) can also provide useful guidance for teaching relevant chemistry concepts. For example, the progression developed by Stevens et al. (2010) describes how students' understandings of atomic structure and the electrical forces governing interactions at the nano-, molecular and atomic levels are related and develop over time. These understandings are relevant to reasoning about chemical bonding and interactions that are central to the structures and functions of key cellular molecules, including DNA and proteins.

Conclusions

Molecular biology is a rapidly growing research field with wide-reaching impacts on many aspects of daily life. Scientific literacy demands on understanding molecular biology ideas have also grown as individuals encounter molecular biology-related technologies and issues more ubiquitously and frequently. However, education in molecular biology, especially at the K–12 level, is lagging far behind the research (Howitt et al., 2008).

On the one hand, there are a variety of challenges associated with learning molecular biology (Tibell & Rundgren, 2010); on the other hand, there is no way to bring into the classroom all the complexity and nuanced understanding of this faced-paced field. As educators and education researchers, we are therefore faced with a choice of identifying the core ideas in the field that are generative and would provide students with conceptual leverage with which to reason about new molecular biology phenomena, technologies and issues they may encounter throughout their lives.

In this chapter, we have taken a stab at identifying these ideas, as well as when and how to teach them, given known learning challenges in this domain. One of our key messages is that we need to refocus biology education on powerful frameworks that can help students understand key molecular mechanisms, such as protein interactions, rather than the specific details of molecular structures and pathways. Engaging students in inquiry and sense-making using these frameworks to explore a variety of meaningful molecular biology phenomena is more likely to engender students who appreciate molecular biology and can successfully engage in personal and civic decision making in their everyday encounters with molecular biology.

References

Corcoran, T., Mosher, F. A. & Rogat, A. (2009). *Learning Progressions in Science: An Evidence-Based Approach to Reform*. New York, NY: Columbia University, Teachers College, Center on Continuous Instructional Improvement.

Dreyfus, A. & Jungwirth, E. (1990). Macro and micro about the living cell: which explains what? In P. L. Lijnse, P. Licht, W. de Vos & A. J. Vaarlo (Eds.), *Relating Macroscopic Phenomena to Microscopic Particles*. Utrecht, The Netherlands: CD-Press, 108–118.

Duncan, R. G. (2007). The role of domain-specific knowledge in generative reasoning about complicated multi-leveled phenomena. *Cognition and Instruction*, 25(4), 271–336.

Duncan, R. G. & Reiser, B. J. (2007). Reasoning across ontologically different levels: students' understandings of molecular genetics. *Journal of Research in Science Teaching*, 44(7), 938–959.

Duncan, R. G., Rogat, A. & Yarden, A. (2009). A learning progression for deepening students' understandings of modern genetics across the 5th–10th grades. *Journal of Research in Science Teaching*, 46(6), 655–674.

Dupré, J. (2009). It is not possible to reduce biological explanations to explanations in chemistry and/or physics. In F. J. Ayala & R. Arp (Eds.), *Contemporary Debates in Philosophy of Biology*. Oxford: Wiley-Blackwell, 32–47.

Elmesky, R. (2013). Building capacity in understanding foundational biology concepts: A K–12 learning progression in genetics informed by research on children's thinking and learning. *Research in Science Education*, 43(3), 1155–1175.

Flores, F., Tovar, M. E. & Gallegos, L. (2003). Representation of the cell and its processes in high school students: an integrated view. *International Journal of Science Education*, 25, 269–286.

Harrison, A. G. & Treagust, D. F. (1996). Secondary students' mental models of atoms and molecules: implications for teaching chemistry. *Science Education*, 80(5), 509–534.

Howitt, S., Anderson, T., Costa, M., Hamilton, S. & Wright, T. (2008). A concept inventory for molecular life sciences: how will it help your teaching practice? *Australian Biochemist*, 39(3), 14–17.

Huang, P. C. (2000). The integrative nature of biochemistry: challenges of biochemical education in the USA. *Biological Education*, 28, 64–70.

Jacobson, M. J. (2001). Problem solving, cognition, and complex systems: differences between experts and novices. *Complexity*, 6, 41–49.

Kozma, R. (2003). The material features of multiple representations and their cognitive and social affordances for science understanding. *Learning and Instruction*, 13, 205–226.

Lewis, J. & Kattmann, U. (2004). Traits, genes, particles and information: re-visiting students' understandings of genetics. *International Journal of Science Education*, 26(2), 195–206.

Marbach-Ad, G. (2001). Attempting to break the code in student comprehension of genetic concepts. *Journal of Biological Education*, 35, 183–189.

Marbach-Ad, G. & Stavy, R. (2000). Students cellular and molecular explanations of genetic phenomena. *Journal of Biological Education*, 34, 200–210.

Morange, M. (2000). *A History of Molecular Biology*. Cambridge, MA: Harvard University Press.

Roberts, D. A. (1982). Developing the concept of "Curriculum Emphases" in science education. *Science Education*, 66(2), 243–260.

Roseman, J., Caldwell, A., Gogos, A. & Kurth, L. A. (2006). Mapping a coherent learning progression for the molecular basis of heredity. Paper presented at the Annual Meeting of the *National Association of Research in Science Teaching*, San Francisco, CA.

Rotbain, Y., Marbach-Ad, G. & Stavy, R. (2006). Effect of bead and illustration models on high school students' achievement in molecular genetics. *Journal of Research in Science Teaching*, 43, 500–529.

Rotbain, Y., Marbach-Ad, G. & Stavy, R. (2008). Using a computer animation to teach high school molecular biology. *Journal of Science Education and Technology*, 17(1), 49–58.

Shea, N. & Duncan, R. G. (2013). From theory to data: refining a learning progression. *Journal of the Learning Sciences*, 22(1), 7–32.

Smith, C., Wiser, M., Anderson, C. W. & Krajcik, J. (2006). Implications for children's learning for assessment: a proposed learning progression for matter and the atomic molecular theory. *Measurement*, 14(1&2), 1–98.

Stevens, S. Y., Delgado, C. & Krajcik, J. S. (2010). Developing a hypothetical multidimensional learning progression for the nature of matter. *Journal of Research in Science Teaching*, 47(6), 687–715.

Stieff, M. & Wilensky, U. (2003). Connected chemistry—incorporating interactive simulations into the chemistry classroom. *Journal of Science Education and Technology*, 12(3), 285–302.

Thörne, K. & Gericke, N. M. (2014). Teaching genetics in secondary classrooms: a linguistic analysis of teachers' talk about proteins. *Research in Science Education*, 44(1), 81–108.

Tibell, L. A. E. & Rundgren, C. J. (2010). Educational challenges of molecular life science: characteristics and implications for education and research. *CBE-Life Sciences Education*, 9, 25–33.

Todd, A. & Kenyon, L. (2016). Empirical refinements of a molecular genetics learning progression: the molecular constructs. *The Journal of Research in Science Teaching*, 53(9), 1385–1418.

van Mil, M., Boerwinkel, D. & Waarlo, A. (2013). Modelling molecular mechanisms: a framework of scientific reasoning to construct molecular-level explanations for cellular behaviour. *Science & Education*, 22(1), 93–118.

van Mil, M., Postma, P., Boerwinkel, D. J., Klaassen, K. & Waarlo, A. J. (2016). Molecular mechanistic reasoning: toward bridging the gap between the molecular and cellular levels in life science education. *Science Education*, 100(3). doi:10.1002/sce.21215

Venville, G. J. & Treagust, D. F. (1998). Exploring conceptual change in genetics using a multidimensional interpretive framework. *Journal of Research in Science Teaching*, 35, 1031–1055.

Wilensky, U. & Resnick, M. (1999). Thinking in levels: a dynamic systems approach to making sense of the world. *Journal of Science Education and Technology*, 8, 3–19.

5

CELL BIOLOGY

William F. McComas

Introduction

> I could exceedingly plainly perceive it to be all perforated and porous, much like a Honey-comb, but that the pores of it were not regular . . . these pores, or cells, . . . were indeed the first microscopical pores I ever saw, and perhaps, that were ever seen.
>
> Robert Hooke (1665), Observation XVIII in *Micrographia*

These words from *Micrographia*, one of the most important books in the history of biology, probably represent all that most students know about the development of the idea of the cell; life processes are contained in tiny boxes, a fact discovered by a seventeenth-century scientist (Figure 5.1). As we will see, this story is an oversimplification, and, as is often the case, the historical details are far more interesting.

Before proceeding it will be useful to define the cell in a modern sense, so we turn to Bruce Alberts (1989, p. 485), one of the co-authors of a very long-lived textbook detailing the molecular biology of cell.

> A cell is . . . nothing less than a special collection of complex molecules, enclosed by a membrane, that has the very special ability to reproduce itself from the much simpler molecules available in its surroundings. Speaking as a chemist, a cell is a self-replicating collection of catalysts. [Later he adds that to speak of them this way] . . . in no way reduces the beauty and wonder of the living state.

It is hard to say whether the sometimes dry language of science or more evocative poetry provides the better description of the cell but, except for evolution, there

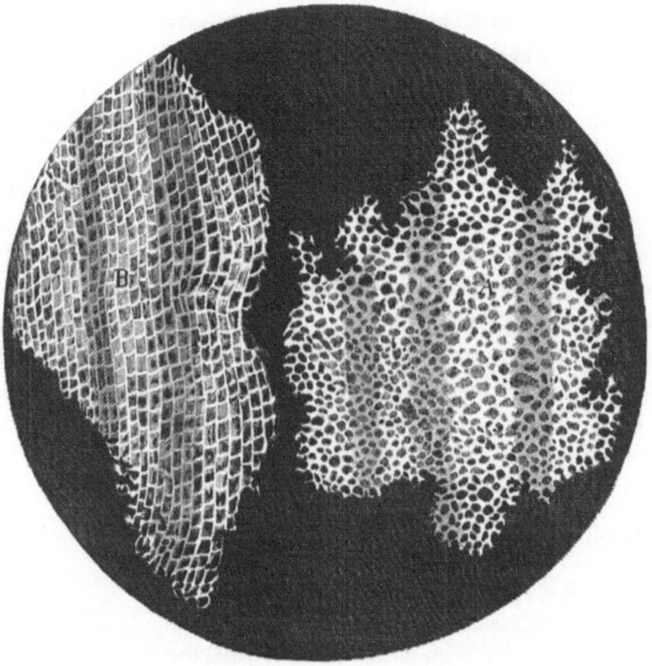

FIGURE 5.1 A drawing of cork cells included in Robert Hookes' *Micrographia* (1665), a classic book illustrating many of the first images visualized through the microscope

is no more fundamental aspect of biology. These tiny packages essentially contain and define what it means to be alive. These microscopic entities may operate individually as single life forms (as with the protists) or in vast numbers, having differentiated to provide highly specialized functions in the tissues and organs of larger organisms. Cells are found in all living things (the status of viruses, which lack cells, is still debated), and it is likely that every science student understands that cells are the basic units and building blocks of structure and function of life itself.

The quote found at the beginning of this chapter is true, but somewhat misleading, as we will see. In 1665, Robert Hooke was the first to gaze through a simple microscope and notice these tiny compartments in a sliver of cork. He named the structures 'cells' after the tiny rooms in which monks reside. However, Hooke knew very little more despite his being credited with a fundamental discovery. For instance, Hooke did not realize that he was looking at cell walls, had no idea that all living things are comprised of cells, did not understand the structure and function of cells, and almost certainly did not consider how these entities might be copied both to replace themselves and to build more involved structures.

Furthermore, he certainly would have been surprised to find that these little rooms define what might be characterized as worlds unto themselves with all the incredible organelles they contain and processes they typically support. Even as the 'father of cell biology', Hooke could have had no idea where his original observation would lead. As Harrish (1999) states, "Hooke did not for a moment suppose that the structures were the residual skeletons of the basic subunits of which all plants and animals were constructed" (p. 1).

With the development of more powerful microscopes and the work of countless cellular biologists, we now know that there are intricate structures (such as mitochondria, chloroplasts, membranes, and specialized vacuoles—at least in eukaryotic cells). These structures and their analogues in prokaryotes (such as bacteria) all support an incredible array of life-sustaining biochemical processes. We now know the cell to be so much more than the tiny 'rooms' noted by Hooke. The cell is "the simplest level of organization the manifests all the features of the phenomenon of life" (Harold, 2001, p. 17). To say it another way, cells are, in effect, highly complex, self-reproducing, living universes.

Cellular Biology: An Educational Perspective

The classic texts in biology for secondary students, starting with the iconic (at least in the U.S.) *Biology for Beginners* by Moon (1921) and similar books of the period, focused on natural history, coupled with what we might label the 'parade of phyla' approach to life science instruction. This consists of multiple chapters, each detailing a group of living things from bacteria through amphibians to mammals, with some similar treatment of plants. This is typically followed by the 'parade of systems', which does the same with topics from nutrition to respiration and the nervous system (reproduction in almost any form was originally absent as a discrete topic) (McComas, 2007). Our area of interest, cell biology, was treated very differently in the early biology texts, which typically featured botany, human physiology, or aspects of natural history. Cells are mentioned only with reference to their discovery and utility in defining single-celled organisms; their anatomy and physiology are not discussed at all.

As biology education developed and, of course, knowledge about the science of biology grew, texts also changed. In two common handbooks for biology teachers, *Educational Biology* (Atwood & Heiss, 1928) and *A Sourcebook for the Biological Sciences* (Morholt et al., 1966), there is a clear cellular trend that continues to this day. More cell-related topics appear in Atwood and Heiss than in earlier guides, and entire chapters of cell-focused material began to appear in biology texts by the mid-1960s, a situation that remains today.

With few exceptions, the 'cell' is the dominant thread that generally directs all introductory biology education. In his review of the history of biology, Francois Jacob, Nobel winner and pioneer molecular biologist, remarked that "with the cell, biology discovered its atom" (1982, p. 121). Wilson (1896), who is often quoted on the importance of the cell, presciently said more than a century ago that the "key to all . . . biological problems must, in the last analysis, be sought in the cell" (p. 1). Therefore, with these facts established regarding the science of biology, I am willing to propose that the cell (along with evolution) has given biological education its pedagogical focus.

A challenge in considering the focus of this present chapter has been to avoid the temptation to conclude that the cell is everything in biology, but I hope to be forgiven for suggesting just that. In reviewing more than a dozen current biology texts designed for secondary education, it seemed that the cell is indeed found almost everywhere, either explicitly mentioned or alluded to implicitly. However, a review of a typical secondary level biology book features basic cell biology in 3–5 early chapters, with the topics featured in Table 5.1 commonly included.

An interesting question regarding this shift from a natural history focus in biology instruction to one that is more cell-focused is why it took so long. Certainly, there are aspects of cell anatomy and physiology that have only relatively recently been discovered and therefore could not logically have appeared earlier, but the delay is particularly puzzling when reviewing the topics in the table of contents in Wilson's (1896) pioneering text in cell biology. Issues well known to scientists at the end of the nineteenth century generally failed to enter many secondary textbooks for decades. A study by Christy (1936) of the most-used biology texts of the period 1907–1935 revealed that the cell was not included on the top ten lists of topics. Hurd's (1961) landmark study of the history of biology instruction in the U.S. shows that of the many goals associated with biology teaching, the cell appeared as a dominant topic only after WWII. Evolution, of course, is the other major notion that should inform biology education, but unlike the cell, it was actively suppressed even into the 1970s thus explaining its absence in the curriculum.

Considering Cell Biology Instruction

As we have found, although in individual classrooms and programs there would be large distinctions in depth and breadth, secondary school cell biology instruction is quite similar in terms of the sub-topics. This level of similarity of goals makes the job of the curriculum designer somewhat easier, but there remains the issue of when these elements might best be taught. In this regard, there is a remarkable tool available called the *Atlas of Science Literacy* (AAAS, 2001). The *Atlas* is essentially a "road map" of when to teach a wide number of science and mathematics subjects, based on a review of students' prior conceptions, and recommendations for what knowledge should be taught first as a foundation for learning related but more advanced topics. The *Atlas* contains, a section on Cell Function and Cells and Organs. Table 5.2 offers a synthesis of these cell-related recommendations for each grade level (AAAS, 2001). The *Atlas* itself should be consulted to visualize the links between elements of cell content, the foundations suggested to support such content, and the links to other science topics (such as chemical reactions, evolution, energy flow) associated with learning about the cell.

This overview of the recommendations in the *Atlas* offers some general assistance as to the order in which key cell topics might best be taught, but this resource does not provide recommendations regarding when, for instance, cellular organelles

TABLE 5.1 Common cell biology topics revealed by a review of several current high school textbooks. Clusters include those topics most frequently linked in a unit of instruction in cell biology

Basic Microscopy (From the perspective of studying the study of the cell)

The Cell Principle

- Prokaryotic and Eukaryotic Cells Compared
- Cells as part of Tissues, Organs, Organ Systems, and Organisms
- Plant and Animal Cells Compared

Structure and Function of Eukaryotic Cells

- Cytoplasm
- Cytoskeleton
- Endo-membrane System (rough and smooth ER)
- Nuclear Envelope
- Nucleus Anatomy and Substructures (chromatin)
- Lysosomes
- Ribosomes
- Mitochondria
- Peroxisomes (single membrane-bound vesicles supporting fatty acid metabolism)
- Chloroplasts (in plants)
- Central Vacuole (in plants)
- Flagella/Cilia

Membranes and Active and Passive Transport (homeostasis)

- Fluid Mosaic Model of the Cell Membrane
- Cellular Junctions (structure and function)
- Passive and Active Transport (ATP)
- Permeability
- Diffusion and Osmosis
- Tonicity (hyper-, hypo-, and isotonic)
- Electrochemical Gradients
- Pumps/Endocytosis/Exocytosis (contractile vacuoles)

Bioenergetics

- Energy and Metabolism
- Heterotrophs vs. Autotrophs
- Potential and Kinetic Energy
- Glycolysis
- Anabolism (small to large; energy required) and Catabolism (large to small; energy is released)
- Citric Acid Cycle
- Oxidative Phosphorylation (in mitochondria)
- Aerobic vs. Anaerobic respiration
- Energy Transport Chains
- Fermentation (lactic acid vs. alcoholic)
- Photosynthesis

- Chloroplasts (and chlorophyll)
- Light and Dark Reactions
- Calvin Cycle (reactions that use energy from the light reactions to form glucose)

Cellular Reproduction

- Stages of Mitosis
- Cell Cycle
- Distinctions Between Prokaryotic and Eukaryotic Cells
- Sexual Reproduction and Meiosis (as compared with mitosis)

TABLE 5.2 An overview of cell-related topics found in the *Atlas of Science Literacy* (AAAS, 2001) summarized from the Cell Function (p. 73 in source) and Cells and Organs (p. 75 in source) maps

Grade Level/Student Age	*Cell-Related Topics Recommended*
Grade K–2 Students who are 5–6 to 7–8 years of age	• Magnifiers can help visualize things impossible to see otherwise • Most things are made of parts • Organisms exchange of food, air, wastes with the environment
Grades 3–5 8–9 to 10–11 years of age	• Some living things are just one cell (but still need to exchange resources with the environment) • Microscopes can help us see cells • Some organisms are single celled; others have many cells (with special functions)
Grades 6–8 11–12 to 13–14 years of age	• Cells contain special structures to engage in exchanges with the environment • All living things are made of cells (one to millions) • Organs and tissues are made of cells • Cell divide to make copies of themselves
Grade 9–12 14–15 to 17–18 years of age	• Cells contain chemical reactions that are impacted by factors such as concentration, pH, temperature, etc. • The properties of molecules can be impacted by the interaction and shape of other molecules • Cells contains specialized sub-units to control waste and energy flow and passing information • Many different molecules (mostly proteins) produced by cells impact processes within and outside the cell • Cells are enclosed in membranes • Proteins have a role in controlling the flow in and out of the cell • Genetic information is encoded in DNA molecules • Different molecules interact to control various cell actions

might be included in educational settings. It is likely, given the complexity of such organelles, that some should be discussed before others. Some guidance on this point would be useful to educators particularly if based upon issues such as learning readiness, student interest, importance and the complexity of ideas to be discussed.

Students' Prior Conceptions Regarding Aspects of Cell Biology

As with all areas of science, there are many well-documented misconceptions in the domain of cell biology. Knowing in advance what students may already think is vital for teachers in the design of effective instructional strategies (Table 5.3). Further, to facilitate diagnosis of misconceptions in advance, many instruments exist, particularly for higher education settings, called concept inventories (CIs) (D'Avanzo, 2008; Libarkin, 2008), including those focused on topics in cell biology. As examples, see Haslam and Treagust (1987) for respiration/photosynthesis, Odom and Barrow (1995) and Fisher et al. (2011) for osmosis/diffusion, and Shi et al. (2010) for issues in molecular cell biology.

A Final Cell Biology Misconception: The Cell 'Theory'

We will end this section on misconceptions with a short excursion into the history of cell biology and its intersection with philosophy. In 1839, zoologist Matthias Jakob Schleiden found differences between plant and animals but determined that they were "fundamentally the same" (Moore, 1993, p. 256). Theodor Schwann offered much the same view about plants from his position as a botanist. Together, they are credited for developing a unified view of the cell with basic tenets that include the notions that 1) all living things are composed of cells and consist of one or more such cells, 2) cells are the basic unit of structure and function in all living things, and 3) that cells come from pre-existing cells (an idea added somewhat later by RudolfVirchow). Harrish (1999, p. xi) concludes, "there can be very few ideas in the history of biology that have had greater influence than what it known as the cell doctrine". Interesting, there is a challenge is regarding what to call this 'idea' of the cell: should it be a theory, doctrine, or even a law?

From a philosophical perspective, laws are generalizations, while theories are explanations for why laws operate in the way that they do (McComas, 2003). In the case of the Schleiden/Schwann idea of the cell, it is clear that their three-part notion is a wonderful generalization. Just reading the three tenets makes this clear. However, these three ideas explain nothing about why cells are the way that they are. Thus, it is clear that "law" or "principle" really is the best name. Curiously, in secondary textbooks the cell idea is almost always misnamed the cell 'theory', while in post-secondary environments the more accurate label cell 'principle' is used. This odd naming issue must be confusing to students and perhaps even contributes to the misconception about the roles of laws and theories.

TABLE 5.3 An overview of common misconceptions associated with cell biology. Those with references were determined through empirical studies. Those lacking references are anecdotal reports appearing in textbooks and on various websites. These have been grouped associated with the principal areas of cell biology that define this aspect of biology instruction

Microscopy

- Anything that is microscopic is approximately the same size, including all cells (although pro- and eukaryotic cells differ greatly in size);
- Molecules and perhaps even atoms can be seen through the microscope (Lee, 1993);
- Anything spherical seen through the microscope is a cell (when in fact many such objects are air bubbles);
- Cells are always too small to see with the unaided eye (this is generally true, but some very large bacteria, amoebae, squid nerve axons, the human ovum, and bird eggs can be seen without a microscope);
- Students confuse cells with atoms and molecules (Arnold, 1983) and/or do not understand the relationship between cells, atoms, and molecules (Sewell, 2002).

Cell Principle and Cell Structure and Function

- Organisms "contain" cells rather than being composed of cells (Driver et al., 1994);
- Cells are two-dimensional (most representations of cells appear 2D, but cells are 3D);
- When single-celled organisms are called 'simple', students think that such cells are prokaryotic (bacteria);
- Single-celled organisms contain organs (perhaps even lungs or intestines) (Dreyfus & Jungwirth, 1988, 1989).

Membranes and Transport (Diffusion and Osmosis)

- Particles stop moving when they are balanced throughout a region (Odom & Barrow, 1995, 2007);
- Particles move because they are crowded (Odom & Barrow, 1995, 2007);
- Particles move with some preference in direction from high to low concentrations (but it is just random) (Zuckerman, 1993);
- Water osmosis at a constant rate (Zuckerman, 1993; Friedler et al., 1987);
- Water moves from hypertonic to hypotonic solutions (Odom & Barrow, 1995);
- Equal concentrations are necessary for osmotic equilibrium to occur (Friedler et al., 1987);
- Osmosis only occurs in living cells placed in hyper- and hypotonic solutions (Odom & Barrow, 1995);
- The central vacuole on plants in a hypertonic salt solution shrinks because salt absorbs the water from the vacuole (Odom & Barrow, 1995).

Bioenergetics

- Plants engage in photosynthesis only (not respiration);
- Respiration is breathing (rather than something at the cellular level) (Bajd et al., 2010);
- The light reaction is just photosynthesis while dark reaction occurs in the dark (night) after the light reaction has occurred or is just respiration by another name (Haslam & Treagust, 1987; Storey, 1989; Amir & Tamir, 1993);
- Plants respire only in the dark (Haslam & Treagust, 1987; Storey, 1989; Amir & Tamir, 1993).

(*Continued*)

TABLE 5.3 (Continued)

Bioenergetics

- Plant cells have chloroplasts and therefore do not have mitochondria (plant cells have both) (Storey, 1991);
- Energy within the cell is used for things like movement and growth (Gayford, 1986; Chabalengula et al., 2012).

Cell Reproduction

- Students think that cells can grow to any size (in fact, the surface to area relationship and the need to move materials in and out of the cell limit the size that can be reached by cells);
- Prokaryotic cells lack a nucleus; therefore, they lack DNA (prokaryotes do not have nuclei but do have DNA);
- DNA is only found in the cell (chloroplasts and mitochondria have their own DNA) (Banet and Ayuso, 2000);
- Thirty-four misconceptions were found in biology preservice teachers related to cell division (most with meiosis), including when DNA replication occurs, when chromosome numbers double, the relationship of centrosome to centriole, and where meiosis takes place (many think it occurs in the sperm or egg cells) (Dikmenli, 2010).

Laboratory Experiences in Support of Cell Biology Instruction

Unlike the basic content of cell biology, the inquiry and/or other hands-on or practical work that students might experience with respect to the cell is more difficult to generalize. This is true because many schools do not typically have students engage laboratory-type work because of limits in available materials. Such experiences are often teacher-determined and rarely specified in curriculum documents or textbooks. Furthermore, in many schools, the potentially dynamic science of biology has been reduced to a non-laboratory lecture class accompanied only by paper and pencil activities to the exclusion of authentic laboratory experiences.

However, in reviewing many sources that offer suggestions for the kinds of hands-on activities that might enhance secondary school biology, there are some trends. Table 5.4 provides a list of the most commonly recommended laboratory experiences to support cell biology instruction at the secondary level. It is encouraging that basic practical work has been developed to support each of the main aspects of cell biology instruction detailed in Table 5.1. Of course, as mentioned, there is no assurance that laboratory activities accompany students' experiences in cell biology education. At the university level, laboratory activities are often more involved particularly in a semester-long cell biology class. As in the secondary school, unfortunately these activities are most frequently confirmatory and based on following specific directions rather than providing opportunities for investigation and experimentation.

TABLE 5.4 Common laboratory (practical) activities often used to support aspects of secondary level cell biology instruction. Note, many cell biology activities are simulations developing and using drawings, ball and stick models, and snap-together chains of beads to represent many of these concepts. Since these do not make use of actual scientific equipment or biological objects, they are not referenced here

- Observing Cells Through the Microscope (usually involves basic microscope technique, specimen preparation, and staining, typically with printed letters of the alphabet, cork slices, onion skin, elodea, various protists, and/or human cheek cells);
- Determining and Comparing Cell Sizes (the issue of magnification);
- Cell Specialization (examining cell types individually or in tissues with prepared slides);
- Cell Size Limitation (the issue of volume to surface area ratios explored with different-sized blocks of gelatin or hard-boiled eggs soaked in dye);
- Osmosis (usually with impermeable and differentially permeable membranes and starch solutions with iodine as an indicator; often accompanied by demonstrations of diffusion);
- Effects of Different Solutions (hyper-, hypo- and isotonic) on cells (typically blood cells, elodea plants, and potato and/or cucumber slice, with the effects judged visually through the microscope and/or qualitatively by look and feel);
- Cellular Respiration (the impact of light, temperature, etc. on germinating peas, which produce CO_2 through the biochemical process of respiration);
- Photosynthesis (the most typical versions of labs involve the freshwater plant elodea under various light and temperate regimes with output measured in CO_2 production; chromatography is often used in introductory biology to separate chlorophyll types);
- Fermentation (glycolysis without oxygen involving yeast and the production of CO_2);
- Stages of Mitosis (either simulated or using onion root tip and/or whitefish blastula prepared slides);
- Cell Cycle (a version of the mitosis laboratory in which students count the number of cells in each mitotic phase with root tip and/or whitefish blastula).

Only rarely is the history of science emphasized in biology instruction, but cell biology has had many contributors and lends itself well to pedagogical exploration through historical re-enactment perhaps accompanied by reading the works of the original contributors. The History of Science Case Study Approach (Klopfer, 1964) includes a unit on *The Cells of Life* (Lind, 1979). Another useful resource book by Gabriel and Fogel (1955) offers an entire section with written reports by Hooke, Dutrochet, Brown, Schwann, Wilson, and Harrison that could frame the basis for laboratory investigations or provide opportunities to read what scientists were thinking at the time of their pivotal discoveries in the life sciences.

Conclusions: Strategies to Improve Cell Biology Instruction

Cell biology and evolution are the two most important concepts in biology and therefore should be major organizing principles in biology education. With respect to the topic of cell biology, there are many resources available to guide instruction. Teachers must understand the basic science and have knowledge of the parameters of the field. Table 5.1 delineates the most important elements involved in the science of cell biology. Of course, it is then important to consider issues of depth and breadth along with learner-readiness. These issues should then be associated with the development of a multi-year instruction plan scaffolding cell biology lessons at increasing levels of sophistication. Suggestions for where specific aspects of cell biology might be taught can be found in the *Atlas of Science Literacy* (AAAS, 2001) as summarized in Table 5.2. Once the goals of instruction are established, educators should take into account what students already know about cell biology. Fortunately, many reports of students' misconceptions are available (Table 5.3).

After considering these initial issues, teachers and curriculum designers can consider *how* to communicate aspects of this fascinating topic in the most engaging way possible. Table 5.4 provides a review of laboratory activities developed to support instruction in cell biology followed by some suggestions for the use of the history of cell biology that might enliven instruction and humanize science teaching.

While it is impossible to synthesize all the suggestions made in the biology education literature about teaching cell biology, it is useful to point out that many specific strategies are offered for teaching specific topics and for teaching cell biology generally. Some particularly useful thoughts include those by Cherif et al. (2016), who suggests role-playing to illustrate the complex nature of the cell; Allen and Tanner's (2002) review of many useful teaching strategies as applied to cell biology; and Veselinovska et al. (2011), Hande and Hande (2009), and Lodish and Rodriguez (2004), who discuss the use of case studies, interactive teaching, and teamwork in teaching cell biology. Allen and Tanner (2003), Yarden et al. (2004), and DiCarlo (2006) specifically advocate concept mapping as an instructional strategy. Tsatsarelis et al. (2000) discuss the role of analogies and metaphors in cell biology, while DebBurman (2002) and Palombi and Jagger (2008) strongly support the use of laboratory activities to support cell biology understanding among students.

Acknowledgment

The author thanks Cameron Simpkins, Lundon Pinneo, and Jenn Oramous for their valuable assistance in preparing this chapter.

References

Alberts, B. M. (1989). Introduction: on the great excitement in cell biology. *American Zoologist*, 29, 483–486.

Allen, D. & Tanner, K. (2002). Approaches in cell biology education. *Cell Biology*, 1, 3–5.

Allen, D. & Tanner, K. (2003). Cell biology and concept mapping. *Cell Biology*, 2, 133–136.

American Association for the Advancement of Science (AAAS). (2001). *Atlas of Science Literacy* (Vol I). Washington, DC: Author.

Amir, R. & Tamir, P. (1993). The "light" and "dark" reactions of photosynthesis: terminology as a source of misconceptions in The *Proceedings of the Third International Seminar on Misconceptions and Educational Strategies in Science and Mathematics*, Ithaca, NY: Misconceptions Trust.

Arnold, B. (1983). Beware the molecell! *Biology Newsletter* (Aberdeen College of Education), 42, 2–6.

Atwood, W. & Heiss, E. (1928). *Educational Biology*. Philadelphia: P. Blakiston's Son and Co.

Bajd, B., Praprotnik, L. & Matyasek, J. (2010). Students' ideas about respiration: a comparison of Slovak and Czech student. *School Health*, 21, 248–251.

Banet, E. & Ayuso, E. (2000). Teaching genetics at secondary school: a strategy for teaching about the location of inheritance information. *Science Education*, 84(3), 313–351.

Chabalengula, V., Sanders, M. & Mumba, F. (2012). Diagnosing students' understanding of energy and its related concepts in biological content. *International Journal of Science and Mathematics Education*, 10(2), 241–266.

Cherif, A., Siuda, J., Jedlicka, D., Bondoc, J. & Movahedzadeh, F. (2016). Not all the organelles of living cells are equal! or are they? engaging students in deep learning and conceptual change. *Journal of Education and Practice*, 7(17), 74–86.

Christy, O. B. (1936). *The Development of the Teaching of General Biology in the Secondary School*. Unpublished dissertation. Nashville, TN: George Peabody College for Teachers (Vanderbilt University).

D'Avanzo, C. (2008). Biology concept inventories: overview, status, and next steps. *BioScience*, 58(11), 1079–1085.

DebBurman, S. (2002). Learning how scientists work: experiential projects to promote cell biology learning and scientific process skills. *Cell Biology*, 1, 154–172.

DiCarlo, S. (2006). Cell biology should be taught as science is practiced. *Nature Review (Molecular Cell Biology)*, 7, 290–296.

Dikmenli, M. (2010). Misconceptions of cell division held by student teachers in biology: a drawing analysis. *Scientific Research and Essay*, 5(2), 235–247.

Dreyfus, A. & Jungwirth, E. (1988). The cell concept of 10th graders: curricular expectations and reality. *International Journal of Science Education*, 10(2), 221–229.

Dreyfus, A. & Jungwirth, E. (1989). The pupil and the living cell: a taxonomy of dysfunctional ideas about an abstract idea. *Journal of Biological Education*, 23, 49–55.

Driver, R., Squires, A., Rushworth, P. & Wood-Robinson, V. (1994). *Making Sense of Secondary Science: Research Into Children's Ideas*. London: Routledge.

Fisher, K., Williams, K. & Lineback, J. (2011). Osmosis and diffusion conceptual assessment. *CBE—Life Science Education*, 10, 418–429.

Friedler, Y., Amir, R. & Tamir, P. (1987). High school students' difficulties in understanding osmosis. *International Journal of Science Education*, 9(5), 541–551.

Gabriel, M. L. & Fogel, S. (1955). *Great Experiments in Biology*. Englewood Cliffs, NJ: Prentice-Hall, Inc.

Gayford, C. (1986). Some aspects of the problems of teaching about energy in school biology. *European Journal of Science Education*, 8(4), 443–450.

Hande, S. & Hande, M. (2009). A handy interactive class for teaching introductory cell biology. *The Journal of Effective Teaching*, 9(1), 48–58.

Harold, F. M. (2001). *The Way of the Cell: Molecules, Organisms and the Order of Life*. New York: Oxford University Press.

Harrish, H. (1999). *The Birth of the Cell*. New Haven, CT: Yale University Press.

Haslam, F. & Treagust, D. F. (1987). Diagnosing secondary students' misconceptions of photosynthesis and respiration in plants using a two-tier multiple-choice instrument. *Journal of Biological Education*, 21, 203–211.

Hooke, R. (1665). *Micrographia: Or Some Physiological Descriptions of Minute Bodies Made by Magnifying Glasses With Observations and Inquiries Thereupon*. London: Martyn and Allersry. Downloaded from https://ceb.nlm.nih.gov/proj/ttp/flash/hooke/hooke.html

Hurd, P. De. (1961). *Biological Education in American Secondary Schools 1890–1960*. Washington, DC: American Institute of Biological Sciences (ERIC 010 991).

Jacob, F. (1982). *The Logic of Life*. New York: Pantheon Books.

Klopfer, L. E. (1964). *History of Science Cases (HOSC)*. Chicago, IL: Science Research Associates.

Lee, O., et al. (1993). Changing middle school students' conceptions of matter and molecules. *Journal of Research in Science Teaching*, 30, 249–270.

Libarkin, J. (2008). Concept inventories in higher education science. A manuscript prepared for the National Research Council Promising Practices in Undergraduate STEM Education meeting, 13–14 October 2008.

Lind, G. (1979). The history of science cases (HOSC): nine units of instruction in the history of science. *European Journal of Science Education*, 1(3), 293–299.

Lodish, H. & Rodriguez, R. (2004). A combination of lecture, problem sets and recitation sections is an excellent way to teach undergraduate cell biology at a high level. *Cell Biology Education*, 3, 202–204.

McComas, W. F. (2003). A textbook case: laws and theories in biology instruction. *International Journal of Science and Mathematics Education*, 1(2), 1–15.

McComas, W. F. (2007). Biology education under the microscope: examining the history and current state of biology education. *The Science Teacher*, 74(7), 28–33.

Moon, T. J. (1921). *Biology for Beginners*. New York: Henry Holt and Company.

Moore, J. A. (1993). *Science as a Way of Knowing: The Foundations of Modern Biology*. Cambridge, MA: Harvard University Press.

Morholt, E., Brandwein, P. F. & Alexander, J. (1966). *A Sourcebook for the Biological Sciences*, 2nd edn. New York: Harcourt, Brace and World, Inc.

Odom, A. & Barrow, L. (1995). Development and application of a two-tier diagnostic test measuring college biology students' understanding of diffusion and osmosis after a course of instruction. *Journal of Research in Science Teaching*, 32(1), 45–61.

Odom, A. & Barrow, L. (2007). Higher school biology students' knowledge and certainty about diffusion and osmosis concepts. *School Science and Mathematics*, 107(3), 94–101.

Palombi, P. & Jagger, K. (2008). Learning about cells as dynamic entities: an inquiry-deriver cell culture project. *Bioscene*, 34(2), 27–33.

Sewell, A. (2002, June). Cells and atoms—are they related? *Australian Science Teachers Journal*, 48(2), 26–30.

Shi, J., Wood, W., Martin, J., Guild, N., Vicens, Q. & Knight, J. (2010). A diagnostic assessment of introductory molecular cell biology. *CBE—Life Science Education*, 9, 453–461.

Storey, R. D. (1989). Textbook errors and misconceptions in biology: photosynthesis. *American Biology Teacher*, 51(5), 271–274.

Storey, R. D. (1991). Textbook errors and misconceptions in biology: cell metabolism. *American Biology Teacher*, 53(6), 339–343.

Tsatsarelis, C., Ogborn, J., Jewitt, C. & Kress, G. (2000). Rhetorical construction of cells in science and in a science classroom. *Research in Science Education*, 30(4), 451–463.

Veselinovska, S. Gudeva, L. & Djokic, M. (2011). Applying appropriate methods for teaching cell biology. *Procedia Social and Behavioral Science*, 15, 2837–2842.

Wilson, E. B. (1896). *The Cell in Development and Heredity*. New York: Macmillan.

Yarden, H., Marback-Ad, G. & Gershoni, J. (2004). Using concept map technique in teaching introductory cell biology to college freshmen. *Bioscene*, 30(1), 3–13.

Zuckerman, J. T. (1993, April 17). Accurate and inaccurate conceptions about osmosis that accompanied meaningful problem solving. Paper presented at the annual meeting of the *National Association for Research in Science Teaching*, Atlanta, GA (ED360176).

6

HUMAN BIOLOGY

Orit Ben Zvi Assaraf and Zohar Snapir

Introduction

The science education community emphasizes the value of engaging in systems thinking in the context of both structure and function. This awareness of systems thinking's relevance has, in turn, led to a growing emphasis on the topic of complex systems in science education. The human body is a complex system that is highly relevant to students, and engaging them in its complexity can therefore promote not only the contextual understanding of the human body but also a more generalized scientific conception of how complex systems operate.

Studies have shown, however, that the complexity of the human body's systemic nature is challenging for students to understand. They have shown, for instance, that high school students have difficulty representing the cellular level of the human body, which restricts their capacity to grasp the body's overall complexity (Ben Zvi Assaraf et al., 2013; Verhoeff et al., 2008). In another example, Zion and Klein (2015) noted that high school students' erroneous perceptions of homeostasis in the human body were limiting their understanding of the body as a complex system. In this chapter, we will review several principles for overcoming the challenge of teaching high school biology students about the human body, and suggest approaches for portraying the human body as a complex system that should make the topic more understandable.

Complex Systems

Complex systems are prevalent in many scientific fields, and at all scales—from the micro scale of a single cell (such as a human fertilized egg) to macro complex systems such as cities or ecosystems (Yoon et al., 2017). In the field of biology, complex systems are found within individual cells, in which complex molecular networks connect signals from the extra-cellular environment to intracellular responses (Chasman et al., 2016). They are also found in individual organisms at the level of physiology, maintaining homeostasis in a variety of different dimensions, like blood sugar and body weight (Rosenbaum & Leibel, 2016). On an even larger scale, interactions between organisms form additional complex systems at the level of biological societies, such as social insect colonies (ants, for example), which are formed by interactions between organisms, each of which is a complex system of its own (Greenwald et al., 2015). Such natural systems are typically dynamic and are often held in states of equilibrium with other interdependent systems, and the interactions within and between them can be unpredictable. Disturbing this web of interconnections can have major implications, as effects cascade across associated networks (Stewart, 2012). Understanding the complexity of natural systems is therefore crucial to a proper understanding of what they are and how they work.

Complex systems are comprised of sub-systems, exhibiting outcomes that arise from the interactions between their structures. These outcomes are the results of causal elements participating in causal event chains that vary in temporal and spatial dimensions (Keynan et al., 2014). Thus, a complex system can be generally described as an entity consisting of a large number of structures, at different levels of organization, in which various processes occur. A complex system functions through the mutual interaction of its various components, which work together to bring about a singular function. The interactions within the system are influenced by feedback loops and are usually nonlinear (Ladyman et al., 2013).

Though biological knowledge of the human body requires familiarity with a wide variety of facts and principles, in the context of systems this variety is customarily centered upon three central system characteristics: hierarchy, homeostasis and dynamism (Tripto et al., 2016). These three characteristics are manifested in all complex systems. The human circulatory system, for example, is a complex system composed of many kinds of cells (representing a micro level of the system), which form the tissues of the organs that work together as a system (at the macro level). The blood is composed of several different kinds of cells suspended in plasma. All of these are components of the system, representing the various levels of its *hierarchy*. The system's functions include cells and plasma, transporting oxygen, carbon dioxide and nutrients, fighting infections and clotting, all representing matter transported both within the system and to other body systems (i.e. its *dynamism*). The blood is circulated in the body through various blood vessels—arteries, veins and capillaries—each of which has a different structure that serves its function, and the heart pumps the blood by creating pressure differentials that drive movement through the blood vessels (Marieb & Hoehn, 2013).

Adding to this complexity, the circulatory system constantly interacts with other systems in the body. These interactions with other body systems are what enables the body's *homeostasis*, which refers both to the maintenance of a stable internal environment and to the regulatory processes (operating via feedback loops) that lead to that stability. What happens in the body during physical exercise, for example, highlights the interactions between the circulatory and the gaseous exchange systems, which are integrated by the nervous system. At the onset of physical exercise, elevated carbon dioxide and H^+ levels serve as signals at the micro level. These signals are processed in the nervous system, which in turn influences the pacemaker in the heart at the macro level, so that the heart beats faster. Blood flow to the working muscles is also increased (Sherwood, 2015).

At the same time, the gaseous exchange system is also influenced by these signals, so that the respiration rate increases. Respiration occurs constantly throughout life, enabling the exchange of oxygen and carbon dioxide in the lungs' alveoli, regulating their levels in their body. The breathing movements are driven by neural circuits in the central nervous system (i.e. the brainstem and the spinal cord). Respiration is under precise control, so gas exchange is matched to the metabolic demand in the body tissues. This is true during resting respiration and also during times of increased metabolism and demand for oxygen, such as during physical exercise (Bell, 2006; Smith et al., 2013). As this example shows, exercise promotes a complex chain of integrated changes that enable the delivery of sufficient oxygen and nutrients to the working muscle. As a result of this integrated activity, during physical exercise more oxygen is delivered to the mitochondria in the cells, and cellular respiration provides enough ATP to fulfil the tissue demand. This example serves as a glimpse into the great complexity and interconnectivity that enables the human body to maintain homeostasis.

Complex systems, such as the human body, are often difficult not only for students to perceive but also for teachers to explain. In the following sections, we will offer several principles for overcoming the challenges encountered while studying complex systems.

Conceptual Representation

The first principle to teaching complex systems like the human body is that it is important to present them within a conceptual framework that addresses, expresses and organizes all of the system's components and the relationships between them. Such conceptual representations can not only help students organize their ideas, but might also make it possible to identify differences in the extent of individuals' system thinking capacity, and of the development of these capacities within each learner.

Several studies have shown that combining such conceptual representation with models promotes deeper systems understanding. This has been demonstrated in the context of a variety of systems, such as ecosystems (Hmelo-Silver et al., 2017), the structure and behavior of honeybees within honeybee system (Danish, 2014) and the human body (Tripto et al., 2016). Reinagel and Speth (2016) have shown that conceptual representation promotes systems understanding at the molecular level as well. They demonstrated that engaging students in building conceptual models to explain how genes determine phenotypes promoted their understanding of this phenomenon. Such an understanding reflects a broader understanding of the connection between the micro and macro levels, and thus of the hierarchy of complex systems. Another study employed visual models that represent scientific phenomena in several complex systems, including the human body, thus promoting students' understanding of dynamism in complex systems (Yoon et al., 2016).

One conceptual framework was proposed by Hmelo-Silver et al. (2017), who modified the earlier Structure Behavior Function (SBF) model (Goel et al., 1996) to create the Components Mechanisms Phenomena (CMP) conceptual framework. This framework entails a perception of all of a system's categories, including the structures (components) within the system and the specific processes and interactions (mechanisms) that occur between them, which lead to the phenomena—the macro scale of processes and patterns within a system. Their study combined the CMP conceptual representation with a curriculum that raises a critical question about a pond ecosystem, encouraging learners to provide mechanistic explanations. The results of the study showed that this approach deepened the participants' systems understandings, as expressed in their improved identification of structures and their ability to provide causal mechanistic explanations in relation to phenomena. This suggests that combining conceptual representation with modeling supports increased systems understanding.

Snapir et al. (2017) used the CMP framework to assess the development of systems understanding of the human body in high school biology students. They adapted and expanded the CMP model by further dividing each CMP dimension into multiple, increasingly complex levels, and evaluated the development of human body systems understanding in high school students. In this longitudinal study, the development of high school biology students' understanding of body systems was manifested by a shift from the basic categories to the more complex ones in each of the CMP model's constituents (Components, Mechanisms, Phenomena), indicating that this framework can serve not only to teach, but also to assess the development of systems understanding.

Alongside the SBF and CMP models, an additional model was developed by Ben Zvi Assaraf and Orion (2005)—the Systems Thinking Hierarchy (STH) model. The STH model proposes that the way students think about and understand a system can be categorized according to eight hierarchical characteristics or abilities, which can be arranged in ascending order of advancement into three sequential levels: (A) analyzing the system components; (B) synthesizing system components; and (C) implementation. The eight characteristics are:

1. Identifying the components and processes of a system (level A).
2. Identifying simple relationships among a system's components (level B).
3. Identifying dynamic relationships within the system (level B).

4. Organizing the system's components, their processes and their interactions within a framework of relationships (level B).
5. Identifying matter and energy cycles within the system (level B).
6. Recognizing hidden dimensions of the system (level C).
7. Making generalizations about a system and identifying patterns (level C).
8. Thinking temporally (level C).

The STH model has been used to assess students' systems understanding in various systems and at various scales, ranging from the macro level of ecological systems (Keynan et al., 2014) to the micro level of genes determining phenotype (Reinagel & Speth, 2016). The STH model has also been used in several studies to assess students' understanding of the human body. Tripto et al. (2017), for example, analyzed students' development of human body systems understanding and showed their progression over the three years of high school biology studies, based on the STH model. The participants of this study showed more complex system models by the end of the study. This was manifested in hierarchy levels ranging from the molecular to the system level, and by an increase in references to dynamic interactions.

These studies indicate that using a conceptual framework as a means of representing the human body can serve as a pedagogical tool for promoting students' understanding of the human body, and for evaluating their progress.

Externalizing Mental System Models

The second principle for teaching complex systems is that students should be given a means of externalizing and examining their own mental system models. Such tools can help both researchers and educators identify and assess patterns and developments in students' thinking. This in turn can provide educators with opportunities to present biological systems to students in ways that improve their skills, develop their meta-cognition and help them apply and adapt their mental models in new contexts (Dauer et al., 2013). Knowing how students organize, store and make use of their biological knowledge can help teachers design teaching that fosters the development of strong cognitive structures throughout the learning process (Dauer & Long, 2015). These structures will serve students as a foundation upon which to build a broader base of biological knowledge (Dauer et al., 2013).

There are several methods for externalizing students' mental models. Reiss and Tunnicliffe (2001) used drawing as means to represent their participants' mental model of the human body. By employing this method with students of various age groups, ranging from 4 to 20 years old, they showed an increase in the students' understanding as their age progressed. More recently, another study examined 8th grade students' understanding of blood circulation. They used a diagram-drawing task as a group activity, revealing students' group models as well as individual models, as the students drew the diagram while engaging in verbal interaction (Lee & Kim, 2014).

Concept maps are another means of externalizing mental models. They are diagrams that indicate interrelationships between concepts and represent conceptual frameworks within a specific domain of knowledge (Novak, 1990). Several science education studies have used concept maps to assess students' conceptual understanding of abstract concepts like hierarchy and homeostasis in the human body (Chang, 2007; Henige, 2012).

Concept maps have been used at various educational levels. Raved and Yarden (2014), for example, used them to evaluate systems thinking competence amongst middle school students in the context of the circulatory system. Concept maps were also used to assess and facilitate knowledge integration in high school biology students (Schwendimann & Linn, 2016) and in undergraduate university students (Southard et al., 2016). Several studies have specifically employed concept maps as a means of externalizing and analyzing high school students' mental models of the human body as a system (Ben Zvi Assaraf et al., 2013; Tripto et al., 2013). Tripto et al. (2017) went beyond examining and demonstrating the use of the concept map as a means of externalizing mental models, using several successive concept maps to discover patterns in students' learning progression over time. In doing so, they also demonstrated that the act of reflectively comparing their own successive concept maps to one another was another means of promoting students' systems understanding.

Another tool for externalizing personal mental models is the Repertory Grid (RG) Technique, based on Kelly's personal construct psychology. Kelly's personal construct psychology states that the world is 'perceived' by a person in terms of the 'meaning' that person applies to it. The bases upon which people develop their personality, attitudes and concepts and perceive reality are their systems of *personal constructs*. Kelly developed a methodology for exploring these systems of personal constructs with repertory grids (RG). This technique is a form of highly structured interviews, assigning relationships to personal constructs and given objects of discourse (Kelly, 1955).

Kelly's repertory grid technique is used in educational research to explore learners' perceptions through the personal constructs they create. It has been used to provide insight about subjects' abilities to identify dynamic relationships within systems, to make generalizations and to identify hidden dimensions of systems. Repertory grids have been used, for example, to track students' understanding of earth systems (Ben Zvi Assaraf & Orion, 2005) and of ecological systems (Keynan et al., 2014). Snapir et al. (2017) used the Repertory Grid as a tool for evaluating the development of systems understanding in the context of the human body, as reflected in the study participants' position within the CMP framework (see above). In the process of creating the repertory grids, the students were asked to explain their perceptions, which also gave them an opportunity to reflect upon and increase their understanding. Repertory grids can thus serve not only as a means of presenting students' understanding of complex systems like the human body, but also to promote systems understanding.

In light of the literature presented in this section, we suggest that using tools such as drawings, concept maps or repertory grids to help students externalize and examine their own systems understanding during the learning process can promote their learning of complex systems, including the human body.

Explicit Guidance

The third principle for addressing the challenge of systems understanding is that students can benefit greatly from explicit guidance in using system language and in drawing clear connections between the various aspects of the complex system. Several studies have shown that explicit scaffolding promotes learning more than implicit instruction, since explicit scaffolding requires less effort on the part of learners in managing and self-regulating their learning (Azevedo et al., 2004; Quintana et al., 2005). Studies also indicate that explicit teaching promotes higher-order thinking strategies. Liu and Hmelo-Silver (2009) showed that without explicit scaffolding, students have difficulty developing systems thinking on their own, and Zion and Klein (2015) noted that without explicit teaching that emphasizes the connection between micro and macro levels in the system's hierarchy, students have difficulty seeing the interactions that make homeostasis possible.

Various studies have shown how explicit teaching helps students draw important connections that are necessary to understanding how complex systems work. Hmelo-Silver et al. (2017), who used an explicit approach to teach phenomena in a complex system, showed that providing students with an explicit mental framework through an intervention that specifically targets thinking about system relations can help them to see more than one link between the mechanism and the component. Following the intervention, the students moved from describing an aquatic ecosystem as a linear predator-prey system to describing an expanded food chain that also included references to symbiotic relationships. Another study showed that explicit teaching encourages students to draw connections between different topics in biology, such as genetics and evolution. Such connections require students to understand the connection between microscopic interactions, such as molecules and genes, and the ways these interactions are manifested at the organism macro level (Tsui & Treagust, 2013). Explicit teaching was also recommended by Kalinowski et al. (2010) and by Speth et al. (2014). Both of these studies refer to students' difficulty connecting genetics at the molecular level to the macro phenomenon of evolution.

Prompting is one type of support that can promote students' metacognitive thinking. Tripto et al. (2016), for example, used the concept maps prepared by the study participants as prompts to guide them in reflecting on their knowledge of the human body as a complex system. This study employed explicit teaching in the form of a metacognitive reflective interview, which is a knowledge integration activity that helps students reflect on their own work and make fuller use of systems language through a series of explicit, specially designed questions. Their interview employed a comparative scaffolding strategy, in which the students compared two concept maps they had created at different points earlier in the learning process. The concept maps served as scaffolding for the students, allowing them to see, verbalize and assess their own thinking as the interview progressed. The results of the study showed that the interview's explicit intervention led to improved expressions of the system patterns 'homeostasis', 'hierarchy' and 'dynamism', thus promoting the students' systems understanding. In light of these findings, we suggest that teachers should provide explicit teaching for specific ideas that encourage students to engage in metacognitive thinking processes.

Reinforcing the Connections Between the System Organization Levels

The fourth principle of teaching the human body is emphasizing the connections between its hierarchical levels. Connecting levels of biological organization is challenging for many students, and the literature has suggested several forms of pedagogical scaffolding to help students make the connections between the organizational levels more coherent, thus promoting their systems understanding.

Verhoeff et al. (2008) claim that systems thinking competence requires the ability to distinguish between different levels of organization. They refer to interrelating concepts at the cellular level as 'horizontal coherence', and to interrelating cellular concepts with higher levels of organization as 'vertical coherence'. Raved and Yarden (2014) showed that focusing on one system and promoting its vertical coherence (between different levels of organization) also improved horizontal coherence (at the same organization level of different system). In their study, they focused on the human circulatory system, explicitly reinforcing the connections between its different hierarchical organizational levels. The participants of the study, 7th grade students, showed not only improved understanding of the dynamic connections between the circulatory system components at different levels of organization (demonstrating improved vertical coherence), but also improved understanding of the connection between the circulatory and the gaseous exchange system, representing vertical coherence.

This approach is consistent with the 'yo-yo' strategy suggested by Knippels (2002), based on the notion that understanding the connection between the micro and macro levels is the basis for understanding the processes and the mechanisms in biological systems. The yo-yo strategy makes a distinction between biological organization levels—from the full-organism level to the micro level—and moves back and forth between the levels to make explicit connections between concepts and phenomena at different levels. A method consistent with this strategy was recently used in a study employing a student-centered approach for improving high school students' vertical coherence in the context of evolution, in which the students were engaged in a lab activity that guided them in thinking across levels (Jördens et al., 2016). Similarly, Van Mil et al. (2013) demonstrated that extending Knippels' yo-yo strategy down to the cellular and molecular level, and engaging students in reasoning across biological levels, can encourage them to draw connections between molecular knowledge and phenomena that occur at the level of cells. For example, Van Mil et al. (2016) examined high school students' mechanistic reasoning between the organism and cellular level. The students were asked to explain the phenomena of disease in the human body by providing a mechanistic explanation at the level of the cell. Thus, the connection between the levels of biological organization can be further emphasized by focusing on the molecular mechanisms that enable high level phenomena.

In conclusion, we suggest that visual models (such as drawings, concept maps or repertory grids), presented within a conceptual framework, can help reveal particular points of difficulty in individual students' system understanding, and thus can also help educators to explicitly guide students in overcoming these obstacles and attaining a more meaningful understanding of the human body as a system. Finally, teachers should be provided with learning materials that explicitly reinforce the connections between the system's different hierarchical organizational levels and the connections between the human body's various sub-systems.

Despite the centrality of complex systems in both the NGSS (2013) and science education research worldwide, in the context of the human body design-based research is still largely absent. Such research is required to bridge the gap between cognitive studies that demonstrate how students understand the human body as a complex system and pedagogical approaches that are designed to promote such understanding.

Acknowledgement

This research was supported by the Israeli Science Foundation Research Grant Application no. 718/11

References

Azevedo, R., Cromley, J. G. & Seibert, D. (2004). Does adaptive scaffolding facilitate students' ability to regulate their learning with hypermedia? *Contemporary Educational Psychology*, 29(3), 344–370.

Bell, H. J. (2006). Respiratory control at exercise onset: an integrated systems perspective. *Respiratory Physiology & Neurobiology*, 152(1), 1–15.

Ben Zvi Assaraf, O., Dodick, J. & Tripto, J. (2013). High school students' understanding of the human body system. *Research in Science Education*, 43(1), 33–56.

Ben Zvi Assaraf, O. & Orion, N. (2005). Development of system thinking skills in the context of earth system education. *Journal of Research in Science Teaching*, 42(5), 518–560.

Chang, S. N. (2007). Externalising students' mental models through concept maps. *Journal of Biological Education*, 41(3), 107–112.

Chasman, D., Siahpirani, A. F. & Roy, S. (2016). Network-based approaches for analysis of complex biological systems. *Current Opinion in Biotechnology*, 39, 157–166.

Danish, J. A. (2014). Applying an activity theory lens to designing instruction for learning about the structure, behavior, and function of a honeybee system. *Journal of the Learning Sciences*, 23(2), 100–148.

Dauer, J. T. & Long, T. M. (2015). Long-term conceptual retrieval by college biology majors following model-based instruction. *Journal of Research in Science Teaching*, 52(8), 1188–1206.

Dauer, J. T., Momsen, J. L., Speth, E. B., Makohon-Moore, S. C. & Long, T. M. (2013). Analyzing change in students' gene-to-evolution models in college-level introductory biology. *Journal of Research in Science Teaching*, 50(6), 639–659.

Goel, A. K., Gomez de Silva Garza, A. G., Grué, N., Murdock, J. W., Recker, M. M. & Govindaraj, T. (1996). Towards design learning environments—I: exploring how devices work. In Fraisson, G. Gauthier & A. Lesgold (Eds.), *Intelligent Tutoring Systems: Lecture Notes in Computer Science*. Berlin: Springer-Verlag, 493–501.

Greenwald, E., Segre, E. & Feinerman, O. (2015). Ant trophallactic networks: simultaneous measurement of interaction patterns and food dissemination. *Scientific Reports*, 5, 12496.

Henige, K. (2012). Use of concept mapping in an undergraduate introductory exercise physiology course. *Advances in Physiology Education*, 36(3), 197–206.

Hmelo-Silver, C. E., Jordan, R., Eberbach, C. & Sinha, S. (2017). Systems learning with a conceptual representation: a quasi-experimental study. *Instructional Science*, 45(1), 53–72.

Jördens, J., Asshoff, R., Kullmann, H. & Hammann, M. (2016). Providing vertical coherence in explanations and promoting reasoning across levels of biological organization when teaching evolution. *International Journal of Science Education*, 38(6), 960–992.

Kalinowski, S. T., Leonard, M. J. & Andrews, T. M. (2010). Nothing in evolution makes sense except in the light of DNA. *CBE-Life Sciences Education*, 9(2), 87–97.

Kelly, G. A. (1955). *The Psychology of Personal Constructs. Volume 1: A Theory of Personality*. New York: WW Norton and Company.

Keynan, A., Ben Zvi Assaraf, O. & Goldman, D. (2014). The repertory grid as a tool for evaluating the development of students' ecological system thinking abilities. *Studies in Educational Evaluation*, 41, 90–105.

Knippels, M. C. P. J. (2002). *Coping With the Abstract and Complex Nature of Genetics in Biology Education: The Yo-Yo Learning and Teaching Strategy*. Doctoral dissertation. Utrecht, The Netherlands: University of Utrecht.

Ladyman, J., Lambert, J. & Wiesner, K. (2013). What is a complex system? *European Journal for Philosophy of Science*, 3(1), 33–67.

Lee, S. & Kim, H. B. (2014). Exploring secondary students' epistemological features depending on the evaluation levels of the group model on blood circulation. *Science & Education*, 23(5), 1075–1099.

Liu, L. & Hmelo-Silver, C. E. (2009). Promoting complex systems learning through the use of conceptual representations in hypermedia. *Journal of Research in Science Teaching*, 46(9), 1023–1040.

Marieb, E. N. & Hoehn, K. (2013). *Human Anatomy & Physiology*. Boston: Pearson Education.

NGSS Lead States. (2013). *Next Generation Science Standards: For States, By States*. Washington, DC: The National Academies Press.

Novak, J. D. (1990). Concept maps and Vee diagrams: two metacognitive tools for science and mathematics education. *Instructional Science*, 19(1), 29–52.

Quintana, C., Zhang, M. & Krajcik, J. (2005). A framework for supporting metacognitive aspects of online inquiry through software-based scaffolding. *Educational Psychologist*, 40(4), 235–244.

Raved, L. & Yarden, A. (2014). Developing seventh grade students' systems thinking skills in the context of the human circulatory system. *Frontiers in Public Health*, 2. doi: 10.3389/fpubh.2014.00260

Reinagel, A. & Speth, E. B. (2016). Beyond the central dogma: model-based learning of how genes determine phenotypes. *CBE-Life Sciences Education*, 15(1), ar4.

Reiss, M. J. & Tunnicliffe, S. D. (2001). Students' understandings of human organs and organ systems. *Research in Science Education*, 31(3), 383–399.

Rosenbaum, M. & Leibel, R. L. (2016). Models of energy homeostasis in response to maintenance of reduced body weight. *Obesity*, 24(8), 1620–1629.

Schwendimann, B.A. & Linn, M. C. (2016). Comparing two forms of concept map critique activities to facilitate knowledge integration processes in evolution education. *Journal of Research in Science Teaching*, 53(1), 70–94.

Sherwood, L. (2015). *Human Physiology: From Cells to Systems*. Belmont, CA: Cengage learning.

Smith, J. C., Abdala, A. P., Borgmann, A., Rybak, I. A. & Paton, J. F. (2013). Brainstem respiratory networks: building blocks and microcircuits. *Trends in Neurosciences*, 36(3), 152–162.

Snapir, Z., Eberbach, C., Ben Zvi Assaraf, O., Hmelo-Silver, C. & Tripto, J. (2017). Characterizing the development of understanding the human body systems in high school biology students—a longitudinal study. *International Journal of Science Education*, 39(15), 2092–2127.

Southard, K., Wince, T., Meddleton, S. & Bolger, M. S. (2016). Features of knowledge building in biology: understanding undergraduate students' ideas about molecular mechanisms. *CBE-Life Sciences Education*, 15(1), p.ar7.

Speth, E. B., Shaw, N., Momsen, J., Reinagel, A., Le, P., Taqieddin, R. & Long, T. (2014). Introductory biology students' conceptual models and explanations of the origin of variation. *CBE-Life Sciences Education*, 13(3), 529–539.

Stewart, M. (2012). Joined up thinking? evaluating the use of concept-mapping to develop complex system learning. *Assessment & Evaluation in Higher Education*, 37(3), 349–368.

Tripto, J., Ben Zvi Assaraf, O. & Amit, M. (2013). Mapping what they know: concept maps as an effective tool for assessing students' systems thinking. *American Journal of Operations Research*, 3(1), 245–258.

Tripto, J., Ben Zvi Assaraf, O., Snapir, Z. & Amit, M. (2016). The 'What is a system' reflection interview as a knowledge integration activity for high school students' understanding of complex systems in human biology. *International Journal of Science Education*, 38(4), 564–595.

Tripto, J., Ben Zvi Assaraf, O., Snapir, Z. & Amit, M. (2017). How is the body's systemic nature manifested amongst high school biology students? *Instructional Science*, 45(1), 73–98.

Tsui, C.Y. & Treagust, D. F. (2013). Introduction to multiple representations: their importance in biology and biological education. In D. F. Treagust & C.Y. Tsui (Eds.), *Multiple Representations in Biological Education*. Dordrecht: Springer, 3–18.

Van Mil, M. H., Boerwinkel, D. J. & Waarlo, A. J. (2013). Modelling molecular mechanisms: a framework of scientific reasoning to construct molecular-level explanations for cellular behaviour. *Science & Education*, 22(1), 93–118.

Van Mil, M. H., Postma, P.A., Boerwinkel, D. J., Klaassen, K. & Waarlo, A. J. (2016). Molecular mechanistic reasoning: toward bridging the gap between the molecular and cellular levels in life science education. *Science Education*, 100(3), 517–585.

Verhoeff, R. P., Waarlo, A. J. & Boersma, K. T. (2008). Systems modelling and the development of coherent understanding of cell biology. *International Journal of Science Education*, 30(4), 543–568.

Yoon, S. A., Anderson, E., Koehler-Yom, J., Evans, C., Park, M., Sheldon, J., Schoenfeld, I., Wendel, D., Scheintaub, H. & Klopfer, E. (2017). Teaching about complex systems is no simple matter: building effective professional development for computer-supported complex systems instruction. *Instructional Science*, 45(1), 99–121.

Yoon, S. A., Anderson, E., Klopfer, E., Koehler-Yom, J., Sheldon, J., Schoenfeld, I., Wendel, D., Scheintaub, H., Oztok, M., Evans, C. & Goh, S. E. (2016). Designing computer-supported complex systems curricula for the next generation science standards in high school science classrooms. *Systems*, 4(4), 38.

Zion, M. & Klein, S. (2015). Conceptual understanding of homeostasis. *International Journal of Biology Education*, 4(1), 1–27.

7

HEALTH AND DISEASE

Jenny Byrne and Marcus Grace

Introduction

Schools have long been identified as an important setting to promote health worldwide (WHO, 1993), but health education in many countries is not regarded as part of the core business of science teachers, even though there is a growing body of evidence of the connections between good educational attainment and positive health outcomes (Brooks, 2014). Teachers are in a prime position to influence and improve the health and wellbeing of their students (Jourdan et al., 2008). This role is especially pertinent for teachers of biology, as specific aspects of health education, such as nutrition and healthy eating, physiology and the impact of exercise, drugs and alcohol education and sex education, fall within their remit. Using health-related issues to introduce aspects of the biology curriculum provides a student-friendly approach, stimulating interest and curiosity in biology topics that may otherwise seem quite remote and abstract—for example, introducing cell division via familiar genetic disorders such as Down Syndrome.

Nevertheless, teachers require appropriate knowledge and skills to teach about health and disease effectively and to manage potentially sensitive issues, especially in the light of increased concerns about obesity, alcohol abuse and risky sexual behaviour among young people. Furthermore, there is increasing recognition that the life course determinants of adult health, including non-communicable (chronic) disease, are acquired in childhood or even before birth, highlighting the significant impact of pre-conceptual and maternal health on adult health and the health of future offspring (Gluckman & Hanson, 2012).

However, health education is rarely compulsory in schools, and in many countries it has a low status. Where it is taught, it is often left to biology teachers, who are themselves not necessarily adequately trained to teach it. For example, in

England training in health is rarely prioritized in teacher training courses and the content and amount of time allocated to health is highly variable, whilst teachers feel unqualified to teach certain aspects of health education, particularly sensitive issues (Byrne et al., 2015).

We regard health-related issues as socio-scientific issues, i.e. issues having a basis in science and a potentially large impact on society (Ratcliffe & Grace, 2003). Given this broad remit, it is not possible to cover all aspects of health and disease in one chapter, so we have focused on a limited number of issues but explored them in more depth.

Conceptualising Health and Disease

The Concept of Health

The concept of health is open to a wide range of interpretation. A marathon runner would perhaps see physical fitness as the epitome of health, whereas an old lady with arthritis who is still able to look after herself and her home would consider herself to be healthy. Health is therefore a relative term and dependent upon many factors such as age, physical and mental states and economic and social status as well as cultural and religious beliefs. The World Health Organization (WHO) considers that health is a fundamental right of every human being and defines health as follows:

> Health is a state of complete physical, mental and social well-being and not merely the absence of disease or infirmity.
>
> *(WHO, 1946)*

WHO has now expanded this definition to include notions of empowerment and the ability to make decisions about one's life as important to health and wellbeing. Figure 7.1 provides a diagrammatic representation of the determinants of health (Dalgren & Whitehead, 1991).

Starting with the personal features of an individual and moving outwards to the different levels of influence on that individual, the diagram illustrates the complexity of the interconnected factors that can affect someone's health. Despite this holistic view of health and wellbeing, health and disease are still often regarded as mutually exclusive, with health equating to the absence of disease.

Naidoo and Wills (2016) discuss a range of theoretical models of health education that can be helpful to plan and develop a teaching programme. The model adopted depends on the aims of the programme, the implicit views of the teacher, the ethos and culture of the school (Pickett et al., 2017) and the wider educational landscape, which includes government policy directives. The model probably most pertinent to educators is the 'education model', which is underpinned by a humanist model of health that presumes all individuals have the innate capability

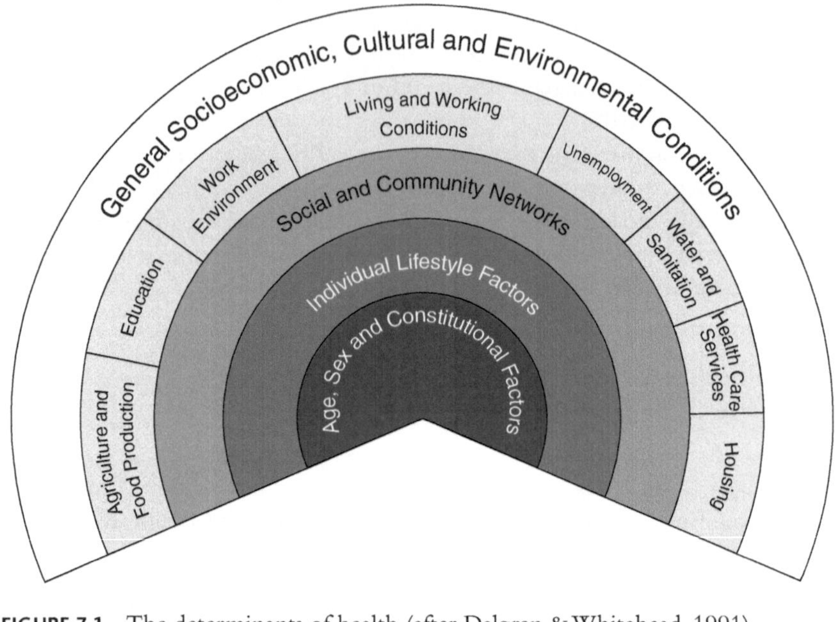

FIGURE 7.1 The determinants of health (after Dalgren & Whitehead, 1991)

to self-actualise in order to reach their full potential given the right conditions. Teaching using the education model focuses on enabling individuals to make informed decisions about their own health and develop attitudes and skills which help them avoid risky health-related behaviours.

The Concept of Disease

Until fairly recently, communicable (infectious) diseases were a priority for WHO, and despite recent outbreaks of communicable diseases such as malaria, Ebola and Zika virus, and continued infection rates of HIV, the global focus is now also on chronic or non-communicable diseases (NCDs). These are related to lifestyle (either a chosen lifestyle or one imposed on the individual) and are the result of a combination of genetic, physiological, environmental and behavioural factors. Current information indicates that NCDs kill 40 million people each year, equating to 70% of all deaths worldwide (WHO, 2017). Each year, 17.7 million people die from cardiovascular diseases, followed by cancers (8.8 million), respiratory diseases (3.9 million) and diabetes (1.6 million). It is therefore important that NCDs are included in a school's health education programme. However, health education lessons need to provide students with more than factual knowledge; they need to support students in becoming health-literate citizens. This is education that facilitates the development of lifelong learning skills to support informed decision making relating to health at individual, family and community levels.

Health Literacy

WHO defines health literacy as "the cognitive and social skills which determine the motivation and ability of individuals to gain access to, understand and use information in ways which promote and maintain good health" (Nutbeam, 1998: 357). Nutbeam (2000) proposes three types of health literacy: i) basic/functional literacy (reading and writing in relation to health); ii) communicative/interactive literacy (cognitive and social skills to extract health information, derive meaning and apply the information to changing circumstances); and iii) critical literacy (skills to critically analyse information and use it to help control changing circumstances). Education for critical health literacy should result in a positive behavioural response to achieve personal empowerment and autonomy in relation to health, and has the potential to facilitate individual or community action. Another important outcome of critical health literacy is improved self-efficacy. Garcia and Mann (2003) tested several social-cognitive models for engaging people in health behaviours and found that models including self-efficacy were better predictors of intentions to engage in positive health behaviours.

Barriers to Teaching About Health and Disease

Health topics can lend themselves perfectly as a context for teaching students about the underpinning biology. They attract media attention; we all have first-hand experience of them and can relate to them accordingly; and young people are often fascinated by diseases, medical conditions and other health matters. So it might seem puzzling as to why health issues are not in fact readily exploited in schools, either within science lessons or as SSIs across the broader curriculum (Lee, 2012). There are actually several reasons, and most relate to the experiences, background education and expectations of biology teachers themselves. We have listed the main reasons below:

- The curriculum and national assessment systems often give priority to acquiring knowledge and understanding of concepts, rather than time for consideration of social and ethical issues. The biology curriculum is very crowded and the number of health topics is numerous, so teachers tend to focus on the biological factual information, although students have indicated that they wish to be consulted about the content of health education and that they want more socio-scientific information and opportunities to discuss health issues other than just the biological information (e.g. Willis et al., 2013).
- There is often a lack of teaching resources reflecting the latest advances in science in a way which is meaningful to school students.
- Some biology teachers do not regard social issues as part of science teaching. Our view is that health issues are multifaceted in nature, and dealing with

the biological concepts in isolation from the social context is unlikely to be effective. However, even teachers more disposed towards a holistic approach are sometimes not certain how to connect the biological concepts with the social context.

- Some teachers are uncertain about where the boundaries lie and what they are 'allowed' to discuss in class, especially regarding sensitive issues such as sex and relationship education and drugs education. They lack knowledge, expertise or confidence to handle controversial issues, which are by their nature associated with large amounts of often incomplete information with no 'correct' answers. Teachers may also be concerned that health issues might relate directly to the lives of some students in the class, demanding additional skills of teaching about sensitive issues. A further constraint of increasing significance is the requirement to gain written consent from all the parents/carers of students in the class prior to teaching about a sensitive issue. This can be time-consuming and present associated logistical problems.

Underlying many of these barriers is low confidence and self-efficacy among pre-service and in-service teachers in teaching health education, due to inadequate support from mentors and senior management in schools.

Even in schools where health education is made a high priority, the learning outcomes may not be as everyone would wish. Evans et al. (2013) noted that in some such schools the prevailing talk among students about healthy eating was in terms of body image, losing weight and 'looking right', rather than about the benefits of a nutritious and balanced diet to improve health and stay healthy. These attitudes and understanding of what it means to be healthy are acquired from family and friends as well as the media and often form "an epistemological obstacle to the acquisition of new scientific knowledge" (Carvalho, 2016: 167).

Biology teachers also sometimes have a poor understanding of the science itself that underpins health issues. This is really not surprising for two main reasons: i) the current trend for biology undergraduates to specialise in one particular field of biology, and many of these fields do not include aspects of human health; and ii) the very difficult challenge of keeping abreast of new scientific research findings and their implications.

One example of a rapidly emerging health-related research field is epigenetics. Consider the following question and—before continuing—consider what answer you would give:

> The risk of getting diseases can be attributed to a combination of our genetics and our environment. Current research indicates that the contribution of our fixed genetic make-up to the risk of adult diabetes and coronary heart disease is:
>
> <10% 10–25% 25–50% 50–75% 75–90% >90%

We asked 56 pre-service postgraduate science teachers this question. Most (57%) answered either 25–50% or 50–75%, and none answered <10%. The answer is actually <10%. To be fair to the teachers, this has only fairly recently been discovered from epigenetics research (Morris, 2012). Epigenetics is an area that has so far rarely found its way into the 'compulsory' school science curriculum in any country. Standard teaching and learning resources for schools talk about some variation within a species being inherited and some being due to the environment, without any consideration of the relative contribution each makes. Among these student science teachers, only 47% had heard the term epigenetics, and only 34% could attempt a definition. However, the concept of epigenetic modification, i.e. that our environment can modify our genes, is hugely important from a health education perspective, because school students (as possible soon-to-be parents) need to be aware that our health-related activities not only affect our own lives, but can also affect the health of our future children. Epigenetic changes control the function of our genes without changing the actual DNA sequence. A key epigenetic modification is DNA methylation, which influences embryonic development and the formation of different cell types and regulates where and when genes are switched on. Recent research (Lilycrop et al., 2017) has indicated that DNA methylation can be affected by a range of environmental factors such as parental health, diet and lifestyle, and this can result in, for example, the child having a greater risk of developing obesity in later life.

Students' Misconceptions About Health and Disease

In taking a constructivist approach to teaching about health and disease, biology teachers need to find out what their students already know and understand about a topic, and ascertain their misconceptions (Driver, 1989). Space prevents a comprehensive coverage of young people's misconceptions about health and disease, but examples include: the belief that antibiotics are effective against viruses as well as bacteria (Cals et al., 2007); the long-term effects of alcohol and cocaine use are similar and akin to the risk of long-term tobacco smoking (Sigelman et al., 2000); and the transmission of HIV is caused by casual physical contact (Osborne et al., 1993). As a more detailed example, we focus below on misconceptions about micro-organisms and disease.

Micro-Organisms and Disease

Medical as well as commercial material and the media often portray micro-organisms as dangerous and potentially fatal. It is therefore not surprising to find that students hold ideas that micro-organisms are harmful, especially to humans, because they cause disease and illness (Byrne, 2011). This pathogenic view of micro-organisms often precludes understanding that diseases, for example, cancer, arthritis or dietary deficiencies, can be non-communicable (Byrne, 2011). This

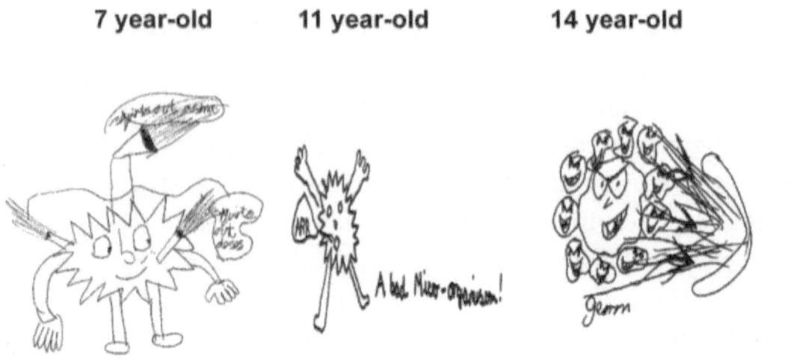

FIGURE 7.2 Drawings illustrating negative views about micro-organisms (Byrne et al., 2009)

exclusively 'exogenous' notion of disease, in which a healthy individual is attacked by micro-organisms and becomes ill, has implications for students being able to comprehend the nature of non-communicable diseases. This may then raise questions about students' ability to understand and appreciate the ethical implications of such techniques as ante-natal screening for genetic disorders (Simonneaux, 2000).

Micro-organisms are often seen as vicious and dangerous; for example, poison, venom, dirt, dust and snakes are commonly depicted by students of all ages. Drawings often show micro-organisms as having teeth (see Figure 7.2) to indicate how easily they could devour parts of the body or infect cells (Byrne et al., 2009).

Alternative, inaccurate ideas about the cause of infection also prevail—for example, the belief that you can catch a cold due to environmental circumstances, such as getting cold, or cold weather, or washing hair and then going outside (Prout, 1985; Byrne, 2011). This 'folk model of infection' seems to be very persistent, and the term 'cold' reinforces the connection with the environment and the common cold (Driver et al., 1994). Consequently, students employ ontological ideas based on everyday experiences.

Infection and Immunity

The notion that illness is inevitable once micro-organisms have entered the body is well established. However, analogies with a pitched battle, in which the body has defence agents to literally fight the intrusive agents in armed combat, are common in attempts to explain both infection and subsequent immune response (Simonneaux, 2000; Byrne, 2011). Antibodies are thought to be particles, elements, white cells or molecules that defend the body against germs and diseases and are thought to be akin to medicines as the same role is attributed to them, albeit as a kind of self-medication. For younger students, recovery from infection is a result of the micro-organisms being physically expelled from the body or through medication that acts as an agent to remove the germs from the body (Byrne, 2011).

Alternative ideas about immunity also are common, with lifestyle factors such as keeping fit, wearing warm clothes or eating properly taking priority over biologic immune responses. However, it is unclear whether these students are referring to an alternative model of health in which lifestyle is important to staying healthy or if they regard these as factors that enhance immune responses. Antibiotics are considered to be curative rather than preventative drugs, although they are also thought to contribute to strengthening the immune system, rather than acting on the micro-organisms. For some students, antibiotics and antibodies are regarded as the same thing (Simonneaux, 1996).

So How Can We Teach About Health and Disease?

The sections above indicate the need for a holistic approach to teaching about health and disease. Due to their interconnectivity, teaching about individual topics separately would be ineffective. A better pedagogical approach is to consider the progressive development of students' life skills, in which they gain age-appropriate accurate knowledge, learn to assess risk and make informed decisions about their behaviour within the context of their everyday lives and experiences, thus increasing their health literacy. Good practice will ideally be a combination of the education model mentioned above and whole school initiatives that support what is happening in the classroom. Table 7.1 provides a worked example about healthy eating to illustrate the progression of skills and knowledge and attitudes.

We would argue that the ultimate goal of health education is to get students to develop heathier and better-informed lifestyles, for themselves, their families and their community—and this requires sustained behavioural change. As already discussed, having increased knowledge and more positive attitudes does not necessarily translate into changed behaviour. So, what do teenagers think about their own health and health-related behaviour? It's a complicated landscape. For instance, Ioannou (2009) reported that some Cypriot teenagers view eating fast foods (as opposed to healthy foods) as a means of expressing their youthful image. Williden et al. (2006) described how most youngsters in New Zealand preferred computers or TV over physical activity, and one-quarter of their parents thought it did not matter what their child ate, as long as they were growing properly. Another barrier to healthy eating that is often overlooked is young people's unfamiliarity with certain kinds of food and food outlets (Pearce et al., 2009).

We carried out a survey of 597 14-year-olds from state schools in the south of England (Grace et al., 2012) and found that two-thirds (67%) believed themselves to be generally in "good health", and 78% described themselves as "fairly" or "very" active. They felt largely in control of their choice of food, with three-quarters (76%) claiming that they choose what they eat "all" or "most" of the time. When asked to consider how their current lifestyle might affect future health,

TABLE 7.1 Healthy eating as worked example of teaching about health and disease in biology

Stage of Education	*Health Education*	*Biology Lessons*	*Whole School*	*Knowledge*	*Skill Development*
Infant (approx. 4–6 years)	The importance of having a balanced diet	Introducing and tasting a range of foods that humans can eat, especially new or novel fruit and vegetables	School lunches and snacks that are nutritionally balanced	Eating the right amounts of different foods	Opportunities to make healthy food choices
Primary (approx. 6–11 years)	The importance of having a balanced diet	Making up a balanced meal using real food examples, models or pictures Exploring what happens to the food that we eat using models, video etc.	School lunches and snacks that are nutritionally balanced Cookery club for students and parents	What constitutes a balanced diet The basic functions of the digestive system	Opportunities to make healthy food choices
Lower secondary (approx. 11–15 years)	Body image, eating disorders and obesity	Small group discussion about the depiction of idealised body types in magazines and newspaper articles and explorations of the nutrition requirements for a healthy body Calculations of energy in a range of meals, e.g. a McDonald's, a school lunch, a packed lunch etc. and energy used up in different forms of exercise	Whole school campaign about body image Fitness and exercise clubs	The function of different food groups to maintain health and the consequences of imbalances in the diet, including obesity, and deficiency diseases Calorific value of different food groups and requirements in a healthy daily diet	Communication and negotiation Decision making about diet and exercise for health Personal organisation to attend fitness clubs
Upper secondary (approx. 15–18 years)	NCDs and health of future offspring	Exploring the effect of non-communicable diseases on physiology Debating the impact of lifestyle factors including poor diet and lack of exercise on the incidence of non-communicable diseases	Promotion of healthy lifestyle	The impact of lifestyle factors on the incidence of non-communicable diseases	Communication and negotiating Assessing risk of an unhealthy lifestyle—now and in the future

72% agreed that the food they eat now will affect their health in the future, but only 35% agreed that the food they eat now could affect the health of any children they may have in the future, with no significant difference between the responses of boys and girls. Of course, students' views will vary around the world, but a simple survey of this kind provides baseline information from which the teacher can plan an appropriate health education programme.

Although there have been worldwide calls for schools to promote health for many years (Tang et al., 2008), school-based health education programmes to tackle issues such as overweight and obesity have only met with modest success (Katz et al., 2008). However, some are showing signs of longer-term achievement, and we conclude by outlining three cases from England.

LifeLab (Grace et al., 2012; Grace & Bay, 2011) helps teenagers to reassure themselves that they can develop more positive attitudes and behaviours in relation to their own health. It consists of: professional development training for the science teachers; lessons in school to explore the science underpinning their own health; a visit to a hospital-based classroom where students carry out practical investigations; and a 'celebration event' for parents and local people where students present posters about their work. There is a structured approach to teaching about sensitive issues, with clear ground rules for students, and the avoidance of any students feeling uncomfortable in the sessions. The whole programme centres around a fictitious character called Chris, and the students investigate Chris's health issues. This enables students to reflect on their own health without discussing themselves explicitly; rather than simply telling students how to live healthier lives, the approach helps them to convince themselves. Randomised controlled trials of LifeLab have shown that students develop positive attitudes and behavioural changes towards their health, which are sustained at least six months later (Grace et al., 2012).

Facts4Life (Bird & Oliver, 2015) is an initiative which aims to tackle 'over-medicalisation' by teaching students about illness, and that most illnesses are not that worrying and do not require individuals to see the doctor or take major drugs. The vast majority of patients who see their doctors do not require major drugs but need reassurance and advice. This project encourages self-help as a part of making students more health literate. It emphasises that being unwell is a normal part of life, so students focus on the symptoms of illness, how illness might track through the body and how we can help ourselves to recover. Early findings from this intervention indicate that the students develop more informed perceptions of health and illness and perceive that they are less reliant on medication and visiting the doctor if they are feeling unwell. A challenge here is to strike the balance between encouraging students to be more resilient and independent while still recognising that there are times when external help is needed.

The Healthy Conversation Skills (HCS) training programme (Lawrence et al., 2016; Black et al., 2012) is another project which is showing signs of success in sustained behaviour change. It is for health care practitioners, but the principles are now being used to train school teachers for work with their students. HCS uses a motivating, student-centred approach to develop skills which support health-related behaviour by employing exploratory conversations, enabling teachers to better understand their students' worlds and empower students to take control of their health issues and thereby increase their sense of self-efficacy. It focuses on training teachers how to develop the skill of asking open-ended 'discovery' questions, which generally begin with 'how' and 'what'. Five core skills are highlighted for the teachers: to be able to identify and create opportunities to hold 'healthy conversations'; to use open-ended 'discovery' questions (which specifically support students to explore their health issues, identify barriers and generate solutions); to reflect on practice; to listen rather than provide information; and to support goal-setting through structured planning.

Conclusions

This chapter has highlighted the multifaceted nature of health issues, the complicated relationship between health and disease and the intricate demands of developing health literate students. This inevitably requires a holistic approach to teaching topics in health, and as much of the content falls within the biology curriculum, it is often biology teachers who are charged with delivering this task.

References

Bird, E. & Oliver, B. (2015). *Facts4Life Primary School Resource: Pilot Evaluation Report. Full Technical Report.* Project Report. University of the West of England. Available at: http://eprints.uwe.ac.uk/27184.

Black, C., Lawrence, W., Cradock, S., Ntani, G., Tinati, T., Jarman, M., Begum, R., Inskip, H., Cooper, C., Barker, M. & Baird, J. (2012). Healthy conversation skills: increasing competence and confidence in front-line staff. *Public Health Nutrition*, 17(3), 700–707.

Brooks, F. (2014). *The Link Between Pupil Health and Wellbeing and Attainment: A Briefing for Head Teachers, Governors and Staff in Education Settings.* London: Public Health England.

Byrne, J. (2011). Models of micro-organisms: children's knowledge and understanding of micro-organisms from 7 to 14 years-old. *International Journal of Science Education*, 33(14), 1927–1961.

Byrne, J., Grace, M. & Hanley, P. (2009). Children's anthropomorphic and anthropocentric ideas about micro-organisms: do they affect learning? *Journal of Biological Education*, 44(1), 37–43.

Byrne, J., Shepherd, J., Dewhirst, S., Pickett, K., Speller, V., Roderick, P., Grace, M. & Almond, P. (2015). Pre-service teacher training in health and well-being in England: the state of the nation. *European Journal of Teacher Education*, 38(2), 217–233.

Cals, J. W. L., Boumans, D., Lardinois, R. J. M., Gonzales, R., Hopstaken, R. M., Butler, C. C. & Dinant, G-J. (2007). Public beliefs on antibiotics and respiratory tract infections: an internet-based questionnaire study. *British Journal of General Practice*, 57(545), 942–947.

Carvalho, G. S. (2016). Current issues in biological education research: the case of health education. In T. Tal & A. Yarden (Eds.), *The Future of Biology Education Research*. Haifa: Israel Institute of Technology, 167–170.

Dahlgren, G. & Whitehead M. (1991). *Policies and Strategies to Promote Social Equity in Health.* Stockholm, Sweden: Institute for Futures Studies.

Driver, R. (1989). Students' conceptions and the learning of science. *International Journal of Science Education*, 11, 481–490.

Driver, R., Squires, A., Rushworth, P. & Wood-Robinson, V. (1994). *Making Sense of Secondary Science: Research Into Children's Ideas.* London: Routledge.

Evans, J., Davies, B., Rich, E. & DePian, L. (2013). Understanding policy: why health education policy is important and why it does not appear to work. *British Educational Research Journal*, 39(2), 320–337.

Garcia, K. & Mann, T. (2003). From 'I Wish' to 'I Will': social-cognitive predictors of behavioral intentions. *Journal of Health Psychology*, 8(3), 347–360.

Gluckman, P. & Hanson, M. (2012). *Fat, Fate and Disease: Why Exercise and Diet are not Enough.* New York: Oxford University Press.

Grace, M. & Bay, J. (2011). Developing a pedagogy to support science for health literacy. *Asia-Pacific Forum on Science Learning and Teaching*, 12(2), Foreword Website: www.ied.edu.hk/apfslt/.

Grace, M. M., Woods-Townsend, K. S., Griffiths, J. B., Godfrey, K. M., Hanson, M. A., Galloway, I., Azaola, M. C., Harman, K., Byrne, J. & Inskip, H. M. (2012). Developing teenagers' views on their health and the health of their future children. *Health Education*, 112(6), 543–559.

Ioannou, S. (2009). 'Eating beans . . . that is a "no-no" for our times': young cypriots' consumer meanings of 'healthy' and 'fast' food. *Health Education Journal*, 68(3), 186–195.

Jourdan, D., Samdal, O., Diagne, F. & Carvalho, G. S. (2008). The future of health promotion in schools goes through the strengthening of teacher training at a global level, *Health Promotion & Education*, 15(3), 36–38.

Katz, G. L., O'Connell, M., Njike, V. Y., Yeh, M. C. & Nawaz, H. (2008). Strategies for the prevention and control of obesity in the school setting: systematic review and metaanalysis. *International Journal of Obesity*, 32(12), 1780–1789.

Lawrence, W., Black, C., Tinatil, T., Cradock, S., Begum, R., Jarman, M., Pease, A., Margetts, B., Davies, J., Inskip, H., Cooper, C., Baird, J. & Barker, M. (2016) 'Making every contact count': evaluation of the impact of an intervention to train health and social care practitioners in skills to support health behaviour change. *Journal of Health Psychology*, 21(2), 138–151.

Lee, Y. C. (2012). Socio-scientific issues in health contexts: treading a rugged terrain. *International Journal of Science Education*, 34(3–4), 459–483.

Lilycrop, K. et al. (2017). ANRIL promoter DNA methylation: a perinatal marker for later adiposity. *Ebio Medicine (Lancet)*, 19, 60–72.

Morris, et al. (2012). Large-scale association analysis provides insights into the genetic architecture and pathophysiology of type 2 diabetes. *Nature Genetics*, 44(9), 981–990.

Naidoo, J. & Wills, J. (2016). *Foundations for Health Promotion*, 4th edn. Amsterdam: Elsevier.

Nutbeam, D. (1998). Health promotion glossary. *Health Promotion International*, 13(4), 349–364.

Nutbeam, D. (2000). Health literacy as a public health goal: a challenge for contemporary health education and communication strategies into the 21st century. *Health Promotion International*, 15(3), 259–267.

Osborne, M., Kistner, J. A. & Helgemo, B. (1993). Developmental progression in children's knowledge of AIDS: implications for education and attitudinal change. *Journal of Pediatric Psychology*, 18(2), 177–192.

Pearce, A., Kirk, C., Cummins, S., Collins, M., Elliman, D. & Connolly, A. M. (2009). Gaining children's perspectives: a multiple method approach to explore environmental influences on healthy eating and physical activity. *Health & Place*, 15(2), 614–621.

Pickett, K., Rietdijk, W., Byrne, J., Shepherd, J., Roderick, P. & Marcus Grace, M. (2017). Teaching health education: a thematic analysis of early career teachers' experiences following pre-service health training. *Health Education*, 117(3), 323–340.

Prout, A. (1985). Science, health and everyday knowledge. *European Journal of Science Education*, 7(4), 399–406.

Ratcliffe, M. & Grace, M. (2003). *Science Education for Citizenship: Teaching Socio-Scientific Issues*. Maidenhead: Open University Press.

Sigelman, C. K., Leach, D. B., Mack, K. L., Bridges, L. J., Rinehart, C. S., Dwyer, K. M., Davies, E. P. & Sorongon, A. G. (2000). Children's beliefs about long-term health effects of alcohol and cocaine use. *Journal of Pediatric Psychology*, 25(8), 557–566.

Simonneaux, L. (1996). *Microbes' Conceptions in France*. Toulouse-Auzeville: ENFA.

Simonneaux, L. (2000). A study of pupils' conceptions and reasoning in connection with 'microbes', as a contribution to research in biotechnology education. *International Journal of Science Education*, 22(6), 619–644.

Tang, K.C., Nutbeam, D. & Aldinger, C. (2008). Schools for health, education and development: a call for action. *Health Promotion International*, 24(1), 68–77.

Williden, M., Taylor, R. W., McAuley, K. A., Simpson, J. C., Oakley, M. & Mann, J. I. (2006). The APPLE project: an investigation of the barriers and promoters of healthy eating and physical activity in New Zealand children aged 5–12 years. *Health Education Journal*, 65(2), 135–148.

Willis, B., Clague, L. & Coldwell, M. (2013). Effective PSHE education: values, purposes and future directions. *Pastoral Care in Education*, 31(2), 99–111.

World Health Organization. (1946). *1946 Constitution*. Geneva: World Health Organization.

World Health Organization. (1993). *The European Network of Health Promoting Schools: Resource Manual*. Copenhagen: WHO Regional Office for Europe.

The World Health Organization. (2017). Noncommunicable diseases. Available at: www.who.int/mediacentre/factsheets/fs355/en/

8

REPRODUCTION AND SEX EDUCATION

Michael J. Reiss

Introduction

Reproduction as a topic in biology has links to many other biology topics[1] and is a key component of any school biology course. Nevertheless, there is considerable variation in the extent to which school biology courses only address mammalian reproduction, sometimes alongside a more superficial treatment of reproduction in flowering plants, or give a broader coverage of the topic. It is argued here that school teaching about reproduction and life cycles should not focus on mammals and flowering plants to the virtual exclusion of other taxa.

The topic of sex education has clear links with that of reproduction, but there are some important differences. For a start, sex education is restricted to humans. It is also important that sex education does not focus only on issues to do with reproduction. The extent to which sex education is taught in schools varies greatly, from country to country and often within countries. It is argued here that while sex education—sometimes called sex and relationships education—should not be treated as being only the preserve of biology teachers; school biology teachers nevertheless have a central role to play.

Students' Misconceptions About Reproduction and Sex Education

There is a large and somewhat diffuse literature about students' misconceptions about reproduction and sex education. As might be expected, misconceptions are more likely to be reported about human reproduction than about reproduction in other species (Scharmann, 1991; Nguyen & Rosengren, 2004). Schussler (2008) showed that children's books about plants not infrequently contain inaccuracies about plant reproduction. Hershey (2005) reported that plant misconceptions, more generally, are common in textbooks, science project books, dictionaries, encyclopaedias, science education journals and educational websites. Yip (1998) found that many students showed problems in relating the time of conception in humans to the condition of the uterine lining, and that they did not appreciate the connections between menstruation, ovulation and the likelihood of a fertilised egg implanting. Misconceptions about contraception are also widespread (e.g. Hamani et al., 2007). More generally, there is a large literature about misconceptions in sex education, particularly for HIV with regards to methods of transmission (e.g. Harvey & Reiss, 1992); there is even a Wikipedia page titled 'Misconceptions about HIV/AIDS'. A number of instruments have been developed to assess knowledge about human reproduction/sex, and so identify misconceptions, an early one being Miller and Lief (1979).

As discussed in this volume's Introductory chapter, the term 'misconceptions' may fail to convey all the ways in which scientifically incorrect views are held about a topic, and this is perhaps particularly the case in regards to reproduction and sex education. Often students, especially young students, are simply ignorant—and this is not something for which they should be blamed. As any biologist knows, the diversity of reproduction in the natural world is immense; it is hardly the fault of students if they have been taught very little of this diversity, or even of the specifics of human reproduction. Moreover, when it comes to sex education, misconceptions or ignorance are sometimes fuelled by prejudice, for example in regards to sexual identities and behaviours.

One example of widespread ignorance, not only among school students but among adults too, is to do with age-dependent female fertility rates in humans (Bunting et al., 2013). It is widely presumed that female fertility rates only drop off once a woman is in her late 30s. The reality, though, is that fertility rates typically drop off from the age of about 32. This is also the case for men, albeit the rate of decline with age is far shallower. Given that more and more women seek nowadays not to have children until they are in their late 20s or early 30s and that if they do not become pregnant, they typically don't seek medical help until they have had more than a couple of years of not conceiving, this biological fact is an important one—and one rarely addressed in school education.

Difficulties in Teaching About Reproduction and Sex Education

One advantage of teaching about reproduction and sex education is that student interest is often high. Nevertheless, there are a number of difficulties in teaching about these topics. One difficulty, especially with regards to human reproduction and associated topics within sex education, is that teachers may feel embarrassed. The solution to this is better teacher education, whether during their initial teacher training or subsequent professional development. Support from senior management within schools is important, too (Buston et al., 2002).

A related difficulty is that in many countries, teachers receive little or no training about sex education. Biology teachers, of course, are nearly always trained to teach about reproduction, both in humans and in other taxa. However, the knowledge and skills to teach sex education well are not the same as the knowledge and skills to teach reproduction well. In particular, good sex education makes use of a range of pedagogical approaches (including role plays and debates) that are less often used in biology teaching. Then there is the fact that teaching about sex education raises issues to do with values to a greater extent than when teaching biology. In schools where religious values are important, attitudes towards sex education may be conservative, for example with regards to homosexuality and abortion. Teachers need to be skilled when teaching sex education in such situations (Halstead & Reiss, 2003). For many teachers, the easier and more pragmatic option may be stick to the biological facts, and this can disadvantage students.

Conceptualising Reproduction and Sex Education

Few students are likely to find it difficult to understand the fundamental concept that reproduction (more precisely, attempting to reproduce either directly or vicariously—via related individuals in cases of kin selection) is a key characteristic of living organisms. Nor are students likely to object to being taught sex education; indeed, surveys repeatedly show that school sex education is desired by the large majority of students and their parents (e.g. Barr et al., 2013). However, there are a number of issues that need to be considered when conceptualising reproduction and sex education.

Diversity in the Natural World

An important contribution that biology can make to students' understandings of the natural world is helping them to appreciate something of the enormous diversity of organisms and their ways of living. This is as true for organisms' life cycles and reproduction as it is for their ecology or any other branch of biology. Hermaphroditic snails were once thought bizarre, jokes of nature (Findlen, 1990). There is great value in students being taught to think about the following issues:

- *Why are some organisms sexual and some asexual?* This question is often included in biology courses, and it is good for students to realise that sexual reproduction enables species to evolve more rapidly to changing environments, because it contributes to the increase of variation within a population. However, students are less often taught the extent to which sexual reproduction is disadvantageous, at least in the short-term, to at least one parent—after all, sexual reproduction is all about passing on only half of one's genetic material to one's offspring. The most likely solution to the evolutionary problem of the evolution of sex is that it enables organisms to deal with the problem of parasites (Hamilton, 1980).

- *How is sex determined?* Students need to be taught, if they don't already know, that (by and large) in humans, females are XX and males XY. However, it is good for students to realise that a great many other sex-determining mechanisms exist. In birds, some reptiles and certain other taxa, sex determination is chromosomal, as for mammals, but females have two different kinds of chromosomes (ZW—equivalent to male mammals being XY) while males are ZZ (equivalent to female mammals being XX). Another chromosome determining system, haplodiploidy, is found in ants, bees and wasps. Here females are diploid, whereas males result from unfertilised eggs and so are haploid. Perhaps the best known non-chromosomal mechanism is when sex is determined by the external temperature (e.g. in some reptiles such as alligators, where eggs are laid and develop externally). Other non-chromosomal mechanisms exist. In clown fish, the large dominant adult in a group is a female with the smaller adults being males.

Morality and Values

Many of the issues that are core to sex education are ones where a moral/values dimension is evident: at what age is it right to have a sexual relationship? Is the notion of consent the be-all and end-all for determining when sexual activity is right? Should someone always be faithful to their sexual partner? Has religion anything positive to contribute to sex education, or is it simply always a constriction on sexual behaviour (e.g. teachings about *in vitro* fertilisation and abortion, and restrictions on any expression of sexuality other than heterosexuality within marriage)? Before deciding how such questions might be answered, it is worth asking whether there is such a thing as a *distinctive* sexual ethics or not. At first the answer may seem obvious. Surely sexual behaviour has its own ethics. People, at different times and in different cultures, argue about the acceptability of polygamy and homosexuality and the age of consent, whether rape can exist within marriage and so on.

However, Igor Primoratz (1999) has argued that sex is morally neutral, so that moral guidance regarding sexual behaviour is provided by the *same* general moral rules and values that apply in other areas:

> Thus adultery is not wrong as extramarital *sex*, but only when it involves breach of promise, or seriously hurts the feelings of the non-adulterous spouse, etc. Prostitution is not wrong as commercial *sex*, but if and when the prostitute is forced into this line of work by the lack of any real alternative. Pedophila is not wrong as adult-child *sex* but because even when the child is

> willingly participating, its willingness is extremely suspect in view of the radical asymmetries of maturity, knowledge, understanding, and power of children and adults. Sexual harassment is not wrong because it is *sexual*, but because it is *harassment*. Rape is not wrong as *sexual* battery, but as sexual *battery*.
>
> *(Primoratz, 1999, pp. 173–174)*

The argument is a powerful one and has much to commend it. After all, if there is something 'special' about sexual ethics, from where does this specialness come? Of course, there are particular ethical questions that it only makes sense to ask in the context of sex, and in that sense, there is a sexual ethics, but in the same way that there is, for example, business ethics and environmental ethics. The issue at hand is whether there is anything *distinctive* about sexual ethics, business ethics and environmental ethics beyond the localised application of more general forms of ethical reasoning—such as the use of the principles of autonomy, the greatest happiness of the greatest number, the avoidance of harm and, as suggested below, the promotion of human flourishing.

It has been argued that there are certain limits to acceptable sexual behaviour, set by the harms caused to others (Halstead & Reiss, 2003). In some cases, such as rape and sexual abuse, these harms are clear-cut; in others, such as visiting sex workers or leaving one sexual partner and changing to another, there are arguments on both sides. What one surely wants is for young people, at the appropriate age, to reflect on and discuss their developing sexual values. There is much, therefore, to be said for them considering such issues. One of the great things about schooling is that teachers are given the authority to promote discussion and get students to think. Schools can therefore add to and shape what children learn from their parents. As Allen (2017) points out, what happens in the sexuality education classroom is key to how and what students learn about sexuality at school. Done well, in a safe environment, teaching can enable students to develop age-appropriate skills rather than being the object of a patronising and fear-driven narrative in which adults try to keep them for as long as possible in a presumed world of childhood innocence. In some countries, sex education is referred to as 'sex and relationships education' or even 'relationships and sex education', acknowledging that much good teaching in this area is concerned with the way we relate to and treat others.

Religion plays an important part, as noted earlier, in the context of morality and values in sex education. Until fairly recently, relatively little had been written in any detail about religious values and school sex education. In recent years, though, there has been an increasing acknowledgement from all sex educators, whether or not they themselves are members of any particular religious faith, that religious points of view needs to be taken into account, if only because a significant number of children and their parents have moral values significantly informed by religious traditions (e.g. Yip & Page, 2013).

Religious believers need no arguments to be voiced in favour of taking religious values seriously, both generally and with particular reference to sexual ethics and behaviour. However, those without a religious faith can often be frustrated at what they perceive as the sexist and heterosexist views of those in religious authority. There is much truth in this. However, things can change and there are pockets of encouragement. Indeed, it is an oversimplification to see religion as always associated with sexual conservatism (Reiss, 2014). More will be gained by working with those of a religious faith than by excluding them. Of course, it important that all those who participate in such working commit to listening and are open to the possibility of change.

Recommendations for Teaching About Reproduction and Sex Education

I concentrate in this section on sex education, which includes human reproduction. There is, of course, great value in students being introduced to the diversity of ways in which organisms reproduce and, as discussed earlier, to the ways in which sex is determined. Examples can (and should) be taken from prokaryotes, fungi, algae, gymnosperms, flowering plants and a number of animal taxa, both invertebrate and vertebrate. Such teaching can also help students better understand the centrality of mitosis and meiosis in reproduction in different taxa. Nowadays, video clips showing the great diversity of sex-determination mechanisms and modes of reproduction in the natural world can easily be sourced on YouTube or elsewhere and are invaluable when teaching.

The Aims of Education and the Aims of Sex Education

It has been argued that there should be two fundamental aims of school education: to enable each learner to lead a life that is personally flourishing and to help others to do so too (Reiss & White, 2013). A central aim of education should therefore be to prepare learners for a life of autonomous, whole-hearted and successful engagement in worthwhile relationships, activities and experiences. In formal school education, this aim involves acquainting students with a wide range of possible options from which to choose both when they are in school and once they leave school. With their development towards autonomous adulthood in mind, schools should provide students with increasing opportunities to choose among the pursuits that best suit them. Young children are likely to need greater guidance from their teachers, just as they do from their parents. Part of the function of schooling, and indeed parenting, is to prepare children for the time when they will need to, and be able to, make decisions more independently.

The notion of human flourishing, for oneself and for others, seems a good basis on which to ground sex education for all students. In 1993, I wrote a paper titled 'What are the aims of school sex education?' (Reiss, 1993). The ones I identified as being the aims of existing sex education programmes—'Stopping girls getting pregnant', 'Reducing the incidence of sexually transmitted diseases', 'Decreasing ignorance', 'Decreasing guilt, embarrassment and anxiety', 'Enabling students to make their own decisions about their sexuality', 'Helping students develop assertiveness', 'Helping students question the present role of women in society', 'Helping students question the present role of men in society' and 'Providing an ethical framework for the expression of sexuality'—can, with some updating and repositioning, each be seen as contributing towards human flourishing.

Since 1993, much has happened in school sex education, however. Encouragingly, these developments have very largely been in the direction of the promotion of human flourishing. There is a greater emphasis on approaches to tackling gender and sexual violence (cf. Maxwell, 2017) as there is now more of an acknowledgement in an increasing number of countries about the prevalence of this, particularly for women and for LGBT (lesbian, gay, bisexual and transgender) youth. As is often pointed out, schools too often are heteronormative. Successful approaches to improving matters include ones based on a notion of inclusion, helping to build safer and more supportive school environments. Ingham (2017) argues that it remains the case that school sex education is expected to be effective in a way that is rarely demanded of other subjects. One problem with this is that too narrow a range of possible outcomes are evaluated; it is easier to count pregnancies than to determine whether young women's capacity for enhanced sexual pleasure has increased.

Good sex education is comprehensive in the sense that it helps students acquire the knowledge, skills and values to make appropriate and healthy choices in their sexual lives. Successful comprehensive sex education should therefore help to reduce rates of sexually transmitted infections and the incidence of unwanted pregnancies; it should also build self-esteem and reduce the likelihood of exploitation. Abstinence-only sex education has a narrower focus, concentrating on reducing the likelihood that young people will engage in sexual intercourse before they are married. A number of studies, especially ones undertaken in the USA, have concluded that whereas comprehensive sex education can meet some of its aims, abstinence-only sex education is generally either ineffective or actually counterproductive (Kirby, 2008; Stanger-Hall & Hall, 2011).

The Role of Biology in Sex Education

Many sex educators view science rather negatively when considering how sex education might best be taught (Reiss, 2017). The 'plumbing and diseases' approach is rightly criticised. As Albury (2017) points out, such a presentation of the science of sex education has a number of shortcomings. One of these is that are students have to seek information on the specifics of sexual practices and relationship skills from other sources; some of these sources are of high quality, but others are not.

However, biology can play a more positive role in sex education. What sex educators are (quite rightly) rejecting is not biology but the rather poor biology that often passes for science when teaching about sex and sexualities. Consider how sex is all too often presented in school biology classrooms. School biology typically examines issues of sex through the lens of reproduction (Reiss, 2007). This automatically tends to assume heterosexuality. Biology is all too often presumed to be a neutral subject, so that many biology teachers in schools continue to teach gender and sex as unquestioned facts. In particular, differences between females and males are often presented as clear-cut and inevitable, and the study of school biology textbooks has shown that they are often sexist and typically ignore lesbian and gay issues (Reiss, 1998). For example, biology textbooks in England for 14–16-year-olds often omit all mention of the clitoris and, when they do refer to it, frequently talk of it in a belittling way as the female's equivalent of a penis. Males are rendered visible, females less so; and the female exists by virtue of comparison with the male. When the possibility of being gay/lesbian is addressed (the furthest that school biology textbooks ever get from heteronormativity), the impression is generally given that this is a sort-of second-best option which the reader may well grow out of.

Closer examination of sex in human biology provides plenty of space for critical reflection and allows for a richer understanding of what it is to be a sexual person. Emily Martin has argued that while menstruation is often viewed in scientific textbooks as a failure (you should have got pregnant), sperm maturation is viewed as a wonderful achievement in which countless millions of sperm are manufactured each day (Martin, 1991). Furthermore, sperm are viewed as active and streamlined, whereas the egg is seen as large and passive, just drifting along or sitting there waiting for sperm to arrive.

It was back in 1948 that Ruth Herschberger argued that the female reproductive components (it is difficult in the scientific discourse around sex to avoid referring to reproduction!) are viewed as somehow being less autonomous than those of the male. The way the egg is portrayed in biology textbooks has been likened to that of the fairy tale *The Sleeping Beauty*, in which a dormant, virginal bride awaits a male's magic kiss. However, biologists have long seen both egg and sperm as active partners. Just as sperm seek out the egg, so the vagina discriminates between sperm, and the egg seeks out sperm to catch. Nevertheless, as Martin pointed out, even when acknowledged, such biological equality is still generally described in a language that gives precedence to the sperm. When the egg is presented in an active role, the image is one of a dangerous aggressor, "rather like a spider laying in wait in her web" (Martin, 1991, p. 498).

Social historical research on sex hormones has also shown how the way that such hormones are presented in textbooks and scientific papers gives messages that go well beyond what the data indicate. Despite the fact that it has been known since the 1920s that each sex contains the 'other's' hormone—so men produce oestrogen and women testosterone—school textbooks typically ignore both this fact and the close chemical similarity between oestrogen and testosterone (Roberts, 2002). Indeed, a different reading of the data to that usually presented in school textbooks—but one more in line with the scientific evidence about the working of sex hormones—is that femaleness and maleness lie on a continuum. Such a model of the consequences of the actions of the sex hormones became common among endocrinologists in the 1940s.

While this model can lead to an essentialist understanding of sexuality and sexual orientation—and it was developed at the same time as a rise in the number of studies of the presumed femininity of gay men—it can also be seen as allowing a far more fluid understanding of sex, accommodating, for example, some forms of intersexuality (cf. Callahan, 2009). The principle of intersexuality dates largely from Magnus Hirschfeld's pioneering work in the first three decades of the twentieth century on sexual difference. By rejecting the discrete categories of male and female, arguing instead that each of us sits on a continuum, Hirschfeld laid the foundation for a radical deconstruction of the sexual binary (Bauer, 2003).

One of the things good teaching can help students to appreciate is the way in which boys/men and girls/women are pressured, respectively, to perform maleness and femaleness, discourses that are structured largely in opposition to each other (cf. Martino & Pallotta-Chiarolli, 2005). My experience of teaching a non-binary version of human sex to students is that many of them find it fascinating; it can help them to see the world, and themselves, in a new light. Indeed, enabling students to see classifications that relate to gender, sex and sexualities less categorically has the potential to lead students to question, even disrupt, other rigid typologies, facilitating the beginnings of an intersectional analysis (cf. Giffney, 2009).

Much of the literature about the 'causes' of sexuality concentrates on gayness, though Lynda Birke, a biologist as well as a feminist and a lesbian, provided a valuable review about lesbianism and over the years has "spent much time and energy refuting the allegations that any social categories (of gender, race or sexuality) are fixed by biology" (Birke, 1997, p. 58). However, as Birke pointed out, there are, of course, a number of reasons for hesitating in entirely rejecting biological notions of sexual orientation/preference. For one thing, some have used such notions politically to argue for gay rights (though this approach is hotly contested—see Schüklenk & Brookey, 1998); more prosaically, it may well yet turn out that there is/are biological bases to at least some people's sexual orientation/preference.

All of which emphasises the importance of biology being taught better in schools for the purposes of sex education than it often is. Much biology teaching is focused around the use of textbooks, yet "teachers can read subtextually and resistantly and can help their students to do likewise. Too rarely are students encouraged to critique their science textbooks; too often are textbooks used as if they contained only unquestionable truths" (Reiss, 1998, p. 148). This is a simple message but one that provides a teacher—and her/his students—with a powerful tool, for it avoids buying into the general assumption of teacher or textbook writer as the expert repository of facts, instead sitting more comfortably with critical and emancipatory understandings of education. This can be more satisfying for teacher and students alike and fits well with an information society that proves students with opportunities to obtain much of the knowledge they want/need to know at the right pace for themselves.

Will Letts (2001) has explored how school science structures and is structured by norms of heterosexual masculinity. Letts' work is particularly valuable as he focused on classroom examples of primary school science—when some might assume that at this age science is fairly neutral (but see the work of the *No Outsiders* Project: DePalma & Atkinson, 2009). He argued that science, including school science, functions as a grand narrative that seduces students and teachers; he concludes:

> As a plan of action, I advocate that school science becomes an active and generative site for critical science literacy. The words 'science literacy' in this phrase are intended quite differently than popular utterances of them have come to mean. 'Science literacy' does not simply mean an intake and consumption of science texts and 'facts', either purposefully or through acts of seduction. I am using critical science literacy to denote something akin to critical media literacy.
>
> *(Letts, 2001, p. 270)*

A school biology classroom for critical biological literacy, at any age, would be one in which the traditional virtues of academic biology—its open-mindedness and refusal to accept tradition on trust—were more widely (reflexively) applied. It would allow young people to think about themselves and their sexualities more meaningfully. It would help those uncomfortable with traditional descriptions of masculinity and femininity to realise that they are not alone in their rejection of such simple dichotomies. All this can be achieved without harming those students who are comfortable with such conventional descriptions. Sadly, such classrooms are still rare. In the long run such teaching, idealistic though this may sound to some, would contribute to making the world a better place both overall and for the many individuals who otherwise feel or find that they do not fit (cf. Britzman, 1995).

Conclusions

Reproduction is rightly considered a core component of school biology. While there is far more to sex education than reproductive biology, biology teachers can play an important role in ensuring that school sex education is effective and contributes to student flourishing. However, good sex education makes different demands on a teacher from those that biology teachers are used to. Accordingly, biology teachers have a right to expect support from senior staff and high-quality training if they are to teach sex education to their students.

Note

1 This chapter on reproduction and sex education has links to the one that precedes it, Chapter 7 on health and disease, and the one that follows it, Chapter 9 on development. There are also links to Chapter 5 on cell biology (because of mitosis and meiosis) and to Chapter 10 on genetics. It has therefore been written so as to complement and avoid substantial overlap with these four chapters.

References

Albury, K. (2017). Section introduction: sexualities, education and the media. In P. Aggleton (Ed.), *Education and Sexualities, Vol I*. London: Routledge, 132–136.

Allen, L. (2017). Section introduction: classroom interactions and processes. In P. Aggleton (Ed.), *Education and Sexualities, Vol I*. London: Routledge, 228–232.

Barr, E. M., Moore, M. J., Johnson, T., Forrest, J. & Jordan, M. (2013). New evidence: data documenting parental support for earlier sexuality education. *Journal of School Health*, 84(1), 10–17.

Bauer, J. E. (2003). Magnus Hirschfeld's doctrine of sexual intermediaries and the transgender politics of (no-)identity. Paper given at the conference Past and Present of Radical Sexual Politics, Amsterdam, 3–4 October 2003. Available at: www.iisg.nl/womhist/radsexpol.html#papers.

Birke, L. (1997). Born queer? Lesbians interrogate biology. In G. Griffin & S. Andermahr (Eds.), *Straight Studies Modified: Lesbian Interventions in the Academy*. London: Cassell, 57–70.

Britzman, D. P. (1995). Is there a queer pedagogy? or, stop reading straight. *Educational Theory*, 45, 151–165.

Bunting, L., Tsibulsky, I. & Boivin, J. (2013). Fertility knowledge and beliefs about fertility treatment: findings from the international fertility decision-making study. *Human Reproduction*, 28(2), 385–397.

Buston, K., Wright, D., Hart, G. & Scott, S. (2002). Implementation of a teacher-delivered sex education programme: obstacles and facilitating factors. *Health Education Research*, 17(1), 59–72.

Callahan, G. N. (2009). *Between XX and XY: Intersexuality and the Myth of Two Sexes*. Chicago: Chicago Review Press.

DePalma, R. & Atkinson, E. (Eds) (2009). *Interrogating Heteronormativity in Primary Schools: The No Outsiders Project*. Stoke-on-Trent: Trentham Books.

Findlen, P. (1990). Jokes of nature and jokes of knowledge: the playfulness of scientific discourse in early modern Europe. *Renaissance Quarterly*, 43(2), 292–331.

Giffney, N. (2009). Introduction: the 'q' word. In N. Giffney & M. O'Rourke (Eds.), *The Ashgate Research Companion to Queer Theory*. Burlington, VT: Ashgate, 1–13.

Halstead, J. M. & Reiss, M. J. (2003). *Values in Sex Education: From Principles to Practice*. London: RoutledgeFalmer.

Hamani, Y., Sciaki-Tamir, Y., Deri-Hasid, R., Miller-Pogrund, T., Milwidsky, A. & Haimov-Kochman, R. (2007). Misconceptions about oral contraception pills among adolescents and physicians. *Human Reproduction*, 22(12), 3078–3083.

Hamilton, W. D. (1980). Sex versus non-sex versus parasite. *Oikos*, 35(2), 282–290.

Harvey, I. & Reiss, M. J. (1992). *AIDSFACTS: Educational Material on AIDS for Teachers and Students*, 4th edn. Cambridge: Daniels Publishing.

Hershey, D. R. (2005). *More Misconceptions to Avoid When Teaching About Plants*. Washington, DC: American Institute of Biological Sciences.

Ingham, R. (2017). Section introduction: intended and unintended effects of sex education. In P. Aggleton (Ed.), *Education and Sexualities, Vol III*. London: Routledge, 2–6.

Kirby, D. B. (2008). The impact of abstinence and comprehensive sex and STD/HIV education programs on adolescent sexual behavior. *Sexuality Research & Social Policy*, 5(3), 6–17.

Letts, W. (2001). When science is strangely alluring: interrogating the masculinist and heteronormative nature of primary school science. *Gender and Education*, 13, 261–274.

Martin, E. (1991). The egg and the sperm: how science has constructed a romance based on stereotypical male-female roles. *Signs: Journal of Women in Culture and Society*, 16, 485–501.

Martino, W. & Pallotta-Chiarolli, M. (2005). *Being Normal is the Only Way to Be: Adolescent Perspectives on Gender and School*. Sydney: University of New South Wales Press.

Maxwell, C. (2017). Section introduction: addressing gender and sexual violence in schools. In P. Aggleton (Ed.), *Education and Sexualities, Vol II*. London: Routledge, 216–221.

Miller, W. R. & Lief, H. I. (1979). The sex knowledge and attitude test (SKAT). *Journal of Sex and Marital Therapy*, 5(3), 282–287.

Nguyen, S. & Rosengren, K. (2004). Parental reports of children's biological knowledge and misconceptions. *International Journal of Behavioral Development*, 28(5), 411–420.

Primoratz, I. (1999). *Ethics and Sex*. London: Routledge.

Reiss, M. J. (1993). What are the aims of school sex education? *Cambridge Journal of Education*, 23, 125–136.

Reiss, M. J. (1998). The representation of human sexuality in some science textbooks for 14–16 year-olds. *Research in Science & Technological Education*, 16, 137–149.

Reiss, M. (2007). Representing the world: difference and science education. In M. Reiss, R. DePalma & E. Atkinson (Eds.), *Marginality and Difference in Education and Beyond*. Stoke-on-Trent: Trentham Books, 61–72.

Reiss, M. J. (2014). Sex education and science education in faith-based schools. In J. D. Chapman, S. McNamara, M. J. Reiss & Y. Waghid (Eds.), *International Handbook of Learning, Teaching and Leading in Faith-Based Schools*. Dordrecht: Springer, 261–276.

Reiss, M. J. (2017). Education and sexualities: the next generation. In P. Aggleton (Ed.), *Education and Sexualities, Vol IV*. London: Routledge, 195–210.

Reiss, M. J. & White, J. (2013). *An Aims-Based Curriculum: The Significance of Human Flourishing for Schools*. London: IOE Press.

Roberts, C. (2002). 'A matter of embodied fact': sex hormones and the history of bodies. *Feminist Theory*, 3, 7–26.

Scharmann, L. C. (1991). Teaching angiosperm reproduction by means of the learning cycle. *School Science and Mathematics*, 91(3), 100–104.

Schüklenk, U. & Brookey, R. A. (1998). Biomedical research on sexual orientation: researchers taking our chances in homophobic societies. *Journal of the Gay and Lesbian Medical Association*, 2(2), 79–84.

Schussler, E. E. (2008). From flowers to fruits: how children's books represent plant reproduction. *International Journal of Science Education*, 30(12), 1677–1696.

Stanger-Hall, K. F. & Hall, D. W. (2011). Abstinence-only education and teen pregnancy rates: why we need comprehensive sex education in the U.S. *PLoS ONE*, 6(10), e24658. https://doi.org/10.1371/journal.pone.0024658.

Yip, A. K.-T. & Page, S.-J. (2013). *Religious and Sexual Identities: A Multi-Faith Exploration of Young Adults*. Farnham: Ashgate.

Yip, D. Y. (1998). Children's misconceptions on reproduction and implications for teaching. *Journal of Biological Education*, 33(1), 21–26.

9

DEVELOPMENT

Kostas Kampourakis and Panagiotis K. Stasinakis

Introduction

To the best of our knowledge, there has been very little recent research in secondary students' conceptions about biological development (e.g. Hiatt et al., 2013). This is quite surprising as there is research on what young children think about, e.g., the development of animals. Interestingly, in those cases of animals whose development includes extensive transformations, known as metamorphoses (e.g., butterflies), it has been found that by the age of 5 years, children may realize that organisms can undergo radical developmental changes (Rosengren et al., 1991). It has also been found that children between 4.5 and 6.5 years old can also provide natural explanations for plant growth, based on features of the external environment, such as whether it has been raining (Christidou & Hatzinikita, 2006).

However, it seems that there is no similar research at the secondary level. Even though the main concepts of development are included in the standards documents, such as the core idea 'LS1.B: Growth and Development of Organisms' in Next Generation Science Standards (2013, p. 145), the impact of the respective processes is not usually discussed in detail. A main reason for this is that the relevant science is relatively recent. It is only during approximately the last 10–15 years that we have come to understand the impact of developmental processes on both individual-level and species-level characteristics. This understanding has emerged from findings in two currently very active research fields: genomics and evolutionary developmental biology (or evo-devo). Thus, the respective findings are relatively new to science educators, and they may therefore feel unprepared to pedagogically manage this topic. However, not discussing the details of development in secondary biology deprives students of an understanding of organisms and life.

Developmental Biology as a Discipline

All students should be expected to be familiar with developmental processes. They have seen themselves and any siblings develop. They may have also made related observations on their pets and plants. However, this does not entail that they understand these developmental processes. For instance, mammals pass through some overall similar developmental stages. After fertilization, an embryo undergoes a number of cell divisions that result in a spherical mass of cells. This then undergoes a tremendous change, resulting in an embryo with bilateral symmetry and three distinct cell layers (ectoderm, mesoderm, endoderm), which develops within an amniotic sac. The initially pluripotent cells of the embryo will gradually differentiate to more specialized cell types, giving rise to the various tissues and organs of the body. Development is quite different in plants. Most of plant development occurs not in the embryo but in the growing plant. All the 'adult' structures of a plant (shoots, roots, leaves and flowers) are produced from localized groups of undifferentiated cells, the meristems (for these and various other examples, see Wolpert, 2011).

Development comprises two major phenomena: growth and differentiation. Growth occurs when the total number of body cells increases due to cell proliferation. At the same time, cells also differentiate and give rise to the various tissues and organs of the body. Both growth and differentiation are determined by local signals and interactions; there is no centralized coordination of development (Davies, 2014, pp. 38–40). Development is characterized both by plasticity and robustness. Developmental robustness is the capacity of individuals of the same species to exhibit the general characteristics of their species irrespective of the environment they live in. Developmental plasticity is the capacity of individuals of the same species with the same genotype to produce different phenotypes during development as a response to local environmental conditions (Bateson & Gluckman, 2011, pp. 4–5, 8). To give one example, *Arabidopsis thaliana* plants have several shared characteristics, such as a stem about 20 cm tall, with a few leaves on it and more forming a rosette at its base. However, when *Arabidopsis thaliana* plants develop while exposed to mechanical stimulation, they develop shorter petioles and bolt compared to plants with the same genotype that were not exposed to mechanical stimulation (Pigliucci, 2005). Developmental plasticity and robustness are two complementary aspects of development that indicate its complexity.

Perhaps contrary to what one might expect, people do not often talk about a theory of development, at least not in the way they talk about a theory of heredity or a theory of evolution. A reason for this is that there is no consensus among experts on this topic. At the philosophical level, a disagreement seems to exist about whether developmental biology can offer such a theory. On the one hand, one can argue that if we consider theories as structured sets of testable explanatory and predictive hypotheses, then establishing such theories is a major goal of contemporary developmental biology; researchers try to explain and predict phenomena, give an important role to testability and aim for unification of the observed phenomena. And even if a single general theory of development does not exist, constructing several such theories is a major goal (Pradeu, 2014). On the other hand, there is the view that a central theory, or a family of theories, is not necessary for structuring knowledge and inquiry in science, and that developmental biology is an example of this. Developmental biology is informed by knowledge from different domains; however, it is not organized by a theory but rather by means of questions. Developmental biologists investigate phenomena such as differentiation, growth, morphogenesis and pattern formation on the basis of multiple questions, which form an anatomical problem-structure for developmental biology (Love, 2014).

Are there reasons to prefer one view over the other philosophical view? Developmental biologists seem to be divided, too. Whereas in evolution and heredity there exist theories and models that are widely applicable, local events and mechanisms are very important in development and produce such a heterogeneity that generalizations are difficult to make and, thus, a general theory of development is difficult to conceive (Arthur, 2014). The importance of local contexts is nicely illustrated by phenomena such as developmental plasticity, already mentioned above, in which the environment affects which one of several phenotypes encoded in DNA will emerge. Another such example is phenotypic heterogeneity, the phenomenon that the same mutation can result in different phenotypes in different individuals; this occurs because genes do not act alone but are rather implicated in complex developmental pathways and networks. However, even though developmental processes are very complex, there exist about 50–60 processes (gene regulatory networks and the process networks that they affect) that underlie development (Gilbert & Bard, 2014). Thus, it could indeed be the case that there are some common principles of development shared by all organisms, which could form a basis for a theory.

The important point from all the above is that it is not self-evident how one could organize development education (compare this to evolution and genetics education, for instance). While considering this, we should note that in biology education we usually have in mind a rather narrow conception of development, which could be described as 'adultocentric' because it focuses on multicellular organisms and on their developmental processes from a fertilized egg towards the adult form. However, this conception does not apply to all organisms; there exist organisms with multigenerational life cycles, as well as ones that consist of more than one genome. The conception of a simple, linear, developmental process does not apply in such cases. Furthermore, many unicellular organisms undergo various processes that might also be described as developmental. All this makes even the definition of what development consists of conceptually demanding (Minelli, 2011).

These considerations highlight some important properties of developmental processes: they are local, context-dependent, complex and variable. Therefore, it may be difficult to teach students about development *per se*. Nevertheless, it is exactly because of these properties that development provides extremely important conceptual tools to better understand genetics and evolution. The underlying idea is the same: adult traits can change due to changes in developmental processes, either within generations (across developmental time) or across generations (across evolutionary time). While students usually learn that genes contain the blueprint for the adult form of an organism, research in genomics supports the conclusion that this is not the case and that several different factors besides genes affect the development of phenotypes. At the same time, although some students find it difficult to understand how large morphological changes take place relatively quickly and thus find hard to understand and accept evolutionary theory, recent research in evolutionary developmental biology supports the conclusion that minor changes in development can result in large morphological changes in adult forms. As has been stated elsewhere, "evolution makes more sense in the light of development" (Kampourakis & Minelli, 2014); this is also the case for genetics (see Kampourakis, 2017). Why development matters for a conceptual understanding of genetics and evolution is explored in the next two sections.

Development and Genetics

Overall, researchers in science education have found that secondary students tend to think of genes as the ultimate determinants of traits. For instance, a study with 383 British students, 14–16-year-olds, found that when they were asked the question, "Why are genes important?", 73% of them responded that genes are involved in the determination of characteristics and only 14% that they are involved in the transfer of information (Lewis & Wood-Robinson, 2000). In another study with 64 15–16-year-old students in the USA, it was found that students tend to map the information included in DNA at higher organization levels rather than that of proteins. In other words, students could map the molecular (genotypic) level directly onto higher levels (which may include that of the phenotype), overlooking the important role of proteins and all other phenomena and mechanisms that result in the production of phenotypes (Duncan & Reiser, 2007).

Research in psychology has also documented people's tendency to infer one's characteristics and behaviors from one's perceived genetic makeup. People may thus tend to view genetically influenced traits as immutable and determined; the relevant genes as being the fundamental cause of the respective trait or condition; and groups that share a genetic trait as being homogeneous and discrete, and to perceive characters or conditions as more natural if they are genetically determined (Dar-Nimrod & Heine, 2011). It has also been found that people tend implicitly (and thus unconsciously) to associate genes with fate concepts (Gould & Heine, 2012). More recent studies have supported the conclusion that people tend to consider a genetic cause as more important than an environmental one, as for instance participants who had read a genetic explanation for obesity ate more in a follow-up task than those who had read a psychosocial explanation, or no explanation at all (Dar- Nimrod et al., 2014).

What genomics research consistently shows is that biological phenomena should be approached holistically by paying attention to phenomena at various levels (cells, tissues, organs) and not only at the level of genes. As already noted, an organism can respond to changing environments through changes in its development and eventually its phenotype. Therefore, genes do not predetermine phenotypes; genes are only implicated in the development of phenotypes. It is indeed the case that changes in single genes can have a big phenotypic impact. But, whereas a change in a gene may even disrupt a whole developmental process, the final outcome is in no way determined by this gene alone (Kampourakis, 2017). Several genes co-contribute to the making of adult phenotypes.

Let us consider a concrete example. The *SRY* (sex determining region Y) gene on the Y chromosome is a gene related to the development of male features. The default developmental outcome for the human embryo is to become a female. The expression of the *SRY* gene is what makes the difference in the outcome because it affects a pathway that guides the development of the male or the female sexual organs. In this sense, *SRY* makes a difference for the development of sex. Embryos carrying the Y chromosome and the *SRY* gene develop testes and a male reproductive system, whereas those not having either the Y chromosome or the *SRY* gene develop ovaries and a female reproductive system (Davies, 2014, pp. 147–151). Interestingly, it has been found that a mutation in the *SRY* gene is enough to make an XY individual develop as a female with underdeveloped reproductive organs (Jäger et al., 1990). It has also been found that a translocation of part of the Y chromosome including the *SRY* gene onto the X chromosome makes an XX individual develop as a hermaphrodite (carrying both male and female organs) (Margarit et al., 2000). One might thus be tempted to state that *SRY* is the gene for sex.

However, if one looks carefully at the details of the process, several proteins (and therefore genes) are involved in the process of sex differentiation. The bi-potential precursor of gonads (testes and ovaries) is established by various proteins, including SF1 and WT1, the early expression of which might also initiate that of SOX9 in both sexes. Beta-catenin can begin to accumulate at this stage, and in XX cells its levels can repress SOX9 production. However, in XY cells, increasing levels of SF1 activate the production of SRY that, along with SF1, enhances SOX9 expression. If SRY activity is weak, low or late, there is no SOX9 expression as beta-catenin levels accumulate and shut it down. In the testis, SOX9 promotes the testis pathway, and it can do so even in the absence of SRY (Sekido & Lovell-Badge, 2009). Therefore, the *SRY* gene does nothing on its own. Sex is the outcome of a complex developmental process that involves several factors. Of course, mutations in this gene can have an enormous impact on the development of an individual. But to understand their effect, one has to consider the whole process of sex development.

It is nowadays clear that genes do not determine characters, but they are implicated in their development. Genes are resources that provide cells with a generative plan about the development of the organism and have a major role in this process through their products. This plan is the resource for the production of robust developmental outcomes that are at the same time plastic enough to accommodate changes stemming from environmental signals. Therefore, in order to understand how characters and diseases arise, it is important to consider developmental processes. Traits develop, and genes have a central role in this process. But to understand this process, we should consider genes in the broader context (cellular, organismal, environmental) in which they exist and in which development takes place (Kampourakis, 2017).

Development and Evolution

As already mentioned, development exhibits both robustness and plasticity. Now, what is interesting is that the plastic responses of organisms to environmental conditions during their development can become stable and inherited by their offspring. In fact, this is how novel, complex phenotypic traits can emerge. Changes in developmental processes can generate novelty and variation that can be selected, and this is how evolution can proceed (Brigandt & Love, 2010). This is the focus of evolutionary developmental biology (usually dubbed as evo-devo). Evo-devo provides us with a new view to understanding the emergence of novel traits in organisms. The evolution of development based on i) the duplication of patterns and processes, from individual genes or genome elements; and ii) the modification from expansion of specific domains and regulatory network interactions to the coordinated modification of organ systems and co-option at multiple levels of biological organization, has led to new developmental patterns and has thus resulted in the evolution of novel traits (Moczek et al., 2015).

This is very important because problems with accepting evolution are often related to explanations of major evolutionary transitions and the origin of complex organs without design (Kampourakis, 2014). Such explanations always refer to a timescale of millions of years that we cannot directly experience. However, simply mentioning that earth sciences suggest that our planet has an age of thousands of millions of years does not make people understand how the human brain or a rose's flower may have evolved. But biological forms, including those with the most complex structures and functions, are also the products of development, a process that anyone can experience in a variety of different organisms. Development is thus easier for people to perceive compared to macro-evolutionary changes (micro-evolution is easier than macro-evolution to perceive and even to observe); in addition, any evolutionary change in the mature, adult form of animals and plants is somehow related to modifications of developmental processes. Therefore, understanding how a developmental process can be modified with particular evolutionary consequences could be a useful context for teaching about evolutionary explanations of organisms. Evo-devo is about both how developmental processes evolve and how changes in these processes impact the evolution of organismal traits. Therefore, learning about development is essential to understanding evolution and the integration of development and evolution (Love, 2013a, 2013b).

In other words, by shifting attention from the large differences between adult phenotypes to the relatively minor differences between the underlying developmental processes, we can go beyond the popular population view of evolution that strictly focuses on natural selection, i.e. on the *survival of the fittest*, and look at the *arrival of the fittest*, i.e. at the mechanisms producing the phenotypic variation that is the raw material for natural selection (Gilbert, 2003). This requires an understanding that development is not the simple, invariant and uniquely determined expression of the genotype, even though the variation that can be produced in a population is (probably always) not entirely random but rather developmentally biased in favor of some alternatives. Organisms have an inherent potential in their genetic material that drives the development of some traits but not others. The respective properties or processes are described as developmental constraints (Arthur, 2004). These properties or processes limit the potential outcomes of evolution. However, evolution can also be 'constraint-breaking'. Rare beneficial mutations make possible the development of very different forms. This can make possible the emergence of a new species that can occupy a previously unoccupied niche. This process of niche-filling evolution can take place whenever there is a vacant niche and a new (mutant) form can occupy it. This process does not depend on selection alone but also, and primarily, on those mutations that can make the difference in bringing about new developmental outcomes (Nei, 2013, pp. 186–187).

To understand this, it is important for students and their teachers to realize that differences in development may contribute to evolutionary change. Such changes are described as developmental repatterning (Arthur, 2004, 2011), and there are four main kinds:

1. *Differences in the timing of developmental events (heterochrony)*. One example is the change of relative timing of egg hatching and segment formation in arthropods. In members of the order Geophilomorpha, segment formation ends before egg hatching, whereas in members of the order Lithobiomorpha segment formation ends long after that. As a result, in the former group the full number of segments has been reached at hatching, whereas in the latter group this number increases over the course of a year after that (Arthur, 2011, pp. 95–97).
2. *Differences in the spatial location of developmental events (heterotopy)*. An example of heterotopy is the morphology of flatfish, the head of which is partially rotated in relation to the rest of the body, resulting in both eyes ending up on the same side of the skull. In this case, the embryos have bilateral symmetry;

however, during development one of the eyes moves across the head and ends up on the same side as the other (Arthur, 2011, pp. 107–109).

3. *Differences in the type of developmental events (heterometry).* One example is the increased brain size of humans compared to chimpanzees. We are roughly 1.25 times larger than chimpanzees, but our brains are 2.7 times larger than theirs (Arthur, 2011, pp. 122–123).
4. *Differences in the amount of activity in developmental events (heterotypy).* The production of hemoglobin S (HbS) in humans instead of the normal hemoglobin A (HbA) is a case of heterotypy (Arthur, 2011, pp. 135–136).

Students do not need to learn the details of these processes. But understanding through some concrete examples that changes in the timing, positioning, amount or type of expression of genes may produce novel characters could help them to also understand that minor changes in development can result in large morphological changes.

The above does not imply that evo-devo concepts are easy to understand. Indeed, Hiatt et al. (2013) identified several conceptual difficulties. Some difficulties related to the understanding of biology in general (26%). Among these, they found misconceptions such as design teleology (50%), misuse of terms (30%), essentialism (10%) and others (anthropomorphism, negative connotations, personification—10% in total). Other difficulties were due to lack of an understanding of development (41%). Among these, they found that students did not mention development (87%) and other misconceptions (such as that a single gene affects a single trait, that gene products are organismal phenotypes, that the environment is irrelevant to phenotype or that HOX genes are the only regulatory genes—13% in total). Several other difficulties related to misunderstandings of evolution (17%). Among these, they found misconceptions such as that characteristics that are not used are lost (21%), the inheritance of acquired traits (19%), that a lack of selection results in stasis (8%), attributing adaptation to individuals, thus exhibiting a lack of understanding of population-level processes (17%), that older clades are more morphologically diverse (9%) and others (all evolution results in adaptation, all evolution is gradual, there is a perfect phenotype, positive natural selection is the only mechanism of evolution, selection acts on genes and not the phenotype, wrong definitions of selection—26% in total). Finally, there were difficulties related to evo-devo (8%). Among these, they found misconceptions such as that changes in gene expression result only from mutations in the gene (39%), that gene expression evolves only when genes appear or disappear (27%), that phenotypic change can only result from a gene appearing or disappearing (25%) and that only closely related species have conserved traits (9%).

However, what we are suggesting here is not to teach evo-devo *per se*. Rather, we suggest using concepts of evo-devo to address students' misconceptions about evolution. There is a lot of research that provides evidence that secondary students—and even their teachers (Stasinakis & Athanasiou, 2016)—tend to explain evolution in scientifically incorrect terms, e.g., by attributing adaptation to purpose and design in nature even after evolution teaching (e.g., Kampourakis & Zogza, 2008, 2009). If students cannot understand how large morphological change is possible, or if anti-evolutionists argue that large morphological change is impossible, then these are problems that science education should address—and evo-devo provides the conceptual tools for this purpose. This is why evo-devo is a conceptually sound and potentially effective strategy to overcome people's difficulties in understanding the evolutionary origin of complex organismal traits.

Teaching Genetics and Evolution From a Developmental Perspective

Genetics has generally been taught from a gene-centric perspective, whereas evolution has generally been taught from a population genetics perspective. Students learn, e.g., how a population of individuals with a trait A can evolve to a population of individuals with a trait B. The usual representation of this evolutionary change is that As change to Bs, but it is also possible to find representations as simplistic as A→B. Just think of the typical examples of how, e.g., the Galapagos finches with large beaks evolved from ancestors with shorter beaks, or—even worse—how 'a' finch with a short beak can evolve to 'a' finch with a large beak. But even if the examples presented to students are about populations evolving some new trait, there is still the problem that the take-home message for students is that an adult phenotype can evolve to a different adult phenotype. In the case of short-beaked and large-beaked finches, this may be easy for students to grasp. But how can they conceive that whales and dolphins have evolved from mammalian quadruped ancestors? This is where development should enter the scene.

The first issue is for students to understand that all adult forms develop from embryonic forms, and that all traits are the outcomes of complex developmental processes. Take, for example, the large beak of the Galapagos finches of species *Geospiza magnirostris*. Developing a beak in birds entails developing a three-dimensional structure that can change along any of the growth axes, depending on the differential expression of genes along molecular pathways (Abzhanov et al., 2006). It might be intuitive for students to think that there exists a specific 'gene for' these beaks, and that any finch with a particular allele would have the same beak, notwithstanding other differences. This is where a teacher's intervention with a developmental perspective could make a difference. If the teacher explained the general process of the development of a beak, without getting into too much detail, it might be possible for students to start realizing that minor variations in the developmental processes might produce a variety of beaks of similar but not exactly the same size and shape. With this in mind, students might then also realize that a change in development, e.g., because of change in DNA that affected the expression of those genes, might significantly affect the outcome. For instance, the over-expression of one gene might result in a significantly larger

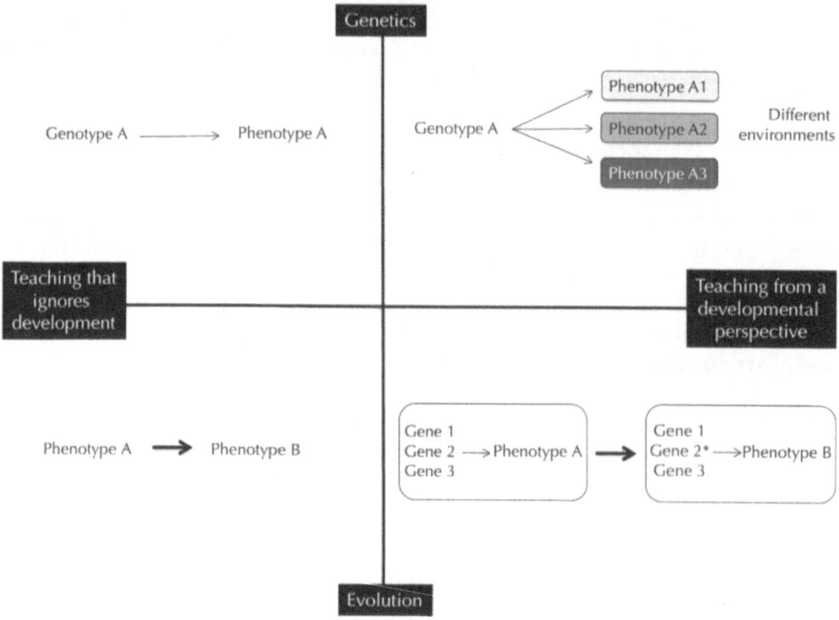

FIGURE 9.1 A developmental reconceptualization of genetics and evolution teaching. The asterisk after 'Gene 2' indicates that there was a change in the expression of the gene and not in the gene itself

beak in the adult form. Now, if this new variant was heritable and if the new phenotype provided an advantage in a particular environment, it might be favored by natural selection and become prevalent in a population. Thus, a population of large-beaked finches could have evolved from a population of small-beaked ones because of a change in development.

This developmental reconceptualization of genetics and evolution teaching is not difficult to do, and it might yield significant learning outcomes. The important condition is to replace the simplistic representations on the left of Figure 9.1 with those on the right of the same figure. To achieve this, the teacher would need to teach about development along with genetics and evolution.

Implications for Secondary Biology Education

Given all the above, we conclude by suggesting concepts of development that secondary students should become familiar with during their biology teaching:

- *Development*: The processes of growth, change and transformation that organisms undergo during their life cycles. Development includes changes that entail increasing order, through the differentiation and integration of specialized parts.

- *Developmental robustness*: The capacity of individuals to develop to an adult form that exhibits the general characteristics of a species irrespective of the environment they live in.
- *Developmental plasticity*: The capacity of individuals of the same species with the same genotype to exhibit phenotypic variation depending on local conditions.
- *Developmental constraints*: Any properties or processes that limit or facilitate evolutionary change by biasing the possible outcomes of development, thus making specific types of variation available for natural selection.
- *Developmental repatterning*: Changes in the timing, positioning, amount or type of expression of a DNA sequence, implicated in some developmental process, which produce novel characters.

We suggest that these concepts should be introduced in secondary biology teaching. Students should first learn more about development and about how it produces phenotypes in a robust manner. Then, teaching about developmental plasticity could address students' misconceptions about genes as the ultimate determinants of traits, whereas teaching about developmental repatterning could help students understand how large morphological changes are possible. Both genetics and evolution would make more sense in the light of development.

References

Abzhanov, A., Kuo, W. P., Hartmann, C., Grant, B. R., Grant, P. R. & Tabin, C. J. (2006). The calmodulin pathway and evolution of elongated beak morphology in Darwin's finches. *Nature*, 442, 563–567.

Arthur, W. (2004). *Biased Embryos and Evolution*. New York: Cambridge University Press.

Arthur, W. (2011). *Evolution: A Developmental Approach*. Oxford: Wiley-Blackwell.

Arthur, W. (2014). General theories of evolution and inheritance, but not development? In A. Minelli & T. Pradeu (Eds.), *Towards a Theory of Development*. Oxford: Oxford University Press, 144–154.

Bateson, P. & Gluckman, P. (2011). *Plasticity, Robustness, Development and Evolution*. Cambridge: Cambridge University Press.

Brigandt, I. & Love, A. C. (2010). Evolutionary novelty and the Evo-Devo synthesis: field notes. *Evolutionary Biology*, 2, 37, 93–99.

Christidou, V. & Hatzinikita, V. (2006). Preschool children's explanations of plant growth and rain formation: a comparative analysis. *Research in Science Education*, 36(3), 187–210.

Dar-Nimrod, I., Cheung, B., Ruby, M. & Heine, S. (2014). Can merely learning about obesity genes affect eating behavior? *Appetite*, 81, 269–276.

Dar-Nimrod, I. & Heine, S. J. (2011). Genetic essentialism: on the deceptive determinism of DNA. *Psychological Bulletin*, 137(5), 800–818.

Davies, J. A. (2014). *Life Unfolding: How the Human Body Creates Itself*. Oxford: Oxford University Press.

Duncan, R. G. & Reiser, B. J. (2007). Reasoning across ontologically distinct levels: students' understandings of molecular genetics. *Journal of Research in Science Teaching*, 44(7), 938–959.

Gilbert, S. F. (2003). Opening Darwin's black box: teaching evolution through developmental genetics. *Nature Reviews Genetics*, 4, 735–741.

Gilbert, S. F. & Bard, J. (2014). Formalizing theories of development: a fugue on the orderliness of change. In A. Minelli & T. Pradeu (Eds.), *Towards a Theory of Development*. Oxford: Oxford University Press, 129–143.

Gould, W. A. & Heine, S. J. (2012). Implicit essentialism: genetic concepts are implicitly associated with fate concepts. *PLoS ONE*, 7(6), e38176. doi:10.1371/journal.pone.0038176

Hiatt, A., Davis, G. K., Trujillo, C., Terry, M., French, D. P., Price, R. M. & Perez, K. E. (2013). Getting to evo-devo: concepts and challenges for students learning evolutionary developmental biology. *CBE Life Sciences Education*, 12, 494–508.

Jäger, R. J., Anvret, M., Hall, K. & Scherer, G. (1990). A human XY female with a frameshift mutation in the candidate testis- determining gene SRY. *Nature*, 348, 452–454.

Kampourakis, K. (2014). *Understanding Evolution*. Cambridge: Cambridge University Press.

Kampourakis, K. (2017). *Making Sense of Genes*. Cambridge: Cambridge Universiy Press.

Kampourakis, K. & Minelli, A. (2014). Evolution makes more sense in the light of development. *The American Biology Teacher*, 76(8), 493–498.

Kampourakis, K. & Zogza, V. (2008). Students' intuitive explanations of the causes of homologies and adaptations. *Science & Education*, 17(1), 27–47.

Kampourakis, K. & Zogza, V. (2009). Preliminary evolutionary explanations: a basic framework for conceptual change and explanatory coherence in evolution. *Science & Education*, 18(10), 1313–1340.

Lewis, J., Leach, J. & Wood-Robinson, C. (2000). All in the genes?—young people's understanding of the nature of genes. *Journal of Biological Education*, 34(2), 74–79.

Love, A. C. (2013a). Interdisciplinary lessons for the teaching of biology from the practice of evo-devo. *Science & Education*, 22, 255–278.

Love, A. C. (2013b). Teaching evolutionary developmental biology: concepts, problems, and controversy. In K. Kampourakis (Ed.), *The Philosophy of Biology. A Companion for Educators*. Dordrecht: Springer, 323–241.

Love, A. C. (2014). The erotetic organization of developmental biology. In A. Minelli & T. Pradeu (Eds.), *Towards a Theory of Development*. Oxford: Oxford University Press, 33–55.

Margarit, E., et al. (2000). SRY gene transferred to the long arm of the X chromosome in a Y-positive XX true hermaphrodite. *American Journal of Medical Genetics*, 90(1), 25–28.

Minelli, A. (2011). Development, an open-ended segment of life. *Biological Theory* 6, 4–15.

Moczek, A. P., et al. (2015). The significance and scope of evolutionary developmental biology: a vision for the 21st century. *Evolution & Development*, 17(3), 198–219.

Nei, M. (2013). *Mutation-Driven Evolution*. Oxford: Oxford University Press.

Pigliucci, M. (2005). Evolution of phenotypic plasticity: where are we going now? *Trends in Ecology & Evolution*, 20(9), 481–486.

Pradeu, T. (2014). Regenerating theories in developmental biology. In A. Minelli & T. Pradeu (Eds.), *Towards a Theory of Development*. Oxford: Oxford University Press, 15–32.

Rosengren, K. S., Gelman, S. A., Kalish, C. W. & McCormick, M. (1991). As time goes by: children's early understanding of growth in animals. *Child Development*, 62(6), 1302–1320.

Sekido, R., & Lovell-Badge, R. (2009). Sex determination and SRY: down to a wink and a nudge? *Trends in Genetics*, 25(1), 19–29.

Stasinakis, P. K., & Athanasiou, K. (2016). Investigating Greek biology teachers' attitudes towards evolution teaching with respect to their pedagogical content knowledge: suggestions for their professional development. *Eurasia Journal of Mathematics, Science and Technology Education*, 12(6), 1605–1617.

Wolpert, L. (2011). *Developmental Biology: A Very Short Introduction*. Oxford: Oxford University Press.

10

GENETICS

Niklas Gericke and Charbel N. El-Hani

Introduction

This chapter is about teaching genetics and will focus on the classical didactic questions: what to teach? How to teach? And, when to teach? The 'what' question revolves around the important issue of didactic transposition, that is, the transformation of genetics from academic science into the genetics of school biology. A central question to be addressed is the relationship between classical (Mendelian) genetics and molecular genetics, and their relation to the gene concept. We will discuss these issues in light of what is usually found in school textbooks and problematize current teaching traditions. It might be the case that the way we teach genetics today is a reason for the well-documented learning difficulties in genetics. Then, we provide suggestions from educational research about how to teach genetics in secondary biology by including a nature of science perspective and building on 'big ideas' of genetics through a learning progression.

Genetics Education and the Problem of Didactic Transposition

Genetics teaching and learning is an important topic in school biology for two main reasons. First, genetics is a cornerstone of any biology course in lower as well as upper secondary biology curricula. Without genetics, it is impossible to explain the hereditary aspects of life; genetics is also fundamental for explaining evolution, a central theory in biology. Second, comparative studies have shown that genetics is perceived as the most difficult topic to teach and learn in biology, first in a study by Johnstone and Mahmoud (1980) that was partly confirmed twenty years later by another study (Bahar et al., 1999). Subsequently, biology education research has identified many specific learning difficulties and alternative understandings not in line with a scientific understanding of genetics (Gericke & Smith, 2014).

The reasons for the perceived learning difficulties are manifold. Knippels (2002) found the following claims in educational literature: domain-specific vocabulary and terminology; mathematical content of Mendelian genetics tasks; complex description of cytological processes; the order of topics presented in the biology curriculum, which generally separates meiosis from genetics; and, finally, the complex nature of genetics due to concepts and processes belonging to different organizational levels. Since then, several other reasons have been added to the list, but all the suggested learning difficulties can be related either to the 'inherent' complexity of the subject matter itself or to the way the subject matter is transformed and presented to students.

Genetics is nowadays a molecular science, and in that sense the subject matter is not more or less complex than other molecular sciences (see Chapter 4 on molecular biology in this volume). However, genetics research predates molecular biology. Genetics had a long history before it went molecular in 1953, when Watson and Crick proposed a model for the structure of DNA. In school biology this older version of genetics, called classical or Mendelian genetics, still comprises a major part of the curriculum.

Hence, a main problem to tackle in genetics education is how to relate and integrate older classical genetics with modern molecular genetics when teaching. This is very difficult since research within philosophy of science has shown that the ways the gene—the most fundamental concept of genetics—is understood and described in classical and molecular genetics are so fundamentally different that they are quite difficult to relate, particularly in a way that students can properly grasp.

This problem is even more accentuated through the process of didactic transposition. The transition from knowledge as a tool to be put to use to knowledge as something to be taught and learnt is what Chevallard (2007) has termed the didactic transposition of knowledge. Didactic transposition concerns the changes that a body of knowledge (and its uses) has to undergo in order to be taught and learnt in school. Evidently, the concepts of genetics were originally not designed to be taught, but to be used in the practice of science. Educational literature in the last ten years has shown that the way scholarly knowledge of genetics, including classical and molecular genetics, is transposed into school biology is a major learning obstacle for students' learning (Gericke & Smith, 2014).

The Different Faces of the Gene in Classical and Molecular Genetics

To understand the underlying epistemological problem when genetics content is transposed into school biology, we first need to take a short historical detour through how the understanding of the gene concept evolved in the biological sciences. The term 'gene' was introduced by Wilhelm Johannsen in 1909. Johannsen called an unobservable hereditary unity that produces the manifest character a 'gene'. The gene was first introduced instrumentally, as a calculation device, not in realist terms as a 'real thing'. 'Gene' was a theoretical term with no accompanying hypothesis about its material nature.

In classical genetics, in roughly the first half of the twentieth century, genes were regarded as any differences in the gametes related to variant traits in the organisms that were independently inherited. They were instrumental unobservable 'difference makers', defined by reference to specific constant relations between observable phenotypic characters (Falk, 1986). Genes were used to explain differences in phenotypes in terms of differences in genotypes (Waters, 1994). Initially inferred from the phenotype, genes were determinants of phenotypic differences in instrumental terms only.

Realist views about genes gained momentum since the 1930s, in a fruitful tension with instrumentalist interpretations, which deepened our understanding of gene structure and function (Falk, 1986). After Watson and Crick's double helix model, genes were reconceptualized as pieces of DNA in realist terms. They became referentially definite (Burian & Kampourakis, 2013) in the form of the 'classical molecular gene concept' (Neumann-Held, 1999), in which a gene was defined as an uninterrupted unit in DNA encoding a single functional product, either a single polypeptide chain or RNA molecule, with clearly delimited beginning and end, and a single function manifested in its encoded product. Genes were not inferred from phenotypes anymore. They were well-defined nucleotide sequences, corresponding to 'open reading frames'. With this inversion of the conceptual relation between genes and phenotypes, linked to a shift from instrumentalist to realist views, the classical molecular concept updated the interpretation of genes as units that had been pervasive since classical genetics.

A determinist parlance about the gene-phenotype relationship, which was warranted from an instrumental perspective, could not hold in realist terms, as it ignored gene-gene and gene-environment interactions. The combination of the instrumental determinism of classical genetics and the realist image of the molecular gene is an important source of misconceptions. While teaching genetics, it is important to distinguish between classical deterministic perspectives, acceptable only in instrumental terms, and molecular realist perspective on genes, only acceptable in non-deterministic terms. The lack of such a distinction has important consequences, since it reinforces genetic determinism, an overstated belief in the power of genes, i.e., the idea that single genes cause phenotypical characteristics without any influence of environmental factors. Although some traits fit this model, most traits are influenced by several genes in interaction with the environment. This is a more scientifically legitimate model to teach in school biology.

Conceptualizing Instrumental and Realist Views on Genes for Genetics Education

Moss' (2003) distinction between 'gene-P' and 'gene-D' is helpful in conceptualizing the distinction between instrumentalist and realist views on genes. Gene-P is the gene determinant of phenotypic differences, with no requisites concerning specific molecular sequences or biological processes resulting in the phenotype. It is the gene interpreted from an instrumental perspective, roughly going back to classical genetics. Gene-D is a developmental resource indeterminate with respect to the phenotype, a molecular entity that is a unit of transcription and/or translation, providing templates for protein or RNA synthesis. Gene-D is embedded into complex gene networks, and its product is embedded into complex interaction networks involved in physiology and development. Genes can be fruitfully conceived in those two different ways, but nothing good results from the conflation of gene-P and gene-D (Moss, 2003).

Gene-P is related to the expression 'gene for', often used in scientific and public understanding, as when we talk about a stretch of DNA as 'a gene for cystic fibrosis'. If interpreted in realist terms and applied to the individual organism, the claim that a DNA segment is a 'gene for a trait' is worrisome: the instrumental nature and proper domain to apply the concept—to explain phenotypic variation in a population in terms of genotypic variation—are neglected. When instrumental and molecular gene concepts are conflated, genetic deterministic views often ensue, as when complex multifactorial traits such as aggressiveness or intelligence are merely attributed to genes. This does not mean, however, that it makes no sense to talk about 'genes for' traits. This expression can be properly used if one is aware that it is an abbreviated way of referring to a relation between phenotypic variation and genetic variation, which can lead to hypotheses about causal relations between genetic *loci* and trait development. 'Gene for' is shorthand for "a locus in which sequence variation causes a difference in phenotype, all other things being equal" (Knight, 2007, p. 298).

The instrumental interpretation of genes is not just a remnant from a time when we did not know much about the genetic material. It is in use today (Orgogozo et al., 2016), showing that a definite material counterpart is not necessary for the gene to play a part in knowledge production. The instrumental nature of this way of understanding genes underlies its utility, since to simplify the genotype-phenotype relationship paves the road to studies associating genotypic to phenotypic variation. But when research on phenotype-genotype mapping shifts from association studies to testing functional hypotheses about the contribution of genomic regions to the phenotype, research and, accordingly, teaching should move from genes-P to genes realistically interpreted, i.e., genes-D.

Let us consider an example well known from secondary school teaching: the 'gene for blue eyes'. Using a simplifying deterministic assumption, we can easily calculate the proportion of blue- and brown-eyed people in a pedigree. But we would gain nothing in predictive power by considering the complex developmental processes connecting genetic *loci* to traits.

However, if we look at the 'gene for blue eyes' in realist terms, we cannot locate an allele situated in a well-defined *locus* as material counterpart. A series of alleles affects iris pigmentation, in the loci HERC2 and OCA2, but also in at least ten other *loci* and through complex gene interactions (e.g., Liu et al., 2010). The 'gene for blue eyes' corresponds to an expression like $(\mathrm{HERC2} \wedge \mathrm{OCA2} \wedge \mathrm{G3} \vee \mathrm{G4} \wedge \mathrm{G5} \wedge, \ldots, \vee \mathrm{G10} \wedge \mathrm{Gn})$. Surely, this is not a material entity to which the concept refers. This does not deny the utility of the instrumental gene-P concept, but means that to understand the causal role of *loci* involved in phenotype development we cannot rely just on association studies using gene-P. We need to move towards mechanistic models well-supported by evidence, considering the proximal role of DNA sequences, going beyond the instrumental gene-P, into the domain of gene-D.

While teaching about genes, it is important to distinguish between gene-P and gene-D. Moreover, if the expression 'gene for' is used, we should explicitly relate it to the description of associations between phenotypic and genetic variations at the population level, not at the individual level. The distinction is important because the limits of applying an instrumental interpretation, talking about 'genes for traits', are typically neglected, reinforcing genetic determinism.

Conceptualizing Genes Through Models in Genetics Education

Gericke and Hagberg (2007) developed a fine-grained reconstruction of how the gene concept, with a focus on the functional aspects, has been understood historically in the science of genetics. Five historical models were identified: *the Mendelian model, the classical model, the biochemical-classical model, the neoclassical model,* and *the modern model.* The first three models relate to classical genetics and an instrumentalist view of the gene, though some aspects of the classical and biochemical-classical models include realist views as well. The neoclassical model relates to a realist view, while the modern model accounts for the complexities of contemporary genetics—for instance, the fact that a single DNA segment can generate several gene products, or that gene expression can be modified by epigenetic markers, which can be altered in response to the environment or experience.

In these models, the different ways the gene has been understood through history are outlined (as discussed earlier). This is referred to as conceptual variation. The aspects of the gene concept where conceptual variation was found relate to: *the structure and function relation to the gene*; *the relationship between organization level and definition of gene function*; *the 'real' approach to define the function of the gene*; *the relationship between genotype and phenotype*; *the idealistic versus naturalistic relationships in the models*; *the reduction explanatory problem*; and *the relationship between genetic and environmental factors*.

In a systematic analysis of Brazilian, Swedish, and English-language high school science textbooks from six countries, Gericke et al. (2014) found a similar distribution in the use of the different models identified by Gericke and Hagberg (2007) (Figure 10.1).

It seems that a similar textbook discourse on genes is found in different countries with an emphasis on classical genetics and considerable attention given to molecular genetics, while neglecting the complexities of contemporary genetics (the modern model). Moreover, a conflation of attributes of different models is common in the textbooks, i.e., conceptual explanations are taken out of one model and integrated into another, leading to the creation of hybridized teaching models in textbooks, which amounts to a substantial part of the textbook descriptions.

But what about the conflation between deterministic views and realist views of molecular genes? These were found to be quite common in textbooks from all countries, since the molecular explanation, aligned with the classical molecular concept, is often hybridized and conflated with the instrumental view, as in gene-P, as exemplified in this quote: "Their [nucleotide base] sequence along the DNA-molecule constitutes the information about different traits in the organism. The part of a DNA-molecule which determines a certain trait is called a gene" (Henriksson, 2007, p. 28). This lends support to genetic determinism: the idea that single genes cause direct effects at the phenotypic level.

The textbooks did not address any epistemological underpinnings of classical or molecular genetics, making it impossible for students to understand that there are different theoretical frameworks for genetics. It can be assumed that this way

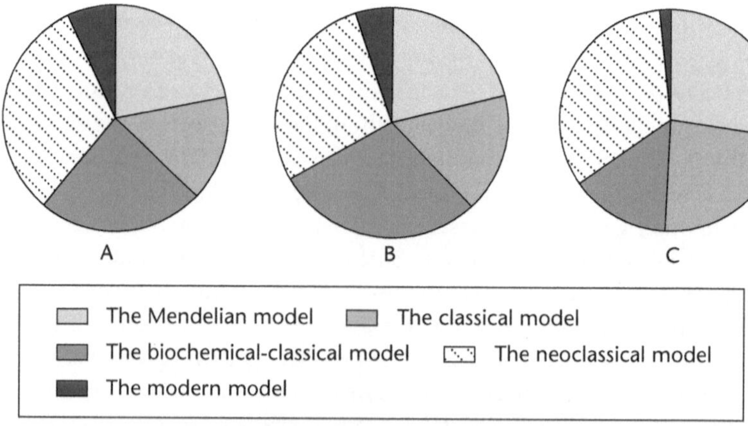

FIGURE 10.1 Distribution of gene function models in high school textbooks (in percentage). A: Swedish textbooks; B: English-language textbooks (Australia, Canada, England, U.S.); C: Brazilian textbooks

of portraying genetics in textbooks has great importance for teaching practices based on previous research on textbooks as curriculum enactor (Nelson, 2006).

As shown here, the body of knowledge we call 'genetics' is very complex, with different concepts and models used for explaining how genetic systems work (El Hani, 2007). If we are using the current curricula, which integrate Mendelian and molecular genetics unreflectively, the learner must conceptualize multiple and different ways of understanding genetics, as shown by the textbook analysis of multiple models accounting for the gene concept. To be able to do so, students must have insight into science as a way of knowing, how it is built—i.e., understand the epistemology of science. Epistemology is the study of the nature of knowledge, justification, and rationality of various beliefs. How we can know what a scientific claim is and how the body of scientific knowledge is built are important epistemological issues. In the school science curriculum, the epistemological aspects of content knowledge are often included into the concept of Nature of Science (Lederman & Lederman, 2014).

Models in science can be used to describe science as a way of knowing (Lehrer & Schauble, 2006). At the same time, models incorporate content knowledge within themselves. One cannot be genuinely understood without the other. If students are equipped with an epistemological understanding (i.e., Nature of Science), they are more likely to be able to discern different models and concepts and understand that they may be separate human products describing genes and that they may operate very differently. It is, for example, important to understand that only few genetic models can predict causally, while most models used in practice, such as gene technologies, only predict in a probabilistic way the outcomes of gene function at an individual level (Kampourakis, 2017). Hence, an epistemological understanding is needed to evaluate the explanatory power of a specific model. The described frameworks in this chapter, dealing with instrumental vs. realist views of genes, gene-P vs. gene-D, and the five historical models, are tools, with different degrees of detail that a teacher can use to make the complex conceptual structures of genetics explicit and visible to students.

Students' Alternative Understandings of Genetics

Most studies in genetics education research show that students of all ages have a tendency to use oversimplified causal explanations instead of biochemical terms or processes when explaining genetic phenomena (e.g., Lewis & Kattmann, 2004; Duncan & Reiser, 2007). Hence, students tend to fall back on descriptions and explanations from classical genetics rather than draw on molecular genetics. In a study in which students were asked to refer to and explain the genetic descriptions within their own textbooks, they used to a large degree concepts from classical genetics and conflated them with molecular genetics when referring to textbook sections addressing the latter (Gericke, et al., 2013). In such explanations, genes are viewed as inherited particles transferred from one generation to the next and, also, as the sole determinants of characteristics with inherent actions, i.e., the gene is thought of as a physical object that takes action in an unalterable way in the organism (e.g., Venville & Treagust, 1998). At the heart of the issue is the 'macro-micro' problem, i.e., that students must understand that genetics concepts and phenomena can be explained at multiple levels of organization: *molecular, cellular, individual*, and *population levels* (Knippels, 2002; Duncan & Reiser, 2007). Related to the problem of discerning organization levels are students' difficulties in distinguishing between genotype and phenotype, between genes and genetic information, and between alleles and genes.

Thus, the most frequent view found among students seems to be that genes are considered, from a realist perspective, to be physical particles and the sole determinants of phenotypic traits (as in 'the gene FOR blue eyes'), ignoring that gene expression is interactive and moderated by other factors—the fact that other components of the genome, cell, and environment impact phenotype as well (Shaw et al., 2008). It seems that students embrace deterministic explanations and tend to reduce classical genetics to molecular genetics.

Deterministic understandings of genetics typically focus on a one-to-one relationship between genes, proteins, functions, and traits, as if particular traits or diseases were generally related to a single gene. This is problematic, not only because it is a simplistic notion and an explanatory model for explaining gene function scientifically that is valid in a few cases, but also because such notions may lead to the naturalistic fallacy, i.e., the claim that what is natural (in this case genetically predetermined) is inherently good or right (Nelkin & Lindee, 2004).

A consequence of the naturalistic fallacy may be that characteristics of specific social group members are perceived as immutable, natural, and homogeneous, which in turn may induce prejudice and stereotypical ideas regarding those social groups, say, related to gender and race (Dar-Nimrod & Heine, 2011). Hence, beyond being a knowledge problem, an overly simplistic understanding of genetics may also support the development of problematic values among students. Besides being able to understand modern molecular genetics, including the probabilistic notion of gene function and its interaction with other genes and the environment, it is also important to be able to cope with personal decision-making in modern medicine, where individualized treatment based on genetic testing and genetic screening is becoming more common. In such a context, a simplistic causal understanding of genes may lead to bad decision-making when interpreting such tests (Bartol, 2014).

The reason why students tend to adopt simplistic notions going back to classical genetics when they are taught molecular genetics is not fully known. However, from didactical research we know that a widespread problem in science education is that once a student has internalized a conception, it is difficult to change her view (Duit & Treagust, 2003). Students more easily take in and accept intuitive ideas in line with their everyday thinking. It is perhaps more intuitive for students to understand genes as active agents that cause characteristics in a one-to-one fashion, as in gene-P, in contrast to understanding them as molecular templates that are resources used in protein or RNA synthesis, as in gene-D. However, it must be noted that we also invest a lot of time teaching Mendelian genetics in school biology (Dougherty et al., 2011) that are in line with these simplistic notions. Hence, the way we teach genetics might be part of the problem. In the next section, we explore alternative ways of teaching that address those potential learning problems.

Teaching Guidelines and Learning Progressions in Genetics Education

How can we teach genetics in a way that counteracts students' simplistic and deterministic understandings of genetics? In what ways can teaching facilitate students' learning of molecular genetics, multifactorial traits, and the implications of gene-environment/experience interaction to gene expression and function? In the research literature, several learning progressions have been proposed for genetics education. Learning progressions are models of how student learning develops over a period of several grades. They depart from consideration of students' prior knowledge and are built toward the goals set by the curriculum for a specific subject.

The learning progression for modern genetics proposed by Duncan et al. (2009) is perhaps the most extensive in the literature, spanning grades 5–10. They emphasized three key aspects of teaching and learning modern genetics: (1) the big ideas in modern genetics and the knowledge and abilities that students should master by the end of compulsory education; (2) the progression of learning that students are expected to make over several grades; and (3) the identification of learning performances and the development of assessments for the proposed progression.

The learning progression is built on the idea that knowledge of three integrated conceptual models, as identified by Stewart et al. (2005), is necessary in order for students to truly understand genetic phenomena. The first model is the genetic model, which deals with the patterns of inheritance observed when organisms reproduce sexually and the probabilities with which different patterns are likely to occur (sometimes referred to as classical or transmission genetics). The second model is the meiotic model, which relates to the cellular processes underlying gene recombination, sorting, and transfer from one generation to the next. The third model is the molecular model, which deals with the mechanisms that link genes to their biological outcomes. All three models are important in teaching, and are demanding for students to comprehend. Therefore, as already argued in this chapter, integrating these models is a very demanding task. The genetic model withholds an instrumentalist view of genes, as in the gene-P concept, while the other two are related to realist notions, as in the gene-D concept.

The learning progression by Duncan and colleagues is organized around two questions: (1) how do genes influence how we, and other organisms, look and function? and (2) why do we vary in how we, and other organisms, look and function? These questions are answered when the following big ideas of genetics are taught in various ways from grade 5–10:

(A) All organisms have genetic information that is hierarchically organized.
(B) The genetic information contains universal instructions that specify protein (primary) structure.
(C) Proteins have a central role in the functioning of all living organisms and establish the connection between genes and traits.
(D) All cells have the same genetic information but different cells use (express) different genes.
(E) Organisms transfer their genetic information to the next generation through reproduction.
(F) There are patterns of correlation between genes and traits and there are certain probabilities with which these patterns occur.
(G) Changes to the genetic information can cause changes in how we look and function (phenotype), and such variation in DNA can serve as a way to identify individuals and species.
(H) Environmental factors can interact with our genetic information.

This list of big ideas builds on many research studies identifying topics of importance for overcoming learning difficulties and counteracting simplistic understanding of genetics.

Another important issue for teaching concerns the inclusion of Mendelian genetics in the curricula, and the importance given to it, as discussed earlier, when we intend to refrain from having students develop simplistic understandings of genetics (as in deterministic views of genetics). Dougherty (2009) called for 'inverting the curriculum', beginning with a presentation of common qualitative traits instead of simple Mendelian ('monogenic') traits (as in (F) in the learning progression above) to address the common student misconception that most human traits fall under the latter category. Many studies propose similar ideas, i.e., that Mendelian and classical genetics should be given less importance in the curricula, and that it is better to start the teaching with molecular genetics, gene-D concepts, and polygenic inheritance. Recently, a study by Jamieson and Radick (2017) has shown that such inverted teaching can have positive effects on learning outcomes and inhibit the development of genetic deterministic understandings among students. However, as argued by Smith and Gericke (2015) we should not forget that Mendel and Mendelian genetics are an important part of the modern culture that students face, and therefore it also needs to be addressed in the curriculum from a literacy perspective. Moreover, it is useful to use the Mendelian model, with the gene-P concept, in genetics teaching, since it is both a simplified model of genetics and plays a role in research, but it needs to be used explicitly and with caution.

Conclusions

In this chapter, we have shown that the body of knowledge we refer to as genetics has a very complex conceptual structure. The reason for this is the historical development of different research programs within the science of genetics, generating different theoretical frameworks. For a scientist, this is generally unproblematic since they can intuitively move between the frameworks. However, in the didactic transposition to biology education, the body of knowledge is changed as it enters the biology curricula and textbooks, and some things are also lost. The different theoretical frameworks are dissolved and re-contextualized. During the re-contextualization processes into school genetics, the concepts from the different frameworks (gene-P/gene-D; instrumentalist/realist views; models of genetics) are sometimes hybridized and conflated. Epistemological aspects of science are also often left out in school genetics, making it difficult for novice learners, such as secondary school students, to interpret and make meaning of school genetics.

Teachers can counteract the effects of these learning barriers by using an epistemologically informed perspective in teaching, making those different conceptual frameworks explicit to students. Instead of conflating the classical and molecular genetics, it is possible to make them explicit in teaching and actively use them as teaching tools in the classroom. If these recommendations are accompanied by a teaching approach including the big ideas of genetics, as outlined in the learning progression discussed earlier, there are good possibilities that students will evolve a more advanced and contemporary understanding of genetics, contrary to deterministic conceptions applied in realist terms, so often reported in the educational literature.

References

Bahar, M., Johnstone, A. H. & Hansell, M. H. (1999). Revisiting learning difficulties in biology. *Journal of Biological Education*, 33(2), 84–86.

Bartol, J. (2014). Re-examining the gene in personalized genetics. *Science & Education*, 22(10), 2529–2546.

Burian, R. M. & Kampourakis, K. (2013). Against "genes for": could an inclusive concept of genetic material effectively replace gene concepts? In K. Kampourakis (Ed.), *The Philosophy of Biology: A Companion for Educators*. Dordrecht: Springer, 597–628.

Chevallard, Y. (2007). Readjusting didactics to a changing epistemology. *European Educational Research Journal*, 6(2), 131–134.

Dar-Nimrod, I. & Heine, S. J. (2011). Genetic essentialism: on the deceptive determinism of DNA. *Psychological Bulletin*, 137(5), 800–818.

Dougherty, M. J. (2009). Closing the gap: inverting the genetics curriculum to ensure an informed public. *American Journal of Human Genetics*, 85(1), 6–12.

Dougherty, M. J., Pleasants, C., Solow, L., Wong, A. & Zhang, H. (2011). A comprehensive analysis of high school genetics standards: are states keeping pace with modern genetics? *CBE-Life Sciences Education*, 10(3), 318–327.

Duit, R. & Treagust, D.F. (2003). Learning in science—From behaviourism towards social constructivism and beyond. In B. J. Fraser & K. G. Tobin (Eds.), *International Handbook of Science Education*. Dordrect, Netherlands: Kluwer Academic Publishers, 3–25.

Duncan, R. G. & Reiser, B. J. (2007). Reasoning across ontologically distinct levels: students' understanding of molecular genetics. *Journal of Research in Science Teaching*, 44(7), 938–959.

Duncan, R. G., Rogat, A. D. & Yarden, A. (2009). A learning progression for deepening students' understandings of modern genetics across the 5th-10th Grades. *Journal of Research in Science Teaching*, 46(6), 655–674.

El-Hani, C. N. (2007). Between the cross and the sword: the crisis of the gene concept. *Genetics and Molecular Biology*, 30(2), 297–307.

Falk, R. (1986). What is a gene? *Studies in the History and Philosophy of Science*, 17(2), 133–173.

Gericke, N., Hagberg, M., Santos, V. C., Joaquim, L. M. & El-Hani, C. N. (2014). Conceptual variation or incoherence? textbook discourse on genes in six countries. *Science & Education*, 23(2), 381–416.

Gericke, N. M. & Hagberg, M. (2007). Definition of historical models of gene function and their relation to students' understandings of genetics. *Science & Education*, 16(7–8), 849–881.

Gericke, N. M., Hagberg, M. & Jorde, D. (2013). Upper secondary students' understanding of the use of multiple models in biology textbooks—the importance of conceptual variation and incommensurability. *Research in Science Education*, 43(2), 755–780.

Gericke, N. & Smith, M. U. (2014). Twenty-first-century genetics and genomics: contributions of HPS—informed research and pedagogy. In M. R. Matthews (Ed.), *International Handbook of Research in History, Philosophy and Science Teaching*. Dordrecht, Netherlands: Springer, Vol. I, 423–467.

Henriksson, A. (2007). *Biologi kurs A* (2nd ed.). Malmö: Gleerups Förlag.

Jamieson, A. & Radick, G. (2017). Genetic determinism in the genetics curriculum—an exploratory Study of the effects of Mendelian and Weldonian emphases. *Science & Education*, 26(10), 1261–1290.

Johnstone, A. H. & Mahmoud, N. A. (1980). Isolating topics of high perceived difficulty in school biology. *Journal of Biological Education*, 14(2), 163–166.

Kampourakis, K. (2017). *Making Sense of Genes*. Cambridge, UK: Cambridge University Press.

Knight, R. (2007). Reports of the death of the gene are greatly exaggerated. *Biology & Philosophy*, 22(2), 293–306.

Knippels, M. C. P. J. (2002). *Coping with the Abstract and Complex Nature of Genetics in Biology Education—The Yo-Yo Learning and Teaching Strategy*. Utrecht, Netherlands: CD-β Press.

Lederman, N. G. & Lederman, J. S. (2014). Research on teaching and learning nature of science. In N. G. Lederman & S. K. Abell (Eds.), *Handbook of Research on Science Education*. New York & London: Routledge, Taylor & Francis Group, Vol. II, 600–620.

Lehrer, R. & Schauble, L. (2006). Cultivating model-based reasoning in science education. In R. K. Sawyer (Ed.), *Cambridge Handbook of the Learning Sciences*. New York: Cambridge University Press, 371–388.

Lewis, J. & Kattmann, U. (2004). Traits, genes, particles and information: re-visiting students' understandings of genetics. *International Journal of Science Education*, 26(2), 195–206.

Liu, F., et al. (2010). Digital quantification of human eye color highlights genetic association of three new loci. *PLOS Genetics*, 6, e1000934.

Moss, L. (2003). *What Genes Can't Do*. Cambridge, MA: The MIT Press.

Nelkin, D. & Lindee, S. M. (2004). *The DNA Mystique: The Gene as a Cultural Icon*. 2nd ed. New York, NY: Freeman.

Nelson, J. (2006). Hur används läroboken av lärare och elever? *NorDiNa*, 2(2), 16–27.

Neumann-Held, E. (1999). The gene is dead—long live the gene: conceptualizing genes the constructionist way. In P. Koslowski (Ed.), *Sociobiology and Bioeconomics: The Theory of Evolution in Biological and Economic Thinking*. Berlin: Springer, 105–137.

Orgogozo, V., Peluffo, A. E. & Morizot, B. (2016). The "Mendelian gene" and the "molecular gene": two relevant concepts of genetic units. *Current Topics in Developmental Biology*, 119, 1–26.

Shaw, K. E., Horne, K.V., Zhang, H. & Boughman, J. (2008). Essay contest reveals misconceptions of high school students in genetics content. *Genetics*, 178(3), 1157–1168.

Smith, M. U. & Gericke, N. M. (2015). Mendel in the modern classroom. *Science & Education*, 24(1), 151–172.

Stewart, J., Cartier, J. L. & Passmore, C. M. (2005). Developing understanding through model-based inquiry. In *How Students Learn: Science in the Classroom*. Washington, DC: The National Academies Press, 515–565.

Venville, G. J. & Treagust, D. F. (1998). Exploring conceptual change in genetics using a multidimensional interpretive framework. *Journal of Research in Science Teaching*, 35(9), 1031–1055.

Waters, C. K. (1994). Genes made molecular. *Philosophy of Science*, 61(2), 163–185.

11

PLANT BIOLOGY

Dawn L. Sanders and Dan Jenkins

Introduction

John King (1997) has observed that "without green plants, the earth would be a stark, largely lifeless place" (p. 215). Consequently, how green plants contribute to life on earth needs to make sense to its human citizens. In this chapter, we draw attention to the importance of "reading the story" (Bromham, 2008) of plant contributions to life on earth through the approach of 'big ideas' (Harlen et al., 2015; Matthews, 2009; Uno, 2009) in biological science. The core concept directing our chapter is the role narratives can play in students' experiences of science education (Bromham, 2008; Avraamidou & Osborne, 2009), and the professional development of teachers. We suggest that realigning the position of plants from isolated characters in school biology to central actors working across a storied landscape of a 'big idea' could radically alter students' conceptions of plant science, and the essential role plants play in the wider world, particularly in addressing some of the current global challenges, such as climate change and food security. We believe that it is necessary, when teaching, to highlight the 'big ideas' of biology to our students as a synoptic narrative. Thus, in this chapter, the role of photosynthesis in plants as enabling life on earth is our example big idea, as affirmed by the following authors:

> Photosynthesis is a fundamental process for life on earth, with many biologists rating it *the* most important natural process.
>
> *(Matthews, 2009 p. 929)*

> We are currently cultivating one-quarter of the total land area, and we are currently, consuming, in one way or another, 40% of all photosynthetic activity.
>
> *(Walker, 2012, pp. 91–92)*

In an educational context, Uno (2009) cites the dual issues of biology textbooks not using plants and trainee science teachers receiving little plant biology education in their training. We suggest a conceptual and didactical gap can arise as a result of this absence. Moreover, perceptions of plants as passive bystanders to environmental change—Beerling's "silent witnesses" (2007, p. 1)—can further diminish student and teacher appreciation of the complex world in which plants exist, no less so than when teaching and studying photosynthesis.

Energy for Life

Photosynthesis is a complex and somewhat abstract process that is often taught in a series of incremental modules beginning with 'plants can make their own food' in primary school and continuing to advanced level equations which are a "summary of some 30 separate steps" (Reiss, 2011, p. 54), in upper secondary school. Indeed, it is a subject often oriented around a series of textbook equations increasing in complexity as a child moves through their schooling. Unlike the topic 'plant adaptation', which is best taught in the physical presence of plants, photosynthesis can be taught without access to living material. Thus, the process of photosynthesis can be perceived by students as abstract, repetitive and disassociated from wider biological processes and, as such, misconceptions abound. Furthermore, the complexity of photosynthesis, with its connections to physical, chemical and biological processes, can make plants look *difficult*. Although all these processes are seen as *biology*, exploring the ideas behind photosynthesis includes:

- Comprehension and acceptance of significant areas of understanding within physical science processes of energy transfer, excited states of pigments and accessory pigments.
- An understanding of wavelengths of light and chemical processes of the reactions, including reduction of compounds to provide energy 'storage' and the complex biochemistry of an array of carbohydrates.
- An understanding of a suite of organelles, cells, tissues, transport systems, enzymes and proteins and the biochemistry of sugars.

We suggest that these complexities and technicalities, when taught through a biological topic called photosynthesis and led by a content testing curriculum, contribute to the loss of the awe about the truly amazing feat of nature that is photosynthesis (Morton, 2007). Given the critical role of photosynthesis as a "fundamental process underpinning life on earth" (Matthews, 2009, p. 929), we believe student understanding of this process is an important part of scientific literacy and, as such, needs to be experienced in ways that foreground photosynthesis as a central biological process, particularly as "living within our biological means is an important issue that will not go away. Photosynthesis is the only way of providing us with our daily food, and so much else" (Walker, 2012, p. 103). In the next section, we focus on relevant studies of students' misconceptions in this area, as a background to our proposed narrative-based approach to teaching photosynthesis, and plant science more broadly.

Student Misconceptions

Lin and Hu (2003) report their research concerning 12- and 13-year-old student understanding of energy flow in biological systems and the linking between the living and non-living worlds, and between different levels of complexity in biology that they base on work by philosophers of biology. Lin and Hu (2003) identify these levels as "phenomenal knowledge" at a whole organism level, "mechanical knowledge" at a cellular level and "physical knowledge" at a molecular level (see Lin & Hu, 2003, p. 1530). Their research demonstrates that students find it hard to link the different levels of biology, especially using ideas of physical knowledge about energy and matter, in their explanations of photosynthesis and respiration. In addition, Marmaroti and Galanopoulou (2006) show that although some students may answer that light or chlorophyll is needed for photosynthesis, they do not always then link this with photosynthesis occurring during the day or under artificial light, or in the green parts of plants.

Ryoo and Linn (2012) consider energy transformations to be critical to understanding photosynthesis and that, as rather an abstract concept and one that is invisible, it is hard for students to understand in any context and may be new to students who are expected to learn about photosynthesis. They also recognise the problem of textbook presentations that inadequately explain, in writing and diagrams or equations, the role of light energy in the process.

Uno (2009) supports the assertion of Lin and Hu (2003) that coherence of understanding between different levels of organisation are important and that teaching using overarching 'big ideas' could be the way forward, e.g. including evolutionary biology in teaching about photosynthesis so that students understand how, and why, chlorophyll developed and its importance in plants making food. Uno (2009) also acknowledges that students' alternative conceptions can be hard to change and should be investigated at the start of a teaching sequence by asking diagnostic questions (or "concept inventories", p. 1755). These are issues, we suggest, that can be approached through teaching strategies designed to encourage conceptual linking between the overall picture and what makes photosynthesis happen. Thus, our recommendations for teaching respond to Uno (2009) by offering models for teachers to develop "concept inventories" (p. 1755) through "historical vignettes" (Clary & Wandersee, 2015; Wandersee & Roach, 1998) and "concept cartoons" (Keogh & Naylor, 1999; Keogh, et al., 2001) as narrative-based activities working alongside seeing the process in action practicals (Jenkins, 2015). This, we suggest, is an approach with which to develop the knowledge levels outlined by Lin and Hu (2003).

Difficulties in Teaching About Plants

Many of the documented difficulties concerning human attention to plants originate from the perceived *otherness* of their physical form and behaviours, as indicated in the extract below:

> Plants are curious organisms—clearly, they are alive but their apparent lack of any central organization (brain, heart, or nervous system) makes it harder for us to understand how they work. They are slow: usually they look today just like they did yesterday and we have to wait, sometimes a long time, to see changes in them.
>
> *(Van Volkenburgh, foreword in Koller, 2011, p. xiii)*

However, these characteristics also offer opportunities. Numerous cellular processes in plants operate at similar rates to their zoological counterparts providing us with ample opportunities to demonstrate many key biological processes, and indeed experiment upon them as models for living organisms more generally. This is emphasised by their use in common practical investigations in the school laboratory, e.g. investigation of osmotic potential on tissue (potatoes), source of enzymes in rate reactions (leaves) and the source of material for lower secondary introduction to cells using microscopes (onion). Yet, this specific use in practical investigations does not always translate into a broader picture of the contribution plants make to biological science, nor to their role in enabling life on earth. Slingsby (2006), for example, has criticised the role of plants in school biology as being reduced to "victims in a series of photosynthesis experiments that don't always work" (p. 51).

If this bleak experimental landscape is adjoined to limitations in teacher and student engagement with plants (Uno, 2009, Nyberg & Sanders, 2014), an emergent problem for society becomes visible. Understanding the contribution of plants to biological life is vital in an era in which plant extinction numbers are increasing (Willis, 2017) and the climate is changing, much of which is attributed to human activity.

Plant-Blindness

The role of plants in the received curriculum is frequently perceived as separate from animals, and often focused on specific *plant topics*, a classic case being that of photosynthesis. Additionally, the planned curriculum often uses animal-based examples for biological concepts (Nantawanit et al., 2011;, Nyberg & Sanders, 2014). In the context of plant science teaching and learning, Ebert-May and Holt (2014) have observed that "plant blindness exists—we need to know the cognitive underpinnings and instructional design that enables students to 'see' plants" (p. 362). In addition, Balding and Williams (2016) suggest that "it is likely that both biological and cultural factors shape human-plant relationships" (p. 3). In response, we believe that 'plant blindness' is but one of the influences on teacher choice. In the planned curriculum, there are many opportunities for teachers to choose which living material to use, and hence create a less zoo-centric focus. For example: why, in enzyme practical investigations, do teachers use a source of catalase from mammalian liver as opposed to potato or other plant tissue? Or why, in the discussion of medicines from plants, is the focus on the product used in human treatments but the purpose in plants often ignored?

Being rooted in one place requires plants to develop highly sophisticated capacities to respond, communicate and react to external stimuli, physical attack and competition with others, develop mutualistic partnerships and, ultimately, reproduce. In addition, the ability of plants to photosynthesise makes them a critical component in food webs and thus important in discussions of food security. Moreover, the intransient nature of plants makes them vulnerable to climate change impacts in relation to pollination vectors, flowering times and habitat needs. Teachers need to be familiar with the science behind these statements—a challenging familiarity if the teacher has neither studied plant science as their biology major nor experienced in-depth plant science in their teacher-training (Uno, 2009). Furthermore, curriculum frameworks often focus on the zoological side of biology; hence, students might make assumptions that plant science has little relevance to contemporary scientific research. As useful models for experimentation in biology, numerous 'firsts in science' (Sainsbury Laboratory University of Cambridge, 2017) were discovered using plants, including:

- *Cells (1655)* Microscopic observations of thin sections of cork led Robert Hooke to discover cells, the building blocks of life.
- *The principles of genetics (1866)* Gregor Mendel's experiments with pea plants allowed him to formulate the basic rules for genetic inheritance of traits.
- *Transposons (1948)* Barbara McClintock used genetics and observations of maize chromosomes to discover transposons, sometimes called jumping genes. These are bits of DNA that move about the genome and can influence the expression of other genes.
- *Totipotency of cells (1957)* Using tobacco cell cultures, Folke Skoog proved the idea proposed by Gottlieb Haberlandt that adult cells could give rise to all cell types. This is the fundamental basis for stem cell biology.
- *Post transcriptional gene silencing (1990s)* The discovery that extra copies of a gene introduced into the genome can trigger silencing of both the introduced and endogenous gene was made first in petunias. The mechanism involved was later shown to involve small RNAs.

Likewise, recent scientific papers, such as Jennings et al. (2010) and Schöner et al. (2017), offer evidence for, in the case of Jennings et al., competition between a carnivorous plant and a spider for the same prey and, in Schöner et al., a pitcher plant benefitting from transferring the pre-digestion of prey to a bat mutualist. These scientific narratives represent the dynamic interconnectedness of the living world and offer contemporary sources for teachers and their students to examine the often complex ecological relationships between plants and animals.

In the context of science education research, Nantawanit et al. (2011) have found that students enjoy "the secret stories of plants" (p. 850) as active constituents of ecosystems, which in their work focused on plant defence strategies. Furthermore, Ebert-May and Holt (2014) note that "thinking of plants at multiple scales requires understanding of core ideas (evolution and ecology), cross-disciplinary concepts (matter and energy), and use of science practices (models, arguments, and cooperative work)" (p. 361).

Returning now to the teaching of photosynthesis, we suggest that using a combination of practical investigations, historical vignettes and concept cartoons offers teachers and students a synoptic approach that brings together seeing the process in action, talking through "conceptual inventories" (Uno, 2009) and knowledge building in multiple ways (Ebert-May & Holt, 2014).

Recommendations for Teaching About Plants

Photosynthesis: Seeing the Process in Action

An approach central to many teaching schemes covering photosynthesis is using practical experimentation to demonstrate the process in action. Two classic demonstrations of the process are commonplace—testing a leaf for starch and observation of bubbling aquatic pondweeds. These activities enable the development of a sense of awe at the process (Jenkins, 2015) and go some way to demonstrate the abstract equations for the process in action. However, these classic examples have limitations and can introduce misconceptions; below, we discuss alternative practical procedures to demonstrate photosynthesis in action that overcome some of these difficulties.

Testing a Leaf for Starch

Barker and Carr (1989) described how many students do not understand the starch test, struggle to understand the complex mechanics of the test and cannot link the presence of starch to photosynthetic activity. They note that one student thought starch was produced when chlorophyll and iodine mixed together. In addition, Kinchin (2000) discusses use of the starch test as proof of photosynthetic activity as problematic; there are numerous examples where starch is present in plant tissue in areas unrelated to photosynthetic activity (e.g. potato tubers found underground). The above issues are then confounded by acknowledgement that the point of demonstrating the presence of *starch* as proof of photosynthetic activity doesn't meet the standard equations for photosynthesis presented to students showing *sugars* not *starch* as one of the end products.

Bubbling Pondweed

This demonstration of photosynthesis in action focuses on the other main product in the generalised equation for photosynthesis, oxygen. However, as many educators attempting this classic demonstration of photosynthesis (first developed by Jan Ingen-Housz in 1779; Ray & Beardsley, 2008) discover, reliability of this method as a quantitative measure of the rate of photosynthesis is full of problems. Mostly these are down to the selection of an appropriate pondweed to 'bubble' reliably and on demand (see Jenkins, 2015).

Both bubbling pondweed and the starch test can be useful as pupil assessment tools; however, Eldridge (2004) notes that when used to demonstrate photosynthesis, these practical investigations do "little to improve student understanding of the whole process" (p. 37). Therefore, discussion with the student is needed on the abstract elements of this as a demonstration of photosynthesis in action. The convention when teaching the evolution of oxygen as a product of photosynthesis involves leaves and stomatal pores in a scenario operating in the open air. However, we introduce an added level of complexity transposing this process into an aquatic plant, which is inverted to collect gas from the cut stem, all operating under water. Through this practical there is potential to introduce a disconnect between the site of photosynthesis discussed with students (leaves) and the collection of the products (from an inverted stem, underwater). The source of carbon dioxide (normally from the air) is often not mentioned, and the fundamental importance of the process (capture of energy and carbon storage products) may be lost to the student. Recognising that dissolved carbon dioxide (present as carbonic acid) is the source of carbon for photosynthesis in aquatic plants enables a discussion with students on the opportunities to measure both reactants and products (Ray & Beardsley, 2008). pH can be used as a proxy for carbon dioxide measured using an indicator present in the water and products (bubbles of gas produced) measured during photosynthesis.

Observing Photosynthesis and Respiration in Plant Material

The misconceptions that photosynthesis is the opposite of respiration and that respiration happens at night and photosynthesis during the day are common. To help explore these misconceptions, several teaching schemes introduce using hydrogen carbonate indicator to demonstrate changes in level of carbon dioxide accompanying respiration and photosynthesis.

Sections of aquatic plants can be put into vials of pH sensitive indicator. In the presence of sufficient light, over time this demonstrates the 'use' (i.e. decrease in concentration) of the reactant carbon dioxide. This is visualised by a colour change in the indicator. In low light levels or in darkness, the same system demonstrates the effects of respiration, increasing carbon dioxide levels (see, for example, Eldridge, D., n.d.).

Using the indicator also provides an opportunity to discuss and introduce the compensation point between photosynthesis and respiration in terms of net carbon dioxide levels. The concept of this balance helps learners understand that respiration is always occurring in plants.

Eldridge (2004) introduced a new procedure to observe and quantify photosynthetic rate in green algae and suggested

> the use of immobilised green algae as photosynthetic organisms which are easy to manipulate. Students may then determine the rate of carbon dioxide absorption by the algae using hydrogen carbonate indicator and a colorimeter or colour scale. In addition, the importance of carbon dioxide as a requirement for photosynthesis is emphasised in an implicit way, as students are in effect taking measurements of how much carbon dioxide is taken up by the algae, which causes the change in the colour of the indicator.
>
> *(p. 38)*

This procedure has been widely adopted by many teachers across the UK and was incorporated into the Strategies for Assessment of Inquiry Learning in Science project materials (www.sails-project.eu) as a more reliable investigation to quantifiably investigate photosynthesis in action. It overcomes the core issue of unreliability of bubbling found when using pondweeds. The algae gain protection from being immobilised in alginate so, unlike a cut section of pondweed, 'algal balls' will continue 'working' in the system for several hours. Once immobilised, the algal balls can be kept for future investigations, allowing a focus on photosynthesis or respiration. Guidance on preparing and using immobilised algal balls as initially outlined by Eldridge (2004) can be found at www.saps.org.uk/algalballs.

Photosynthesis: A Narrative Approach

We agree with Matthews (2009) in his foregrounding of photosynthesis as a 'fundamental process' in biological science; moreover, this is a topic in which students struggle, as previously outlined. Therefore, in this section we focus on narrative-based ways to complement seeing the process in action through practical experiments. Both of the suggested narrative methods attempt to overcome misconceptions held on the topic of photosynthesis using student-centred approaches, which work to make explicit alternative conceptions held by students. By attempting to draw out students' own views on the chemistry, physics and biology concepts involved, these approaches begin to tackle the complexities noted by Lin and Hu (2003), and are influenced by narrative-based approaches to science education (e.g. Avraamidou & Osborne, 2009).

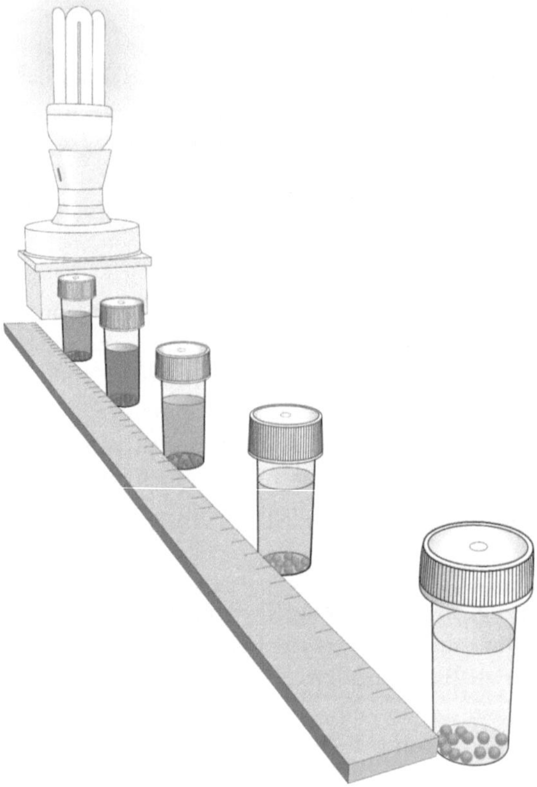

FIGURE 11.1 Immobilised algal balls in hydrogen carbonate indicator

Credit: National Centre for Biotechnology Education, University of Reading. Drawn by Dean Madden

Historical Vignettes

In this section, we discuss the use of 'historical vignettes' (Clary & Wandersee, 2015; Wandersee & Roach, 1998) in the context of the experimental history of photosynthesis as a narrative-based model with which to record, in a teaching context, students' alternative conceptions concerning photosynthesis.

Example: *'Where does the wood come from?'*

Using historical work carried out by Jean Baptiste van Helmont in the sixteenth century, students examine their views on 'where does the wood come from?' After reading an extract from van Helmont's diary, students discuss his interpretations. Four typical student responses to van Helmont's experiment are given in the speech bubbles in Figure 11.2.

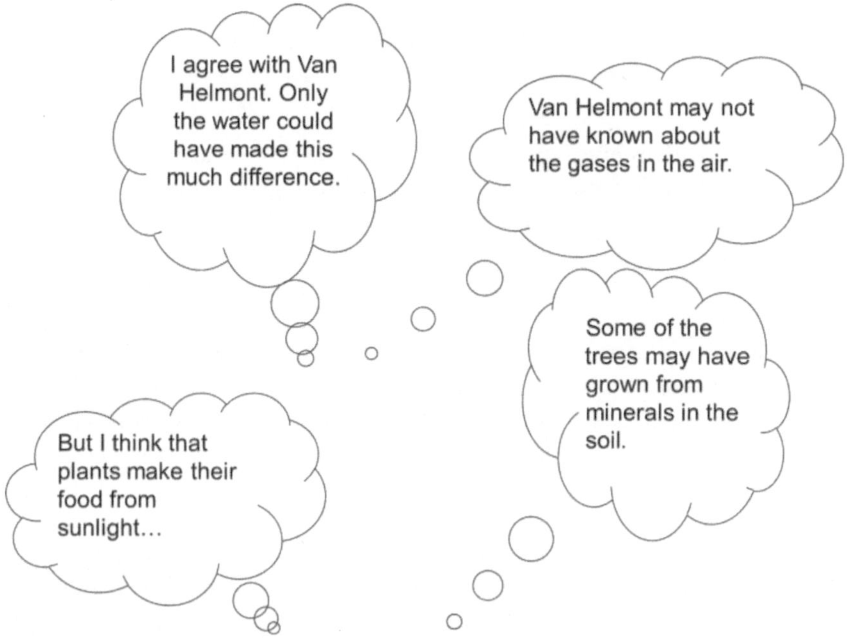

FIGURE 11.2 Where does the wood come from? Eldridge, D. (n.d.). *Where does the wood come from?* Retrieved from www.saps.org.uk/secondary/teaching-resources/134

Credit: Science & Plants for Schools, University of Cambridge.

Students are given cards as a stimulus to talk in structured groups about each of the statements and examine data and pictures of more recent experiments. They summarise their thoughts and feed back to the rest of the class. Throughout the discussion it is important to introduce and reinforce the idea that an increase in mass (biomass) is good evidence that photosynthesis has taken place and that carbon dioxide from the air contributes to this.

By acknowledging the biological, chemical and physics processes involved in photosynthesis, this teaching scheme attempts to recognise some of the abstract elements of photosynthesis. The work scheme continues to explore how carbohydrates are made and helps teachers redress the commonly held misconception that the mass of a plant comes from the soil. It is worth noting that van Helmont believed water was the source of the accumulation of mass in a plant, not carbon dioxide. Discussion of this with students presents another opportunity to acknowledge and explore alternative conceptions, their presence and importance throughout the history of science.

Using historical characters alongside original student thinking enables students to find that their conceptions of the seemingly counter-intuitive and abstract process of photosynthesis are justified. By exploring the historical development of our understanding of photosynthesis, alternative conceptions can be voiced and appropriate conceptions then scaffolded by the teacher working alongside their students, for:

> Not only does history facilitate understanding of the nature of science for our students, but it also provides historical analogies of modern topics, enabling our students to develop useful research skills, effective scientific argumentation, and better scientific understanding.
>
> *(Clary & Wandersee, 2015, p. 329)*

Van Helmont is not the only historical character to be used in these 'historical vignettes'. Matthews (2009) has written at length on the attributes of working with Joseph Priestley and his photosynthesis investigations and concludes that "Priestley is an under-utilised figure in science education" (p. 955). He goes on to note:

> Unfortunately, Priestley's contribution to the modern understanding of photosynthesis is seldom mentioned in school curricula. This is a pity as his role was pivotal, and students can very easily be led through many of the same steps that he took.
>
> *(Matthews, 2009, p. 955)*

Matthews' (2009) statement suggests the use of historical vignettes alongside replications of Priestley's investigations, without, we suggest, causing distress to a mouse.

Concept Cartoons

One of the biggest challenges for teachers of all subjects is how to bring student voice into lessons in ways which a) are meaningful and authentic and b) help students to voice their uncertainties as well as express their "islands of expertise" (Crowley & Jacobs, 2002). Concept cartoons are a response to this challenge:

> Concept cartoons extend the range of pedagogical strategies available to teachers. They present learners with a set of alternative ideas about a scientific concept in visual form. They are used mainly in the classroom to support teaching and learning in science by generating discussion, stimulating investigation and promoting learner involvement and motivation.
>
> *(Keogh et al., 2001, p. 137)*

Within a carnivorous plant education programme, which one of us developed in 2012, the students are asked for their views on plant nutrition, using Figure 11.3 as a stimulus for student reflection to aid teacher diagnostics or "concept inventories" (Uno, 2009, p. 1755). The student speech and thought bubbles are

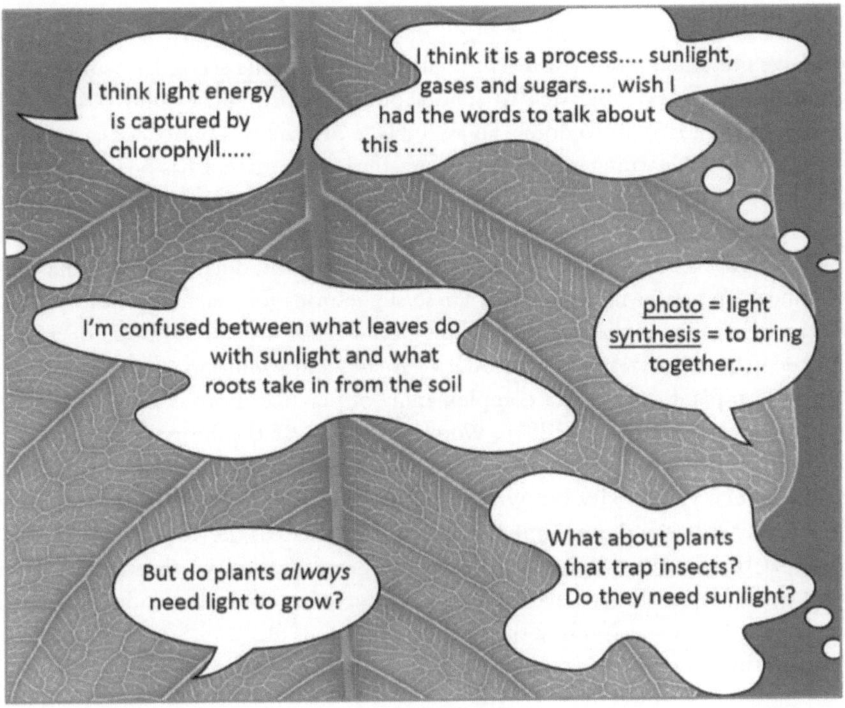

FIGURE 11.3 Concept cartoon from *Murderous Plants*

Credit: The Charles Darwin Trust. Reproduction Permission Linnean Learning; (c) Dawn Sanders and Karen Goldie-Morrison.

drawn, in part, from the research of Driver et al. (1994) on secondary students' understandings of photosynthesis. More specifically, students can connect with the represented thoughts, theories and ideas. The notion behind this multimodal representation is to support student confidence in publicly discussing their internal uncertainties when represented as an external voice in a teaching aid, and to fulfil Keogh et al.'s (2001) recognition that

> an effective auditing strategy will both diagnose the students' level of understanding of science concepts and help them begin to reconstruct their understanding.
>
> *(p. 138)*

Thus, the concept cartoon, used in this way, can act as a proxy for internal dialogues, previous learning experiences and tentative ideas, in order to decentre individual performativity and instead facilitate group discussion, aided by the teacher's didactical moves.

Conclusions

We have presented a case for teaching plant science in the secondary curriculum using the 'big ideas' approach to school biology affirmed by Uno (2009) and Harlen et al. (2015). In so doing, an assemblage of teaching approaches, in which student-centred learning complements practical experiments, has been suggested for one strand of plant science, namely the process of photosynthesis. Our teaching recommendations are a response to a body of research literature on student misconceptions and common didactical moves, and the documented limitations of standard practical investigations. These suggestions for teaching photosynthesis can be readily applied to other biological science arenas, examples of which include ecosystems, respiration, modern genetics and evolution.

Modern plant science is a complex arena of practices from computer modelling to field-work (Jahren, 2016; Woollard, 2017). At the heart of this science lies an incredibly diverse kingdom of life forms. School biology cannot hope to replicate such complexity, but what it can do is provoke awe and wonder in the presence of plants and the significant biological processes they underpin. We suggest that teaching that combines seeing the process in action through practical work with historical narratives and concept cartoons will assist students to 'read the story' (Bromham, 2008) of the vital contributions plants make to life on earth, through their "noble and important services" (Hales, 1727, p. 324).

Acknowledgements

Thanks to Dr Indira Banner for her input to the student misconception section of this chapter.

References

Avraamidou, L. & Osborne, J. (2009). The role of narrative in communicating science. *International Journal of Science Education*, 31(12), 1683–1707.

Barker, M. & Carr, M. (1989). Photosynthesis—can our pupils see the wood for the trees? *Journal of Biological Education*, 23, 41–44.

Balding, M., & Williams, K.J.H. (2016), Plant blindness and the implications for plant conservation. *Conservation Biology*, 30, 1192–1199.

Beerling, D. (2007). *The Emerald Planet: How Plants Changed Earth's History*. Oxford: Oxford University Press.

Bromham, L. (2008). *Reading the Story in DNA*. Oxford: Oxford University Press.

Clary, R. M. & Wandersee, J. H. (2015). The history of science in the science classroom: the past is the key to the future in science education. *Earth Sciences History*, 34(2), 310–332.

Crowley, K. & Jacobs, M. (2002). Building Islands of expertise in everyday family activities. In G. Leinhardt, K. Crowley & K. Knutson (Eds.), *Learning Conversations in Museums*. New Jersey: Lawrence Erlbaum, 333–356.

Driver, R., Squires, A., Rushworth, P. & Wood-Robinson, V. (1994). *Making Sense of Secondary School Science: Support Materials for Teachers*. London: Routledge.

Ebert-May, D. & Holt, E. (2014). Seeing the forest and the trees: research on plant science teaching and learning. *CBE—Life Sciences Education*, 13, 361–362.

Eldridge, D. (2004). A novel approach to photosynthesis practicals. *School Science Review*, 85(312), 37–45.

Eldridge, D. (n.d.). *Photosynthesis—a Survival Guide*. Available at: www.saps.org.uk/secondary/teaching-resources/134.

Hales, S. (1727). *Vegetable Staticks, or, an Account of Some Statical Experiments on the Sap in Vegetables: Being an Essay Towards a Natural History of Vegetation*. London: Innys.

Harlen, W., Bell, D., Devés, R., Dyasi, H., Fernández de la Garza, G. Léna, P., Millar, R., Reiss, M., Rowell, P. & Yu, W. (2015). *Working with Big Ideas of Science Education*. Trieste: The Science Education Programme (SEP) of IAP.

Jahren, H. (2016). *Lab Girl*. New York: Alfred A. Knopf.

Jenkins, D. (2015). Plants: an ideal living material for teaching science. *School Science Review*, 97(358), 52–56.

Jennings, D., Krupa, J., Raffel, T. & Rohr, J. (2010). Evidence for competition between carnivorous plants and spiders. *Proceedings of the Royal Society B: Biological Sciences*, 277, 3001–3008.

Keogh, B. & Naylor, S. (1999). Concept cartoons, teaching and learning in science: an evaluation. *International Journal of Science Education*, 21(4), 431–446.

Keogh, B., Naylor, S., de Boo, M. & Feasey, R. (2001). Formative assessment using concept cartoons in initial teacher training in the UK. In R. Duit (Ed.), *Research in Science Education: Past, Present and Future*. Rotterdam: Kluwer Academic Publishers, 137–142.

Kinchin, I. M. (2000). Confronting problems presented by photosynthesis. *School Science Review*, 81(297), 69–75.

King, J. (1997). *Reaching for the Sun: How Plants Work*. Cambridge: Cambridge University Press.

Koller, D. (2011). *The Restless Plant*. Harvard: Harvard University Press.

Lin, C. &Hu, R. (2003). Students' understanding of energy flow and matter cycling in the context of the food chain, photosynthesis, and respiration. *International Journal of Science Education*, 25(12), 1529–1544.

Marmaroti, P. & Galanopoulou, D. (2006). Pupils' understanding of photosynthesis: a questionnaire for the simultaneous assessment of all aspects. *International Journal of Science Education*, 28(4), 383–403.

Matthews, M. (2009). Science and worldviews in the classroom: Joseph Priestley and Photosynthesis. *Science & Education*, 18, 929–960.

Morton, O. (2007). *Eating the Sun: How Plants Power the Planet*. New York: Harper Perennial.

Nantawanit, N., B. Panijpan, and P. Ruenwongsa. 2012. Promoting students' conceptual understanding of plant defense responses using the fighting plant learning unit (FPLU). *International Journal of Science and Mathematics Education*, 10(4): 827–864.

Nyberg, E. & Sanders, D. (2014). Drawing attention to the 'green side of life'. *Journal of Biological Education*, 48(3), 142–153.

Ray, A. & Beardsley, P. (2008). Overcoming student misconceptions about photosynthesis: a model- and inquiry-based approach using aquatic plants. *Science Activities*, 45(1), 13–22.

Reiss, M. (Ed.) (2011). *Teaching Secondary Biology*. London: Hodder.

Ryoo, K. & Linn, M. (2012). Can dynamic visualizations improve middle school students' understanding of energy in photosynthesis? *Journal of Research in Science Teaching*, 49(2), 218–243.

Sainsbury Laboratory University of Cambridge. (2017). Available at: www.slcu.cam.ac.uk/outreach/resources/posters/firstinplants.

Schöner, C. R., Schöner, M. G., Grafe, T. U., Clarke, C. M., Dombrowski, L., Tan, M. C. & Kerth, G. (2017). Ecological outsourcing: a pitcher plant benefits from transferring pre-digestion of prey to a bat mutualist. *Journal of Ecology*, 105, 400–411.

Slingsby, D. (2006). The future of school science lies outdoors. *Journal of Biological Education*, 40(2), 51–52.

Strategies for Assessment of Inquiry Learning in Science (SAILS) project. Available at: www.sails-project.eu.

Uno, G. (2009). Botanical literacy: What and how should students learn about plants? *American Journal of Botany*, 96(10), 1753–1759.

Walker, T. (2012). *Plants: A Very Short Introduction*. Oxford: Oxford University Press.

Wandersee, J. H. & Roach, L. M. (1998). Interactive historical vignettes. In J. J. Mintzes, J. H. Wandersee & J. D. Novak (Eds.), *Teaching Science for Understanding: A Human Constructivist View*. San Diego: Academic Press, 281–306.

Willis, K. (Ed.) (2017). *State of the World's Plants 2017*. Report. London: Royal Botanic Gardens, Kew.

Woollard, A. (2017). Interview: Dame Ottoline Leyser. *The Biologist*, 64(4), 16–19.

12

ENERGY, PHOTOSYNTHESIS, AND RESPIRATION

Ute Harms and Uwe Bertsch

Introduction

All living systems depend on a concept scientists refer to as *energy*. In the nineteenth century, energy was introduced as a scientific concept in physics. Subsequently, this concept expanded into chemistry and biology, where it was applied to explain disciplinary phenomena. Today, energy is a central model of the life sciences to explain the evolution, existence, and functioning of life at all system levels, e.g. the cell, the organism, or the ecosystem. Without energy, there is no life. Living systems are open and can absorb and release energy, but they cannot produce or destroy it, though energy can temporarily be stored. But to establish or to maintain the order of living systems, (thermal) energy must be released to the environment. This leads to an increase of disorder in their surroundings that scientists call *entropy*. Roughly speaking, energy flows through open systems, ending up in thermal energy that cannot be used by organisms for biochemical processes any longer. Thus, to maintain life on Earth, a continuous energy input is a requisite.

The phenomenon of 'life' first appeared about 3.5 billion years ago. Then, simple organisms arose that possessed the physicochemical properties to absorb energy from their environment, to use it to synthesize their own complex structures from simple starting molecules, and to carry out functions such as movement and reproduction. These two strands of (i) capturing energy and (ii) processing energy for life functions are the two basic metabolic processes in biology: (i) photosynthesis and (ii) respiration.[1] The concepts *energy*, *photosynthesis*, and *respiration* are key to explaining the existence and evolution of life and thus are fundamental for teaching and learning biology.

Photosynthesis and respiration are both mandatory content in secondary biology education. The respective anabolic and catabolic biochemical processes merge changes of matter and energetic processes. In photosynthesis, cyanobacteria, algae, and mainly green plants capture light energy and use it to synthesize energy-rich biomolecules. Organisms obtain energy-rich molecules such as glucose either through their own process of photosynthesis, or by the intake of energy-rich molecules in food. The release of this energy in a controlled way inside cells is called respiration. Whilst the role of *matter* in both processes becomes concrete—as, for example, by an organism growing or losing weight—the shared *energetic* processes are abstract. In traditional biology education, the role of energy in these contexts is hardly made explicit and if so only marginally. But in conjunction with the introduction of new science standards and curricula in many countries, the overarching importance of energy in science has been strengthened (e.g. Germany: KMK, 2005; USA: NRC, 2012; see Table 12.1). As a disciplinary core idea energy is introduced as a basis for learning and revision of different content (e.g., photosynthesis and respiration), aiming to stimulate the coherence and depth of understanding within each discipline, in this case the life sciences (NRC, 2012).

TABLE 12.1 The concepts energy, photosynthesis, and respiration (highlighted) as proposed by the *Next Generation Science Standards* for the life sciences through primary and secondary level (NGSS, 2013; Opitz et al., 2015)

Content Area	*Grade*	*Content*
Organization for matter and energy flow in organisms	3–5	"Food provides animals with the materials and energy they need for body repair, growth, warmth, and motion. Plants [. . .] obtain energy from sunlight."
	6–8	"Plants use the energy from light to make sugars through photosynthesis. Within individual organisms, food is broken down through a series of chemical reactions that rearrange molecules and release energy."
	9–12	"Through cellular respiration, matter, and energy flow through different organizational levels of an organism as elements are recombined to form different products and transfer energy."
Cycles of matter and energy transfer in ecosystems	6–8	"Food webs model how matter and energy are transferred among producers, consumers, and decomposers as the three groups interact within an ecosystem."
	9–12	"Photosynthesis and cellular respiration provide most of the energy for life processes. [. . .] At each link in an ecosystem elements are combined in different ways and matter and energy are conserved."

Biology teaching aims to support students to develop conceptual knowledge. Many studies, mainly from cognitive psychology, have shown that misconceptions hinder the reconstruction of scientific concepts. Misconceptions are defined here as incomplete or incorrect explanations of scientific concepts that derive from everyday experiences or previous teaching (e.g., Vosniadou & Skopeliti, 2013). If they stay undetected and if they are not made explicit during teaching, they impede conceptual change and thus the development of valid scientific concepts (Bransford et al., 2000). During the last three decades, many scholars have elaborated misconceptions on energy, photosynthesis, and respiration. These stretch across more or less all ages and can even be found among science teachers. The knowledge of these misconceptions is fundamental for developing teaching programmes and learning opportunities for energy, photosynthesis, and respiration.

Students' Misconceptions About Energy, Photosynthesis, and Respiration

Energy is a word from our everyday life. Everybody 'uses energy'; we are asked to 'save energy'; and when we are doing sports we 'lose energy' that we can 'replace' by eating. So, everybody—including young children—develops everyday connotations of energy that often differ from its scientific meaning (e.g., Jin & Wei, 2014; Needham, 2014). In recent decades many scholars—mainly from physics education—have investigated students' energy conceptions. The findings stretch from kindergarten to the end of secondary level and reveal a continuous limited scientific understanding of this abstract concept (e.g., for kindergarten: van Hook & Huziak-Clark, 2008; for primary: Liu & Ruiz, 2008; for lower secondary: Trumper, 1993; for upper secondary: Finegold & Trumper, 1989; Duit & Kesidou, 1988). Dissipation (degradation) and conservation of energy seem to be especially difficult to comprehend (e.g., Goldring & Osborne, 1994; Liu & McKeough, 2005). Students hold the misconception that energy isn't dissipated when flowing through an ecosystem but is recycled in biodegradation and afterwards absorbed by plants (Beals et al., 2012; Wernecke et al., 2018a). This reflects the idea that energy is moving in a cycle like substances do (Hirca & Akdeniz, 2008).

Zangori and colleagues (2017) found that students mixed up matter and energy when asked to use mathematical representations to argue for the cycling of matter and flow of energy among organisms in an ecosystem. Students believe that energy can be recycled (e.g., by decomposers) and reused (e.g., by plants) in an ecosystem (e.g., Lancor, 2014). Even though secondary level students name the sun as the source of energy for plants, they sometimes also put forward soil, water, or air. Barak and colleagues (1999) showed that students from the lower secondary level may believe that water, vitamins, oxygen, and air are sources of energy for animals. Evidently, students lack the ability to distinguish between substances that are vital and those that also deliver energy (Boyes & Stanisstreet, 1991). Özay and Öztas (2003) reported the same findings from their questionnaire study with

9th graders. These results are also supported by Leach and colleagues (1996), who found that many 16-year-old students use energy, substance, nutrition, and nutrients synonymously.

Misconceptions also exist in regards to the topic of energy flow in ecosystems (Wernecke et al., 2018a). For example, students describe plants as organisms that use the soil as an energy source. This notion, that scientists had even before the metabolic process of photosynthesis was elucidated (Métioui et al., 2016), may be held by students from the secondary level (Simpson & Arnold, 1982) and even teachers (Beals et al., 2012) nowadays. As in biology class, energy is considered in open systems; students show severe difficulties in comprehending that for the whole system of the biosphere the law of energy conservation—which they learned in physics—also applies (Barak et al., 1997). They argue that energy is lost or destroyed (Opitz et al., 2017a, 2017b). Perhaps this misconception is so frequent because the principle of energy conservation in open systems cannot be experienced by the senses. Although university students can reproduce a definition on energy at the beginning of their studies, they are hardly able to apply it to biological problems, and they perceive energy in physics as something different from energy in biology (Chabalengula et al., 2011). This was interpreted by the authors as a vitalistic biological understanding. In addition, elementary teachers (e.g., Trumper et al., 2000), pre-service biology students and biology teachers (e.g., Chabalengula et al., 2011) show limited understanding of energy in biological contexts.

In contrast to energy, there are hardly any connotations of the word photosynthesis from everyday life, as photosynthesis is a technical term. However, many scholars have revealed a variety of misconceptions that students of all ages and adults show (Parker et al., 2012). For example, in the context of plant nutrition, students map their experience of eating (i.e., the intake of matter) to plant nutrition, which leads to the scientifically inadequate idea that plants get the lion's share of their food directly from the environment, not by synthesizing it themselves through photosynthesis (Leach et al., 1996). Misconceptions on photosynthesis can be categorized into three groups: (i) the role of light; (ii) the role of the gases and substances involved; and (iii) the interplay of photosynthesis and respiration.

The role of light in photosynthesis is unclear for many students. They believe that sunlight plays a role as a substance that is built into the plant (Eisen & Stavy, 1988). Misconceptions also occur as to the question of where exactly in the plant photosynthesis takes place. In addition, the function of chlorophyll in the light reactions is unclear (Marmaroti & Galanopoulou, 2006). A frequent misconception that clearly shows students have difficulty grasping the complexity of photosynthesis is the notion that the carbon dioxide going into the plant is directly changed into the oxygen that the plant releases as a product of photosynthesis (Parker et al., 2012). Students are unclear about the fact that the carbon dioxide plants absorb through their stomata is (i) the starting point for the synthesis of glucose molecules (using indirectly the energy from the Sun transformed into ATP and $NADPH^+$ molecules during the light reactions) and (ii) the only material that contributes to the increase of plant mass (Eisen & Stavy, 1988). Hence, students don't comprehend that indirectly—via the synthesized glucose molecules—the carbon dioxide is the foundation for energy storage in the plant (Köse, 2008), which humans use by eating vegetables and by using fossil fuels.

In this context, it comes to the fore that students have problems grasping the carbon transformation processes that are at the heart of photosynthesis and respiration (Zangori et al., 2017). Accordingly, misconceptions of photosynthesis and respiration are strongly interlinked (Parker et al., 2012). Brown and Schwartz (2009) and Wilson and colleagues (2006) showed that students think photosynthesis needs the input of carbon dioxide by humans. Amir and Tamir (1994) revealed students' idea that plants do not respire at all. The authors also showed that students believe photosynthesis provides the plant directly with energy. Students often understand respiration to be a pure gas-exchange event. As they believe that photosynthesis is the opposite of cellular respiration, it is also viewed as gas exchange, or how plants 'breathe' (Amir & Tamir, 1994). Respiration is often understood as being the opposite of photosynthesis, because some reactants of photosynthesis (carbon dioxide, water) are the products of respiration, whilst oxygen, a reactant of respiration, is a product of photosynthesis. Students do not seem to comprehend the differences between the processes in chemical pathways and their location in the plant (Eisen & Stavy, 1988). Furthermore, students hold the misconception that respiration only occurs in the dark (e.g., Marmaroti & Galanopoulou, 2006).

Difficulties in Teaching About Energy, Photosynthesis, and Respiration

The various misconceptions revealed about energy, photosynthesis, and respiration indicate the demanding challenge for teaching them. But what are the fundamental difficulties? In our view, they can be ascribed to two salient features of the three concepts: their *non-concrete, abstract nature*, and their *high complexity*.

Difficulties Related to the Concepts' Non-Concrete, Abstract Nature

Energy, photosynthesis, and respiration are not directly visible and hence are difficult to grasp. Accordingly, teaching these concepts needs tools that render them tangible. Such tools mainly are the language (spoken and written) that the biology teacher uses in class; pictures, texts, and mathematical equations in biology textbooks; and conventional experimental protocols, as well as hands-on activities. Relating to the depictive representations, we have little evidence about the particular features that foster understanding of energy, photosynthesis, and respiration and do not generate or strengthen misconceptions. Wernecke and colleagues (2016) analysed the representation of energy in biology textbooks. They discovered that flow diagrams with arrows indicating sequences are frequently used to illustrate energetic aspects of photosynthesis, matter cycles, and ecological energy flows. Their abstractness is often alleviated by concrete elements such as schematic drawings, realistic drawings, or photographs. Representations with such multiple layers are especially demanding: students must both relate the concrete elements to the corresponding things in the real world and decode the meanings of symbols such as arrows (Roth et al., 2005).

Another difficulty is that learners often just skip depictive representations and thus fail to exhaust their full potential (Cromley et al., 2010). Also, to exploit them an adequate pre-knowledge must be available for and activated by the students (Kragten et al., 2015), something that teachers often presuppose. Another problem for learning from diagrams is sufficient knowledge of diagrammatic conventions (Cheng & Gilbert, 2015), specifically of biological flow diagrams, which are often not taught explicitly to students but are taken for granted by the teacher, especially in higher grades. Regarding descriptive representations (texts, mathematical equations) for teaching abstract scientific concepts and processes like energy, photosynthesis, and respiration, metaphors have been found to be valuable tools (e.g., for the global warming associated with increases in atmospheric carbon dioxide concentrations metaphorically described as 'the greenhouse effect'; Niebert & Gropengießer, 2014). Metaphors can help students to visualize abstract phenomena, motivate them, and make new ideas intelligible (Duit, 1991). However, a meta study by Niebert et al. (2012) revealed that effective instructional metaphors need unambiguous, embodied sources that students understand directly. These requirements are not always fulfilled by common metaphors, causing several difficulties in teaching that are likely to be overlooked. Metaphors used by experts (e.g., teachers) in scientific discourse are often conventionalized and not perceived as figurative language. Also, so-called dead metaphors do not invoke a concrete source domain; and students who are not familiar with the specific scientific meaning of a metaphor may comprehend it differently, meaning that their ideas might be more affected by the source domain (Rundgren et al., 2008).

Although metaphors have the potential to enable communication about abstract concepts like energy, photosynthesis, and respiration and facilitate understanding, studies revealed that students struggle with their adequate usage. For example, Niebert and Gropengießer (2014) elucidated that students and scientists use the same image schemata for understanding the greenhouse effect, but inadequate mapping from source to target domains by students often results in scientifically incorrect conceptions, as the misconception about plant nutrition described earlier shows (Leach et al., 1996). However, an issue that has received little attention so far is whether energy-related metaphors used in educational contexts are adopted and used in the intended way by students. As the understanding of energy is indispensable to deeply understanding photosynthesis and respiration, the usage of metaphors can be assumed to be a source for misconceptions in learning the three concepts in question (Wernecke et al., 2018b).

Difficulties Related to the Concepts' High Complexity

The complexity addressed here can be described as twofold: each of the three concepts itself is complex, and the interrelationship between them represents a second dimension of complexity. Energy flows through different trophic levels of an ecosystem. To build up and maintain ordered structures that are fundamental for life, organisms need to take up and transform energy, either by using sunlight in photosynthesis or through catabolic pathways that oxidize chemical compounds. As energy sources are often limited, the survival and evolutionary success of organisms depends also on the efficiency of an organism's energy transformations. Thus, energy describes biological processes on all system levels from the biosphere down to a single cell. Concurrently, energy manifests in different forms, is transferred from system to system, and dissipates. To comprehend all this poses high challenges for students.

Furthermore, even though the frequent occurrence of the energy concept in biology has the potential to connect topics across different grade levels and thus to support students to reconstruct an integrated understanding of biology (e.g., Bransford et al., 2000), it also may overcharge the students' abilities. This problem might be intensified by the fact that energy is also a crosscutting concept in biology, chemistry, physics, and the engineering subjects (e.g., NRC, 2012). In addition to school science instruction, the energy concept also significantly concerns students' everyday lives, for example, when they speak about nutrition, sports, climate change, or energy 'consumption' and 'saving' (Opitz et al., 2015). In these contexts, different connotations for energy exist simultaneously, with vitalistic conceptions (e.g., 'I have a lot of energy; I think I want to go running!') differing from the scientific concept (i.e., energy as an abstract, conserved quantity); this can make it more difficult for students to learn about energy in biology and science classes (e.g., Lancor, 2014).

Energy can be described as a threshold concept (*sensu* Meyer & Land, 2005) for understanding photosynthesis and respiration. Without conceptual knowledge of energy, students fail to develop a deeper understanding of the latter (Loertscher et al., 2014). For example, regarding photosynthesis, Parker (2012) argues that the amount of information (e.g., inputs and outputs of light-dependent reactions and the Calvin Cycle) leads students to pure memorization instead of higher order thinking. He thus concludes that in order to get a grip on learning photosynthesis, traditional content should be organized around a few themes like matter, energy, and scale. To grasp the meaning of energy, photosynthesis, and respiration, biochemical knowledge is also needed. Lin and Hu (2003) addressed energy flow and matter cycling in 7th grade students. They concluded from their study that even though energy is only explicitly introduced two years later, energy already has an important role for contexts relevant to students in the 7th grade. The textbook emphasis on recall definitions, insufficient teacher knowledge, and students' difficulty with transferring principles from the non-living world to living beings obviously limit integrated conceptual learning on energy and beyond.

Conceptualizing Energy, Photosynthesis, and Respiration

Against the background of the manifold misconceptions found and the difficulties arising from them for teaching and learning biology, we propose a conceptualization of energy that meets its claim of being a disciplinary core idea. Our conceptualization highlights energy and puts photosynthesis and respiration in the context of energy flow. Furthermore, this conceptualization takes into account the crosscutting character of the concept of energy to enhance science learning beyond disciplinary boundaries by building a bridge to the conceptualization of energy in physics. Thus, a bipartite conceptualization of energy arises, subordinating photosynthesis and respiration in a wider (energy) context and integrating the disciplinary and crosscutting character of energy (Figure 12.1). The biological view on energy starts where energy from the universe enters the biosphere as a result of the actions of photosynthetic organisms. Energy then flows through living systems at all levels of complexity, from the ecosystem down to the molecular level. The whole amount of radiation energy fixed by autotrophic organisms in the process of photosynthesis ends up in thermal energy that disperses in the atmosphere. The final end of the flow is the chemical reactions of respiration, where thermal energy is released and low-energy molecules are left over. By conceptualizing photosynthesis and respiration in such a way, student understanding should go beyond just grasping the two basic physiological processes of biology; this should help students to realize their fundamental role for the evolution, existence, and interrelationships of living beings overall. We presume that in this way a deeper, more coherent biological understanding will develop that is needed to deal with and comprehend related socio-scientific issues, e.g. climate change.

At each level of biological organization (in short, 'system level'), there are many educational relevant biological topics that should be put into the context of energy flow—for example, the predator-prey relationship at the ecosystem level or information processing at the level of the nervous system (see Figure 12.1 for further examples). And here the crosscutting character of the energy concept comes into play. We refer to a common framework from physics education that describes energy in four aspects, i.e. (i) energy forms, (ii) energy transfer and transformation, (iii) energy dissipation, and (iv) energy conservation (Neumann et al., 2013). Besides putting the particular biological topics into the framework of energy flow along the biological system levels, these four aspects should—whenever possible—be applied for integrating the biological and physics perspective on

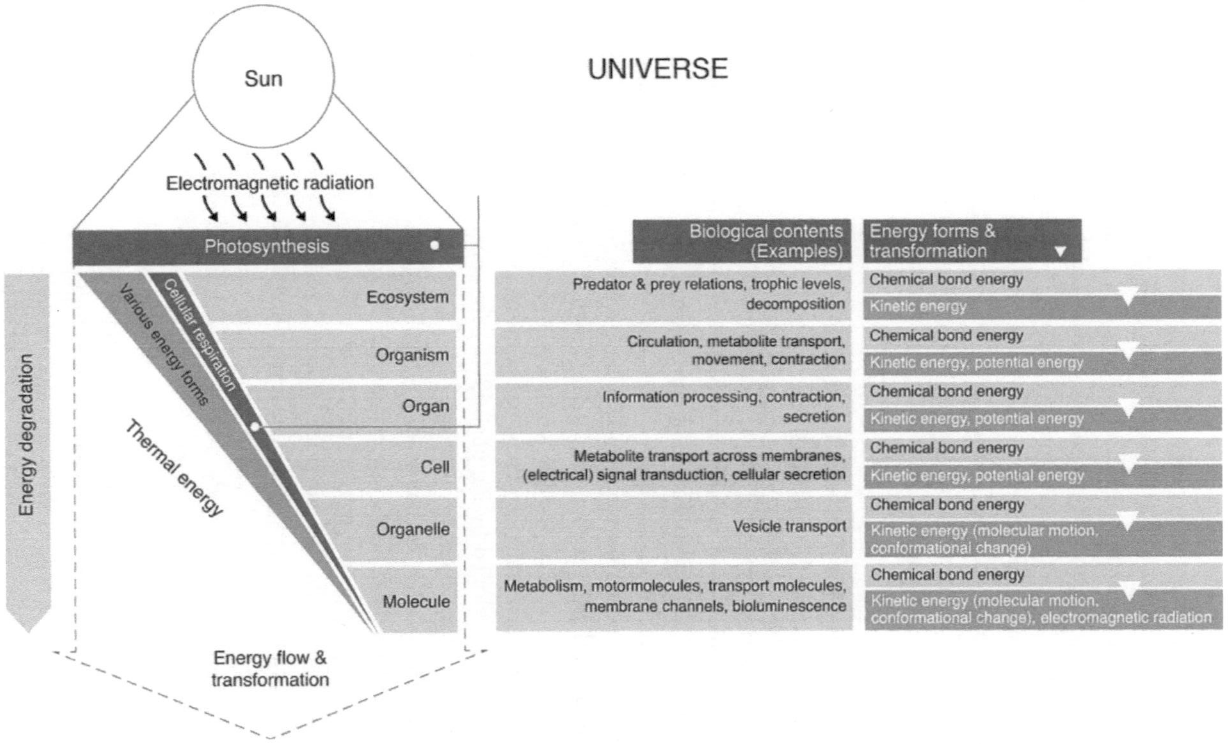

FIGURE 12.1 Interplay of energy flow, forms, transformation, and degradation in biology. The arrow (left) illustrates the aspect of energy conservation

energy. Coming back to the example of predator-prey relationship: discussing predator-prey relationships implies different forms or manifestations of energy like kinetic energy (e.g., the predator has to run to get the prey) and electric energy (regarding the affected neurobiological processes). Energy from the prey is transferred to a higher trophic level (the predator who transforms the chemical energy absorbed from the prey into kinetic energy for its movement or its growth). All the energy transformation processes lead to the dissipation of energy, meaning that in all of them thermal energy is generated that is no longer available for life functions. In a nutshell, we propose a bipartite conceptualization of energy (integrating photosynthesis and respiration) that (i) follows energy through all levels of biological organization, aiming for a coherent understanding of biology through the lens of energy, and (ii) exceeds disciplinary boundaries by linking to its conceptualization in physics.

Recommendation for Teaching About Energy, Photosynthesis, and Respiration

The conceptualization described above focuses on thinking vertically across the different levels of biological organization, and on drawing interdisciplinary relations by applying a conceptualization of energy from physics teaching to reach a horizontal coherence across disciplines. This new view on energy, photosynthesis, and respiration education we introduce here relates to approaches developed for other biological areas such as evolution (Tibell & Harms, 2017) and genetics (Knippels, 2002), where comprehension of the basic concepts also requires vertical and horizontal thinking. Thus, we propose to transfer the learning and teaching strategy developed by Knippels (2002) for genetics teaching to energy, photosynthesis, and respiration education.

Such teaching implies the need to: (i) explicitly distinguish between the different system levels; (ii) interrelate concepts at the same level of organization; (iii) connect concepts from different levels of organization; (iv) switch back and forth between the different concepts and levels to trigger micro-macro thinking; and (v) reflect on a meta level which levels of biological organization have been criss-crossed. Furthermore, relating to Parker et al. (2012), when teaching energy, photosynthesis, and respiration, the role of matter should clearly be separated from teaching the energy story. By this theoretically fine-grained, structured approach, we aim to mitigate the difficulties in learning that arise from the concepts' considerable complexity. Now, empirical studies must show their effectiveness. Regarding difficulties related to the concepts' abstract nature, several studies show that depictive representations are regarded as helpful. For example, van Heuvelen and Zou (2001) found that bar chart representations of work-energy processes help university students to understand energy conservation in physics. It should be investigated whether this finding could be replicated for biological contexts. Similarly, Ryoo and Linn

(2012) showed that dynamic and static depictive representations promote middle school students' understanding of energy transformations in photosynthesis. Wernecke et al. (2018a) transferred the method of learning from mistakes from mathematics teaching to energy education. They derived an incorrect depictive representation of energy flow in an ecosystem from well-known student misconceptions. In an experimental study, this approach showed positive effects on secondary students' conceptual knowledge of energy. Because of the three concepts' abstract nature, schoolbooks and teachers use several metaphors when teaching energy, photosynthesis, and respiration. As Wernecke and colleagues (2018b) show, these can pose severe obstacles for learning. Therefore, teachers should thoroughly scrutinize the language applied when teaching these abstract concepts. Currently, there is a lack of intervention studies testing methods for how to teach energy, photosynthesis, and respiration successfully. However, teaching methods that have empirically been shown to promote understanding of other complex and abstract biological concepts (e.g., evolution) could be used to derive recommendations for teaching the three concepts in question as, for example, learning from adaptive prompts and worked examples (Neubrand et al., 2016; Neubrand & Harms, 2017).

Note

1 The term 'respiration' is used in this text exclusively for cellular respiration.

References

Amir, R. & Tamir, P. (1994). In-depth analysis of misconceptions as a basis for developing research-based remedial instruction: the case of photosynthesis. *The American Biology Teacher*, 56, 94–100.

Barak, J., Gorodetsky, M. & Chipman, D. (1997). Understanding of energy in biology and vitalistic conceptions. *International Journal of Science Education*, 19, 21–30.

Barak, J., Sheva, B., Gorodetsky, M. & Gurion, B. (1999). As 'process' as it can get: students' understanding of biological processes. *International Journal of Science Education*, 21, 1281–1292.

Beals, A. M., McNall Krall, R. & Wymer, C. L. (2012). Energy flow through an ecosystem: conceptions of in-service elementary and middle school teachers. *International Journal of Biology Education*, 2, 1–18.

Boyes, E. & Stanisstreet, M. (1991). Misconceptions in first-year undergraduate science students about energy sources for living organisms. *Journal of Biological Education*, 25(3), 209–213.

Bransford, J. D., Brown, A. L. & Cocking, R. R. (Eds) (2000). *How People Learn: Brain, Mind, Experience and School*. Washington, DC: National Academy Press, 3–23.

Brown, M. H. & Schwartz, R. S. (2009). Connecting photosynthesis and cellular respiration: preservice teachers' conceptions. *Journal of Research in Science Teaching*, 46(7), 791–812.

Chabalengula, V., Sanders, M. & Mumba, F. (2011). Diagnosing students' understanding of energy and its related concepts in biological contexts. *International Journal of Science and Mathematics Education*, 10(2), 241–266.

Cheng, M. M. W. & Gilbert, J. K. (2015). Students' visualization of diagrams representing the human circulatory system: the use of spatial isomorphism and representational conventions. *International Journal of Science Education*, 37, 136–161.

Cromley, J. G., Snyder-Hogan, L. E. & Luciw-Dubas, U. A. (2010). Cognitive activities in complex science text and diagrams. *Contemporary Educational Psychology*, 35, 59–74.

Duit, R. (1991). On the role of analogies and metaphors in learning science. *Science Education*, 75, 649–672.

Duit, R. & Kesidou, S. (1988). Students' understanding of basic ideas of the second law of thermodynamics. *Research in Science Education*, 18(1), 186–195.

Eisen, Y. & Stavy, R. (1988). Students' understanding of photosynthesis. *The American Biology Teacher*, 50, 208–212.

Finegold, M. & Trumper, R. (1989). Categorizing pupils' explanatory frameworks in energy as a means to the development of a teaching approach. *Research in Science Education*, 19(1), 97–110.

Goldring, H. & Osborne, J. (1994). Students' difficulties with energy and related concepts. *Physics Education*, 29(1), 26.

Hirca, N. C. & Akdeniz, F. (2008). Investigating grade 8 students' conceptions of 'energy' and related concepts. *Journal of Turkish Science Education*, 5(1), 75–87.

Jin, H. & Wei, X. (2014). Using ideas from the history of science and linguistics to develop a learning progression for energy in socio-ecological systems. In R. F. Chen, A. Eisenkraft, D. Fortus, J. Krajcik, K. Neumann, J. Nordine & A. Scheff (Hrsg.), *Teaching and Learning of Energy in K-12 Education*. New York, NY: Springer, S. 157–173.

KMK—Sekretariat der ständigen Konferenz der Kultusminister der Länder in der Bundesrepublik Deutschland. (2005). *Bildungsstandards im Fach Biologie für den Mittleren Schulabschluss—Beschluss vom 16.12.2004* [German national education standards concerning biology for middle school graduation. Decision from December 16th 2004].

Knippels, M.-C. (2002). *Coping with the Abstract and Complex Nature of Genetics in Biology Education; the Yo-Yo Learning and Teaching Strategy*. Utrecht: University of Utrecht Press.

Köse, S. (2008). Diagnosing student misconceptions: using drawings as a research method. *World Applied Sciences Journal*, 3, 283–293.

Kragten, M., Admiraal, W. & Rijlaarsdam, G. (2015). Students' ability to solve process-diagram problems in secondary biology education. *Journal of Biological Education*, 49, 91–103.

Lancor, R. A. (2014). Using student-generated analogies to investigate conceptions of energy: a multidisciplinary study. *International Journal of Science Education*, 36(1), 1–23.

Leach, J., Driver, R., Scott, P. & Wood-Robinson, C. (1996). Children's ideas about ecology 2: ideas found in children aged 5–16 about the cycling of matter. *International Journal of Science Education*, 18, 19–34.

Lin, C.-Y. & Hu, R. (2003). Students' understanding of energy flow and matter cycling in the context of the food chain, photosynthesis, and respiration. *International Journal of Science Education*, 25, 1529–1544.

Liu, X. & McKeough, A. (2005). Developmental growth in students' concept of energy: analysis of selected items from the TIMSS database. *Journal of Research in Science Teaching*, 42(5), 493–517.

Liu, X. & Ruiz, M. E. (2008). Using data mining to predict K—12 students' performance on large-scale assessment items related to energy. *Journal of Research in Science Teaching*, 45(5), 554–573.

Loertscher, J., Green, D., Lewis, J. E., Lin, S. & Minderhout, V. (2014). Identification of threshold concepts for Biochemistry. *CBE-Life Sciences Education*, 13, 516–528.

Marmaroti, P. & Galanopoulou, D. (2006). Pupils' understanding of photosynthesis: a questionnaire for the simultaneous assessment of all aspects. *International Journal of Science Education*, 28, 383–403.

Métioui, A., Matoussi, F. & Trudel, L. (2016). The teaching of photosynthesis in secondary school: a history of the Science approach. *Journal of Biological Education*, 50(3), 275–289.

Meyer, J. H. F. & Land, R. (2005). Threshold concepts and troublesome knowledge (2): epistemological considerations and a conceptual framework for teaching and learning. *Higher Education*, 49(3), 373–388.

National Research Council [NRC]. (2012). *A Framework for K-12 Science Education: Practices, Crosscutting Concepts, and Core Ideas*: The National Academies Press.

Needham, R. (2014). Using 'energy ideas' in the teaching of biology. *School Science Review*, 96(354), 74–77.

Neubrand, C., Borzikowsky, C. & Harms, U. (2016). Adaptive prompts for learning evolution with worked examples—highlighting the students between the "novices" and the "experts" in a classroom. *International Journal of Environmental & Science Education*, 11(14), 6774–6795.

Neubrand, C. & Harms, U. (2017). Tackling the difficulties in learning evolution: effects of adaptive self-explanation prompts. *Journal of Biological Education*, 51(4), 336–348.

Neumann, K., Viering, T., Boone, W. J. & Fischer, H. E. (2013). Towards a learning progression of energy. *Journal of Research in Science Teaching*, 50, 162–188.

NGSS (Lead States). (2013). *Next Generation Science Standards: For States, By States*. Washington, DC: The National Academies Press.

Niebert, K. & Gropengießer, H. (2014). Understanding the greenhouse effect by embodiment—analysing and using students' and scientists' conceptual resources. *International Journal of Science Education*, 36, 277–303.

Niebert, K., Marsch, S. & Treagust, D. F. (2012). Understanding needs embodiment: a theory-guided reanalysis of the role of metaphors and analogies in understanding science. *Science Education*, 96, 849–877.

Opitz, S., Blankenstein, A. & Harms, U. (2017a). Student conceptions about energy in biological contexts. *Journal of Biological Education*, 51(4), 427–440.

Opitz, S., Neumann, K., Bernholt, S. & Harms, U. (2017b). How do students understand energy in biology, chemistry and physics? development and validation of an assessment instrument. *Eurasia Journal of Mathematics, Science and Technology Education*, 13(7), 3019–3042.

Opitz, S., Harms, U., Neumann, K., Kowalzik, K. & Frank, A. (2015). Students' energy concepts at the transition between primary and secondary school. *Research in Science Education*, 45(5), 691–715.

Özay, E. & Öztaş, H. (2003). Secondary students' interpretations of photosynthesis and plant nutrition. *Journal of Biological Education*, 37, 68–70.

Parker, J. M., et al. (2012). Exploring undergraduates' understanding of photosynthesis using diagnostic question clusters. *CBE-Life Sciences Education*, 11(1), 47–57.

Roth, W.-M., Pozzer-Ardenghi, L. & Han, J. Y. (2005). *Critical Graphicacy: Understanding Visual Representation Practices in School Science*. Dordrecht: Springer.

Rundgren, C.-J., Hirsch, R. & Tibell, L. A. (2008). Death of metaphors in life science?—a study of upper secondary and tertiary students' use of metaphors in their meaning-making of scientific content. In C.-J. Rundgren (Ed.), *Visual Thinking, Visual Speech—a Semiotic Perspective on Meaning-Making in Molecular Life Science*. Norrköping, Sweden: FontD, 1–24.

Ryoo, K. & Linn, M. C. (2012). Can dynamic visualizations improve middle school students' understanding of energy in photosynthesis? *Journal of Research in Science Teaching*, 49(2), 218–243.

Simpson, M. & Arnold, B. (1982). The inappropriate use of subsumers in biology learning. *European Journal of Science Education*, 4(2), 173–183.

Tibell, L. A. E. & Harms, U. (2017). Biological principles and threshold concepts for understanding natural selection—implications for developing visualizations as a pedagogic tool. *Science & Education*. doi: 10.1007/s11191-017-9935-x

Trumper, R. (1993). Children's energy concepts: a cross-age study. *International Journal of Science Education*, 15(2), 139–148.

Trumper, R., Raviolo, A. & Shnersch, A. (2000). A cross-cultural survey of conceptions of energy among elementary school teachers in training empirical results from Israel and Argentina. *Teaching and Teacher Education*, 16(7), 697.

van Heuvelen, A. & Zou, X. (2001). Multiple representations of work—energy processes. *American Journal of Physics*, 69, 184–194.

van Hook, S. & Huziak-Clark, T. (2008). Lift, squeeze, stretch, and twist: Research-based Inquiry Physics Experiences (RIPE) of energy for kindergartners. *Journal of Elementary Science Education*, 20(3), 1–16.

Vosniadou, S. & Skopeliti, I. (2013). Conceptual change from the framework theory side of the fence. *Science & Education*, 23(7), 1427–1445.

Wernecke, U., Schütte, K., Schwanewedel, J., & Harms, U. (2018a). Enhancing conceptual knowledge of energy in biology with incorrect representations. *CBE – Life Sciences Education (LSE)*, 17(1), 1–11, ar5. doi: 10.1187/cbe.17-07-0133

Wernecke, U., Schwanewedel, J., & Harms, U. (2018b). Metaphors describing energy transfer through ecosystems: helpful or misleading? *Science Education (SciEd)*, 102(1), 178–194.

Wernecke, U., Schwanewedel, J., Schütte, K. & Harms, U. (2016). Wie wird Energie im Biologieschulbuch dargestellt?—Entwicklung eines Kategoriensystems und exemplarische Anwendung auf eine Schulbuchreihe [How is energy represented in biology textbooks?]. *Zeitschrift für Didaktik der Naturwissenschaften*, 22(1), 215–229.

Wilson, C. D., Anderson, C. W., Heidemann, M., Merrill, J. E., Merritt, B. W., Richmond, G., Sibley, D. F. & Parker, J. M. (2006). Assessing students' ability to trace matter in dynamic systems in cell biology. *CBE—Life Sciences Education*, 5, 323–331.

Zangori, L., Peel, A., Kinslow, A., Friedrichsen, P. & Sadler, T. D. (2017). Student development of model-based reasoning about carbon cycling and climate change in a socio-scientific issues unit. *Journal of Research in Science Teaching*. doi:10.1002/tea.21404

13
ECOLOGY

Konstantinos Korfiatis

Ecological Literacy

The concept of ecological literacy refers to the importance of educating citizens so that they not only know quite a bit about ecology, but are also able to critically process that knowledge and apply it for informed decision-making and in environmentally related actions (Mc Bride et al., 2013; Casper & Balgopal, 2017). Knowledge of ecology is essential so that students develop understanding of critical environmental issues such as global warming and loss of biodiversity (Hokayem & Gotwals, 2016). As Jordan et al. (2009) argue, ecological literacy must be part of a citizen's lifetime learning experience, beginning in the primary grades and continuing through to informal adult learning experiences. Ecological literacy must include an understanding of the links between human actions and their subsequent effects on ecosystems (Balgopal & Wallace, 2009). Humans make decisions as individuals and as groups that alter ecosystem structure and functioning at multiple scales. Not all human actions have destructive effects, however. Humans can reduce impacts to native ecosystem structures or improve ecological functioning through restoration (Ruppert & Duncan, 2017). This much depends upon the education of ecologically literate citizens.

Within the above line of reasoning, ecology education shares common grounds with conservation education. Specifically, they both have as a goal the understanding and appreciation of the natural world. Learning about the diversity of living things is an important topic for ecology and conservation as well (Primack, 2012; Allen, 2016). Conservation education, however, is especially focused on the protection of the planet's biodiversity in all its forms (genetic diversity, species, communities and ecosystems diversity). It aims not only to develop people's knowledge of Earth's biological diversity and awareness of threats to biodiversity, but also to increase skills and intentions to act to protect and maintain the genes, species, communities and ecosystems that make up that diversity (Primack, 2012).

Conservation education developed from the realization of conservation biologists that the public affects the success or failure of environmental management efforts. Researchers could spend years designing plans or studying biological processes but fail to achieve conservation goals without adequate public awareness, support and involvement (Jacobson et al., 2015). As in the case of ecology education, it is reported that although people are aware of environmental problems and declare that they care for the environment, their knowledge about conservation is minimal. People's concern for biodiversity is confined to attractive and/or emotionally appealing species (e.g. eagles, panda), while they dislike or are indifferent to others such as invertebrates, despite their crucial role in maintaining life through their role in decomposition processes (Jacobson et al., 2015). Therefore, a good knowledge of ecology lies at the core of conservation education.

What a Student Needs to Know

Ecological literacy has to a large extent to do with the development of knowledge about systems and the development of system thinking skills (McBride et al., 2013). Ecosystems are complex, open systems, and understanding interdependent relationships within them requires systemic reasoning (Hokayem & Gotwals, 2016). Understanding critical environmental issues requires thinking about how different systems work and interact with each other (Long et al., 2014).

Therefore, a framework for ecological literacy includes:

I. An understanding of ecological connectivity (interactions and relationships) and key ecological concepts, including: functional connections within species and between species and the environment; patterns of population growth; and the dynamics of intra-specific and inter-specific relationships (Jordan et al., 2014).
II. An understanding of structure and the function of ecological systems, and of the flow of energy and matter through them. Producers, consumers and decomposers are the main components of an ecosystem structure, and they regulate the flow of matter and energy in the system. The role of each species in an ecosystem is one of the reasons that biodiversity is considered important. Therefore, understanding the structure and functions of ecosystems, as well as the role of each ecosystem's components, is important for understanding the effect of human actions as well. Ruppert and Duncan (2017) present the example of wetlands' ecosystems, where organic nitrogen is converted to atmospheric nitrogen, removing potential toxic wastes that contribute to dead zones in coastal fisheries. Ecosystems, however, have a limited capacity to provide resources and process waste, which is regulated by the flow of

matter and energy within them. If humans extract resources or dump wastes at a rate that exceeds an ecosystem's capacity, that ecosystem will become over-exploited or polluted.

III. 'Systems thinking' also includes ideas about order and organization, equilibrium and change and consequences of system perturbations (Long et al., 2014; Hokayem et al., 2015). The comprehension of these concepts entails an appreciation of feedbacks and constraints, and an ability to explain and predict basic patterns of population dynamics (Jordan et al., 2009). Indeed, feedback loop reasoning is an important aspect of systemic reasoning (Hokayem & Gotwals, 2016). Ecosystems can change rapidly and unexpectedly, but resilient systems are less likely to experience a dramatic change in ecosystem function. Stability in ecosystems is regulated by negative feedbacks. Negative feedbacks keep the condition in a system in a state of 'dynamic equilibrium'. In a predator-prey system governed by the existence of negative feedback mechanisms (usually in the form of density-dependent relationships between predator and prey), for example, when predators eat prey and expand their population, fewer prey become available, keeping the size of the predator population in check. In complex ecosystems, there are many interlocking feedback relationships (e.g., many predator-prey relationships, among others), which help maintain the functioning of an ecosystem over time (Jacobson et al., 2017). The absence of feedback mechanisms, however, produces unpredictable, and often chaotic, modes of system function.

IV. An important aspect of understanding complex systems is to identify patterns at the system level and connect those patterns to behaviors and interactions of constituent components. In many situations, the patterns at the system level are 'emergent', meaning that the patterns are caused by the interactions and collective behaviors of constituent components (Hokayem & Gotwals, 2016).

V. Last, but not least, natural history, i.e. the study of organisms in their environment, is an indisputable component of the science of ecology (Krebs, 2010), and consequently of ecological literacy. According to King and Achiam (2017), an understanding of natural history is vital for resolving issues of localized biodegradation and biodiversity loss, which in turn contribute to global issues of food security, disease control and climate change. We need to know what living organisms are, where they live, what they eat, how they mate and how they die in order to make sense of earth's ecosystems. The first steps towards ecological literacy are for inquirers to become familiar with the natural history of their local and abiotic environments and from this to progress to the acquisition of skills such as posing interesting questions about their own surroundings and consideration of the consequences of various human activities. Only in such a context could an understanding of ecological concepts and content be meaningfully acquired (Lefkaditou et al., 2014).

Students' Misconceptions

Educational research has consistently focused on students' misconceptions in ecology and their lack of comprehension of the structures and functions of ecosystems. Research evidence converges at the conclusion that most difficulties should be attributed to the dynamic and systemic nature of ecological processes combined with the lack of systemic thinking skills necessary for a student to understand ecological processes and human-nature interactions.

Students have difficulties in understanding the kinds of relationships organisms establish between themselves and with the abiotic elements of their environment. It is even argued that student understanding of interactions in ecosystems does not progress much over grade levels, although ecosystems in many countries are already an important topic from early elementary school (Hokayem & Gotwals, 2016). As Jordan et al. (2014) mention, learning about interdependence in ecosystems is difficult. In the case of feeding relations, for instance, and their association with the concept of food webs, it has been repeatedly reported that students tend to think in terms of isolated food chains rather than interconnected and complicated food webs. When tracing the effect of a change or perturbation through the structure of a complex system, students tend to follow this effect along one direction only, away from the locus of perturbation, without considering possible effects in other branches of a food web or in other parts of an ecosystem indirectly linked with the source of the perturbation (Lefkaditou et al., 2014; Allen, 2016). When considering the effects of changing population sizes, children tend to be able to more easily comprehend changes up the trophic levels than down—that is, in the direction producer (plant) → primary consumer (herbivore) → secondary consumer (carnivore) (Demetriou et al., 2009; Gotwals & Songer, 2010). For example, they can appreciate that if there is a fall in a population of rabbits due to disease, then this would mean foxes would starve and begin to die. It seems that a reduction in a prey population causing predator starvation is easier to understand than proliferation of prey as a result of predator decline (Allen, 2017). Precisely because students do not comprehend how changes in a part of a system could affect the rest of the system, they tend to think that a change in one population in a food web could only affect another population if they are related as predator and prey; alternatively, they may think that if something happened to one population in a food web, then the same effect will result for all populations within the web (Demetriou et al., 2009; Gotwals & Songer, 2010; Allen, 2017).

It has even been reported that, in some cases, older primary children 'invented' food sources not actually available in a given food web when asked about the effect on the population of a consumer if its prey is being dramatically reduced (Allen, 2017). According to Allen (2016), this represents an erroneous belief that prevents a correct understanding of sustainability, implying that natural resources are infinite and a species can easily shift from one resource to another. It also has been shown that upper secondary-aged students are willing to skip trophic levels, ignoring the adjacency rule, and have, for instance, a predator able to eat a plant (Allen, 2017).

Students with misconceptions about trophic relations have difficulties in understanding environmental issues. It has been shown, for instance, that students who did not understand the complexity of interrelationships within a food web were not able to explain the effect of an oil spill in an aquatic environment (Hogan, 2000).

Students also appeared to be unware of the factors that regulate natural systems. In the case of herbivores, for instance, they miss identifying both reproduction and food as the factors regulating population in a context of an absence of predators (Hokayem et al., 2015). This example is just an expression of most students' failure to understand how feedback mechanisms regulate a system such as a food web. For example, in a hypothetical predator-prey system of wolves and deer, students usually believe that the wolves will first consume all the deer and then die, since they will have exhausted their feeding sources. Thus, they seem to not understand the existence of regulatory mechanisms in many populations that do not allow them to increase until they exhaust all resources and then go extinct (Lefkaditou et al., 2014).

Students also typically cannot connect different levels of a system (Jacobson et al., 2017). They seldom connect, for example, individual processes of matter transformation and energy transformation at a molecular level with patterns of the cycling of matter and energy flow processes at the level of ecosystems. In another example concerning the water cycle, research has shown that school students do not usually think about water in dynamic, cyclical systems (Gunckel et al., 2012).

The lack of understanding of complex systems functioning and interactions seems to be accompanied by rather teleological explanations, and the inability of students to grasp issues of change and stability (Hovardas & Korfiatis, 2011). As students find it difficult to follow the effects of a disturbance through different branches of a food web or, more importantly, the feedback effects taking place in interactions between populations (Hogan, 2000), some students think that any disturbance in an ecosystem will eventually result in its collapse (White, 2008), while others hold that no matter the kind or the magnitude of the disturbance, the ecosystem will eventually recover to its initial state (Ergazaki & Ampatzidis, 2012).

Ecology Teaching: For Better or for Worse?

Students' understanding of ecosystem processes is critical to biological education. Yet, teaching life systems can be difficult, because systems are dynamic and often behave in a non-linear manner (Jordan et al., 2014). This difficulty in bringing the dynamic nature of ecological phenomena into the classroom results in many deficiencies in teaching: relationships between processes are rarely stressed and ecological issues are rarely taught in ways that highlight systemic reasoning (Jordan et al., 2014; Hmelo-Silver & Azevedo, 2006; Hokayem & Gotwals, 2016).

Therefore, we face a situation where ecology teaching itself often produces misunderstandings and obstacles in students' attempts to comprehend ecology! A characteristic example is the educational approach suggesting that in teaching feeding relationships, students have to comprehend first the concept of a food chain, as a lower anchor of a learning progression, and then to move to the learning of food webs. However, there is enough evidence to show that by starting with the idea of a food chain students become 'addicted' to the idea of linear and teleological thinking, which is later difficult to change (Songer et al., 2009; Demetriou et al., 2009). Indeed, food chains and food pyramids might pertain to a linear approach when these representations are not accompanied by feedback loops in instruction and school textbooks (Hovardas, 2016).

Another criticism concerns the fact that environmental systems are treated by teachers and textbooks alike as static collections of objects and cycles to be memorized, while the focus should be on the dynamic character of those systems. This leads to a simplistic understanding of causality and the systemic nature of interactions taking place within the food web (Demetriou et al., 2009).

Jin et al. (2013) describe how a lack of a systemic approach to the teaching of the carbon cycle in secondary school students undermines the aim of having people understand the processes creating global climate change. They argue that curriculum materials used in secondary schools often address the reactants and products of carbon-transforming processes, without articulating the big ideas of how matter and energy transform. Jacobson et al. (2017) pinpoint that research to date has focused on students' learning about climate ideas (e.g., greenhouse gases, carbon cycle) but not the complex systems ideas (e.g., self-organization, positive feedback, emergence) that can provide a conceptual foundation for understanding climate systems and how humans impact the global environment (Jacobson et al., 2017).

Long et al. (2014) criticize how the teaching of ecological models takes the form of discrete topics that students learn about through narrative descriptions, instead of engaging students in building and reasoning about ecological models (e.g. predicting consequences of environmental changes by explaining the movement of carbon through an ecosystem).

Last, but not least, a growing body of research shows the use of the 'balance of nature' metaphor in educational practice enforces misconceptions, since it promotes teleological ideas (i.e. nature tends to balance) and simple linear causality (i.e. the natural course of systems eventuates in balance and any divergence from balance leads to destruction; Hovardas & Korfiatis, 2011; Ruppert & Duncan, 2017).

A Teaching Suggestion

Despite difficulties, various studies have shown that students are able to understand ecological concepts and grasp ecological processes after proper learning interventions (Hokayem & Gotwals, 2016; Ruppert & Duncan, 2017). Many of these learning interventions are based on the use of simulations. Ben Zvi Assaraf and Orion (2010) and Evagorou et al. (2009) showed that software programs coupled with explicit instruction about systemic reasoning produced successful results in teaching middle school students about ecosystems. Computer-based instructional tools (e.g. STELLA, EcoBeaker, Net-logo and Starlogo) offer many opportunities for students to visualize system connectivity and illustrate ideas about emergent properties, self-organization, loop reasoning and other ecosystem reasoning skills (Gunckel et al., 2012; Long et al., 2014; Jacobson et al., 2017; Jordan et al., 2014; Hokayem et al., 2015; Hokayem & Gotwals, 2016). Furthermore, certain games have proved as effective as computer simulations in helping students understand predator-prey relationships or the effects of disturbances in an ecosystem (Hovardas, 2016).

However, in ecology education, we should not teach models in a decontextualized manner. We should not think of ecology education only as something that happens in computer labs. Authenticity in ecological education has no meaning without field experiences. The natural history part of ecology is important but has not attracted as much research as it should (Lefkaditou et al., 2014).

I argue that it is possible to create an integrated framework for teaching ecology by putting together the insights that ecological educational research offers (Jordan et al., 2014).

In outlining the basic features of an educational proposal for ecological literacy, I find the 'Structure-Behavior-Function' approach to complex systems, proposed by Hmelo-Silver et al. (2007), to be very useful. According to this approach, 'Structure' refers to the elements of a system and the kind of relationships between them (e.g. the different species in the case of a food web and the foraging relationships between them). 'Function' refers to the role of each element in a system (e.g. the role of producers, consumers and decomposers in a food web). 'Behavior' refers to the dynamic mechanisms that yield a reaction or an outcome within a system (e.g. changes caused in a food web after a disturbance). Therefore, the S-B-F approach offers a framework which encapsulates in a simple schema all the ingredients for building learning sequences for ecological literacy.

Structure is the first element to start a learning sequence, because it is the simplest (Hmelo-Silver et al., 2007); in the case of ecology, it can be combined with outdoor learning and the first-hand study of the natural word. The study of the structure of a system is not restricted to the study of its individual parts, but also focuses on the way those parts interact and function in forming the whole. This process can also provide the basis for the comprehension of more abstract ecological concepts.

Demetriou et al. (2009) propose a learning sequence for the understanding of the structure of a food web (see also Allen, 2016, ch. 4). The concept of the food web is a central concept for ecological literacy since trophic relationships comprise the structural framework within which most natural processes take place. It is the topic one has to begin with in order to proceed with subjects such as the flow of nutrients and energy, the structure and function of ecosystems or population dynamics. The curriculum developed in Demetriou et al. (2009) allows students to construct a food web using real data collected from the field and following guidelines for the identification of organisms and their food preferences. The proposed curriculum facilitated the construction of a food web in three steps:

I. Identifying the elements of a system (e.g. by discovering in situ organisms living in an area—a wetland, an urban park or in school's yard—and identifying the species with the use of dichotomous keys).
II. Defining the relationships between elements (e.g. by gathering information from the internet or from a science book on who is eating whom in the specific study site).
III. Constructing the food web: knowing the elements of a trophic structure (i.e. the organisms that live in an area) and the way they are connected to each other (i.e. the food preferences of each species), students are able to proceed to the construction of the food web. To accomplish this, each team of students can start by choosing one species, checking which of its food preferences were found during the field study and connecting them with an arrow in the appropriate direction. They can then continue with other species until they have in their diagram all the species they have recorded. The outcome is a diagram depicting the trophic relations between the species living in the specific ecosystem (Demetriou et al., 2009).

This learning suggestion focuses initially on the visible elements of a system, such as the species living in an ecosystem, and easily understandable relations, such as predator-prey relationships. The emphasis at the beginning of a learning sequence on the visible elements can be more helpful for students while trying to comprehend how a system is constructed and how structure interplays with function and behaviour.

The assistance of computer simulations would be necessary if the learning intervention aims to study the behaviour of the food web, e.g. to study predator-prey relations or competition between species, or consequences of the disappearance of a population, or of the appearance of a new predator etc. Computer simulations offer the necessary means so that ecological processes that are difficult to observe in nature or to study in a laboratory setting (patterns of population growth, seasonal oscillations, flow of energy, etc.) can be successfully discussed and studied in this way (Korfiatis & Tunnicliffe, 2012). Students should be given opportunities to manipulate virtual systems and see how changing one component in a system (e.g. the size of a population) will affect other organisms. This will help them understand that food webs are not static and unchanging but are flexible and will vary in different situations.

The educational process described above has not yet sufficiently captured the examination of the role of organisms in a food web (producers, consumers, decomposers) and especially the role of decomposers which students find difficult to understand. In other words, it does not provide a sufficient educational approach for studying the function of a system and its parts. This is the part that can be covered by an experimental approach to ecological teaching which, it should be noted, will not usually have the hypothesis-testing, single lesson format that is the rule in science education. A 'project-like' approach, taking place over a longer period of time and allowing for multipurpose activities and an open agenda, has been argued as more appropriate for learning about ecology (Korfiatis & Tunnicliffe, 2012). Cultivation of a shrimp habitat in a bottle, keeping terraria or aquaria in classrooms or cultivating plants in pots are exemplary instances of the kind of classroom activities that are appropriate for ecology education.

The educational approach described above presents a different way of looking at inquiry teaching than is typical in science classrooms, and it provides a context for a more rounded education with respect to the practice and epistemology of science (King & Achiam, 2017). It combines the benefits of students experiencing natural history in situ with the involvement in modelling processes and longitudinal experimental processes. Other researchers have also highlighted the effectiveness of a teaching approach that combines place-based education and model-based instruction in teaching ecological topics such as watersheds, pollution and climate change (Ruppert & Duncan, 2017; Jacobson et al., 2017). This educational approach is relevant for conservation education as well, since the latter also pays attention to outdoor and first-hand experiences for helping students understand natural systems and environmental issues (Jacobson et al., 2015). It also favours experiments, simulations, discovery learning and similar techniques that employ the environment as a learning framework.

Conclusion

Ecology education has the difficult task of teaching students the structure and function of the world's ecosystems as well as their interrelations with humanity. The teaching approach outlined above comprises an outdoor component (which allows for developing a knowing of the 'parts' of the system—individual specimens, species, etc.—and to start pondering their role), an experimentation component (which is important for observing and comprehending roles) and a modelling component (as the most effective approach for comprehending the behaviour of the system; Korfiatis & Tunnicliffe, 2012).

The proposed teaching approach emphasizes the role of scientific practices such as observation and comparison, along with modelling construction and experimentation. Many school textbooks and classroom practices tend to assert that the validity of scientific knowledge is judged solely by its agreement with experimentally acquired evidence. We argue that curricula should include a greater explication of the ways in which ecology proceeds. In creating more balanced curricula, learners' perceptions of the sciences as being monolithic and depersonalized will also be countered (King & Achiam, 2017). It is an approach that gives greater emphasis to pluralism and allow learners to explore real problems in their own communities, where situational and local knowledge will emerge in addition to classroom science knowledge (Lefkaditou et al., 2014; Jacobson et al., 2017).

References

Allen, M. (2016). *The Best Ways to Teach Primary Science: Research into Practice*. Maidenhead: Open University Press.

Allen, M. (2017). Early understandings of simple food chains: a learning progression for the preschool years. *International Journal of Science Education*, 39(11), 1485–1510.

Balgopal, M. & Wallace, A. (2009). Decisions and dilemmas: using writing to learn activities to increase ecological literacy. *The Journal of Environmental Education*, 40(3), 13–26.

Ben Zvi Assaraf, O. & Orion, N. (2010). Four case studies, six years later: developing system thinking skills in junior high school and sustaining them over time. *Journal of Research in Science Teaching*, 47, 1253–1280.

Casper, A. & Balgopal, M. (2017). Conceptual change in natural resource management students' ecological literacy. *Environmental Education Research*. doi:10.1080/13504622.2017.1350830

Demetriou, D., Korfiatis, K. & Constantinou, C. (2009). Comprehending trophic relations through food web construction. *Journal of Biological Education*, 43, 53–59.

Ergazaki, M. & Ampatzidis, G. (2012). Students' reasoning about the future of disturbed or protected ecosystems and the idea of the 'balance of nature'. *Research in Science Education*, 42, 511–530.

Evagorou, M., Korfiatis, K., Nicolaou, Ch. & Constantinou, C. (2009). An investigation of the potential of interactive simulations for developing system thinking skills in elementary school: a case study with fifth-graders and sixth-graders. *International Journal of Science Education*, 31, 655–674.

Gotwals, A. & Songer, N. (2010). Reasoning up and down a food chain: using and assessment framework to investigate students' middle knowledge. *Science Education*, 94, 259–281.

Gunckel, K., Covitt, B., Salinas, I. & Anderson, C. (2012). A learning progression for water in socio-ecological systems. *Journal of Research in Science Teaching*, 49, 843–868.

Hmelo-Silver, C. E. & Azevedo, R. (2006). Understanding complex systems: some core challenges. *The Journal of the Learning Sciences*, 15(1), 53–61.

Hmelo-Silver, C. E., Marathe, S. & Liu, L. (2007). Fish swim, rocks sit, and lungs breathe: expert-novice understanding of complex systems. *Journal of the Learning Sciences*, 16(3), 307–331.

Hogan, K. (2000). Assessing students' systems reasoning in ecology. *Journal of Biological Education*, 35, 22–28.

Hokayem, H. & Gotwals, A. (2016). Early elementary students' understanding of complex ecosystems: a learning progression approach. *Journal of Research in Science Teaching*, 53, 1524–1545.

Hokayem, H., Ma, J. & Jin, H. (2015). A learning progression for feedback loop reasoning at lower elementary level. *Journal of Biological Education*, 49(3), 246–260.

Hovardas, T. (2016). A learning progression should address regression: insights from developing non-linear reasoning in ecology. *Journal of Research in Science Teaching*, 53, 1447–1470.

Hovardas, T. & Korfiatis, K. (2011). Towards a critical re-appraisal of ecology education: scheduling an educational intervention to revisit the 'balance of nature' metaphor. *Science & Education*, 20, 1039–1053.

Jacobson, M. J., Markauskaite, L., Portolese, A., Kapur, M., Lai, K. & Roberts, G. (2017). Designs for learning about climate change as a complex system. *Learning and Instruction*, 52, 1–14.

Jacobson, S, McDuff, M. & Monroe, M. (2015). *Conservation Education and Outreach Techniques*, 2nd edn. Oxford, UK: Oxford University Press.

Jin, H., Zhan, L. & Anderson, C. (2013). Developing a fine-grained learning progression framework for carbon-transforming processes. *International Journal of Science Education*, 35(10), 1663–1697.

Jordan, R., Brooks, W., Hmelo-Silver, C., Eberbach, C. & Sinha, S. (2014). Balancing broad ideas with context: an evaluation of student accuracy in describing ecosystem processes after a system-level intervention. *Journal of Biological Education*, 48(2), 57–62.

Jordan, R., Singer, F., Vaughan, J. & Berkowitz, A. (2009). What should every citizen know about ecology? *Frontiers in Ecology and the Environment*, 7, 495–500.

King, H. & Achiam, M. (2017). The case for natural history. *Science & Education*, 26, 125–139.

Korfiatis, K. & Tunnicliffe, S. D. (2012). The living world in the curriculum: ecology, an essential part of biology learning. *Journal of Biological Education*, 46, 125–127.

Krebs, C. (2010). Case studies and ecological understanding. In I. Billick & M. Price (Eds.), *The Ecology of Place: Contributions of Place-Based Research to Ecological Understanding*. Chicago and London: The University of Chicago Press, 283–302.

Lefkaditou, A., Korfiatis, K. & Hovardas, T. (2014). Contextualising the teaching and learning of ecology: historical and philosophical considerations. In M. R. Matthews (Ed.), *International Handbook of Research in History, Philosophy and Science Teaching*. Dordrecht: Springer, 523–550.

Long, T., Dauer, J., Kostelnik, K., Momsen, J., Wyse, S., Bray, S. & Ebert-May, D. (2014). Fostering ecoliteracy through model-based instruction. *Frontiers in Ecology and the Environment*, 12, 138–139.

McBride, B. B., Brewer, C. A., Berkowitz, A. R. & Borrie, W. T. (2013). Environmental literacy, ecological literacy, ecoliteracy: what do we mean and how did we get here? *Ecosphere*, 4(5), 67.

Primack, R. (2012). *A Primer of Conservation Biology*, 5th edn. Sunderland MA, Sinauer Associates.

Ruppert, J. & Duncan, R. (2017). Defining and characterizing ecosystem services for education: a Delphi study. *Journal of Research in Science Teaching*, 54, 737–763.

Songer, N. B., Kelcey, B. & Gotwals, A. W. (2009). How and when does complex reasoning occur? empirically driven development of a learning progression focused on complex reasoning about biodiversity. *Journal of Research in Science Teaching*, 46, 610–631.

White, P. (2008). Beliefs about interactions between factors in the natural environment: a causal network study. *Applied Cognitive Psychology*, 22, 559–572.

14

EVOLUTION

Ross H. Nehm

Introduction

Although evolutionary biology has emerged as one of the most vibrant, powerful and useful scientific disciplines, helping learners understand, accept, and appreciate these attributes continues to be a persistent educational challenge. Indeed, at the same time that evolutionary principles are being used to help identify new pathogens, design new drugs, and conserve endangered species, antiscientific groups are working to eliminate evolution from curricula and promote misleading perspectives designed to foster negative attitudes toward evolution (National Research Council, 2012). The contrast between the contemporary successes of evolutionary science on the one hand, and societal ambivalence or antipathy towards it on the other, is striking. Clearly, evolution education demands further attention from scientists and educators, and while many topics have been shown to be important to this pursuit (e.g., genetics, nature of science, as well as worldviews and religious and cultural factors—see the respective chapters in this volume), this chapter explores the conceptual difficulties inherent to teaching and learning about evolutionary change. Effective teaching requires a deep understanding of student thinking, and so this chapter approaches the challenge from the perspective of students. Readers interested in exploring additional aspects of evolution education are encouraged to consult Kampourakis (2014), Kampourakis and Nehm (2014), Nehm and Kampourakis (2014), and Rosengren et al. (2012).

Student Thinking About Evolutionary Change

Documenting Student Thinking About Patterns and Processes of Evolutionary Change

One of the most fundamental attributes of life on Earth is that it changes over time, albeit at different tempos (Simpson, 1944). Indeed, evolutionary change can be remarkably fast (e.g., molecular evolution of the influenza and HIV viruses) or remarkably slow (e.g., the phenotypic stability of fossils over millions of years; Nehm & Budd, 2008). Given that all lineages across the tree of life are known to evolve, an inevitable question that emerges from these patterns (e.g., the gain and loss of features ranging from ecolocation to limbs) is whether they can be effectively explained by a unifying set of causal processes. Put another way: how do the myriad patterns of change relate to processes of change? Like many other scientific disciplines, the life sciences have used observable patterns (e.g., changes in lizard limb length over time, similarities among DNA sequences) as entry points for exploring and uncovering the hidden processes that account for change. Mutation, heredity, natural selection, genetic drift, and speciation are examples of processes that effectively explain patterns of change across the tree of life (for reviews, see Coyne, 2010; Mayr, 1994).

Assessing student understanding is a remarkably complex task, and entire fields of research are devoted to this pursuit (NRC, 2001). Over the past half century, biology educators have developed, deployed, and refined educational assessment tasks that present students with patterns of evolutionary change or biological differences and prompt students to explain how these differences came to be (for a review of the history of these items, see Ha & Nehm, 2014). Although the tasks from different studies vary in important ways (e.g., Bishop & Anderson, 1990), they are united by the goal of uncovering student thinking about the processes responsible for biological patterns. Recently, these assessment tasks have been updated and standardized to produce the Assessment of Contextual Reasoning about Natural Selection (ACORNS) instrument (Nehm et al., 2012). Using ACORNS instrument items, student explanations for patterns of biological difference can be elicited and explored across different *lineages* (e.g., plants vs. animals), different trait *polarities* (e.g., gain vs. loss of a trait), different taxon *familiarities* (penguin vs. prosimian), different *scales* (within- vs. between-species), and different trait *functions* (e.g., claws vs. fur color). These items provide educators with an expansive range of surface features that can be used to design assessment items and explore student thinking about process.

For example, an educator interested in how a student thinks about the loss of a trait between species of familiar plants would ask: *How would biologists explain how a species of cactus without spines evolved from a species of cactus with spines?* The skeletal structure of an ACORNS item is as follows, and permits substitution of multiple features: "*How would* [A] *explain how a* [B] of [C] [D1] [E] *evolved from a* [B] *of* [C] [D2] [E]?" The item skeleton is fleshed out with specific features depending on the interest of the investigator, such as: A = perspective (e.g., biologists, you), B = scale (e.g., species, population), C = taxon (e.g., plant, animal, bacteria, fungus), D = polarity (e.g., with, without), and E = trait (e.g., useable, static). Students offer explanations to the question in an unconstrained (open-ended) form, which allows them to choose which ideas they view as most relevant (in contrast to forced-choice assessments that include limited options chosen by the instrument developer). The diversity of life offers a range of attributes for students to think about, and assessment tools like the ACORNS allow educators to explore the range of student thinking about biological diversity using items differing in surface features (e.g., plants, trait loss, functional traits).

Open-ended assessment tasks like the ACORNS have revealed how students think about evolutionary patterns and processes across the tree of life. Students utilize a broad array of normative (scientifically accurate) and non-normative (naïve or scientifically inaccurate) ideas when explaining evolutionary change. Non-normative ideas common to student thinking have been thoroughly documented in the evolution education literature (Table 14.1; see Gregory, 2009 for a complete list). Teleological, intentional, and typological thinking, use-inheritance models, evolutionary 'pressure' frameworks, and many other factors have been identified as abundant across different age groups (e.g., middle school to university), geographic regions (e.g., United States, Europe, Asia, South America), and cultures (e.g., Western and Eastern). Consequently, biology teachers throughout the world should anticipate non-normative ideas as common features of students' evolutionary thinking (Table 14.1).

Non-normative ideas are often mixed together with normative scientific ideas (e.g., mutation, heritable variation, differential survival) in students' evolutionary explanations. Studies of large numbers of student explanations (>15,000) indicate that 'pure' naïve or 'pure' normative explanations are the exception. For example, in response to a question about trait change between species, students may explain that the spines of a species of cactus were lost because the cactus species did not need the spines anymore, and so the spines were not passed down to the next generation. Or, students may explain that the environment changed, and so all of the members of the species adapted (acclimated) to the environment over many generations. Although many combinations of normative and naïve ideas are possible, the ideas presented in Table 14.1 have been shown to be very common. ACORNS items provide instructors with a tool for documenting the range of ideas that students utilize when approaching problems about evolutionary change. For educators teaching in English, machine-learning tools have been developed to automatically analyze and describe the composition and structure of students' evolutionary explanations (Moherrari et al., 2014).

Curricular Examples and Student Thinking About Evolution

Although all life changes through time, and patterns of biological difference are ubiquitous across hierarchical levels (e.g., populations, species), student thinking appears to be strongly dependent on *what* is different (or changing) and the *ways* in which it changes (Nehm & Ha, 2011). In other words, student thinking about evolutionary change across the tree of life lacks coherence or stability. Determining which features are responsible for differences in student reasoning requires experimental studies that randomly assign items with single feature manipulation to participants. Although few studies of this nature have been completed, some

TABLE 14.1 Examples of non-normative student ideas used to explain patterns of evolutionary change

Student idea	*Description*
Adaptation as acclimation	Evolutionary change occurs through the gradual adjustment or acclimation of individuals, populations, or species over time in response to changing abiotic or biotic conditions. (This idea is often combined with the inheritance of acquired characteristics.)
Inheritance of acquired traits	Traits acquired during the lifetime of an individual, population, or species are inherited. (This idea is often combined with essentialism and adapt/acclimate and/or use/disuse.)
Trait use/disuse	The use (or lack of use) of character states (traits) directly causes the increase, decrease, gain, or loss of these states in individuals, populations, or species. (This idea is often combined with the inheritance of acquired characteristics.)
Environment as necessary and sufficient cause	Without environmental change, evolution does not (or cannot) occur. Explanations of evolutionary change begin with environmental (biotic or abiotic) change as the necessary and sufficient driver. (This idea is often combined with adaptation as acclimation, use/disuse, and needs and goals.)
Pressures and forces	Environmental and evolutionary pressures and forces exerted on biological units (individuals, species) induce or 'force' them to change (e.g., mutate) or evolve (change traits) in the face of death or extinction. (This idea is often combined with environment as necessary cause and others.)
Intentionality	Biological units (individuals, species) make conscious choices (they learn, decide, fight) to address environmental (biotic and abiotic) challenges in the face of death or extinction.
Essentialism/typology	Aggregate biological units (populations, species) are, at their core, homogeneous and uniform, and characterized by (often implicit, intangible) unchangeable features. Lack of pre-existing variation is associated with this idea.
Needs and goals/teleology	Evolutionary change is initiated and driven by the needs and/or goals of individuals (or collective biological units). Goals and needs originate as a result of biotic and abiotic drivers. (This idea is often combined with environment as cause of trait change.)

generalizations are beginning to emerge. First, controlling for traits and taxa (i.e., keeping them the same), explaining biological differences between individuals in different populations tends to be significantly easier for students compared to explaining differences between different species. Thus, population-level reasoning appears to be easier than species-level reasoning. Second, controlling for traits and taxa, the evolutionary gain of new traits tends to be easier for students to understand and explain compared to the evolutionary loss of traits. Third, explaining evolutionary changes in familiar animals tends to be easier for students than explaining evolutionary change in plants (which tend to never be as familiar to students as animals). Fourth, useable traits appear to invoke more non-normative ideas compared to static traits (e.g., claws vs. fur color). These findings indicate that teachers' choices of curricular examples (i.e., types of change in a lineage) are likely to impact student thinking; each curricular case will tend to evoke different normative and non-normative ideas in learners' minds. Although this situation creates challenges, it can also be used as a powerful tool to advance student understanding of evolution (see 'Teaching About Evolutionary Change' below).

Monitoring Changes in Student Thinking in Response to Instruction

A major goal of evolution education is to improve student understanding; therefore, monitoring changes in student thinking is a necessary aspect of instruction. Monitoring student use of non-normative ideas (Table 14.1; Gregory, 2009) is important but insufficient. Students can combine normative and non-normative ideas together in different ways, and may use non-normative ideas in one situation but not in another (Nehm & Ha, 2011). How, then, should teachers approach the challenge of monitoring student understanding? Growth in student understanding could occur in many different ways, and involve: (1) changes in the *composition* of ideas utilized in problem solving, (2) changes in the *structure* or arrangement of ideas, or (3) changes in the *coherence* or consistency of ideas across different problem types, or across the same problem types in different contexts (e.g., plant gain and animal loss problems) (see Figure 14.1A). Changes in the hierarchical status of ideas across contexts and problem types is also possible. The *composition-structure-coherence* framework illustrates the different ways in which progress in student understanding of evolutionary change can be conceptualized and measured (Nehm & Kampourakis, 2016). Conceptual abstraction (i.e., normative schemas consistently applied across problem contexts) is quite rare and represents the upper anchor of understanding for university undergraduates in their first few years; consequently, teachers should expect learning gains to involve the addition of new normative ideas alongside non-normative ideas *within* specific contexts.

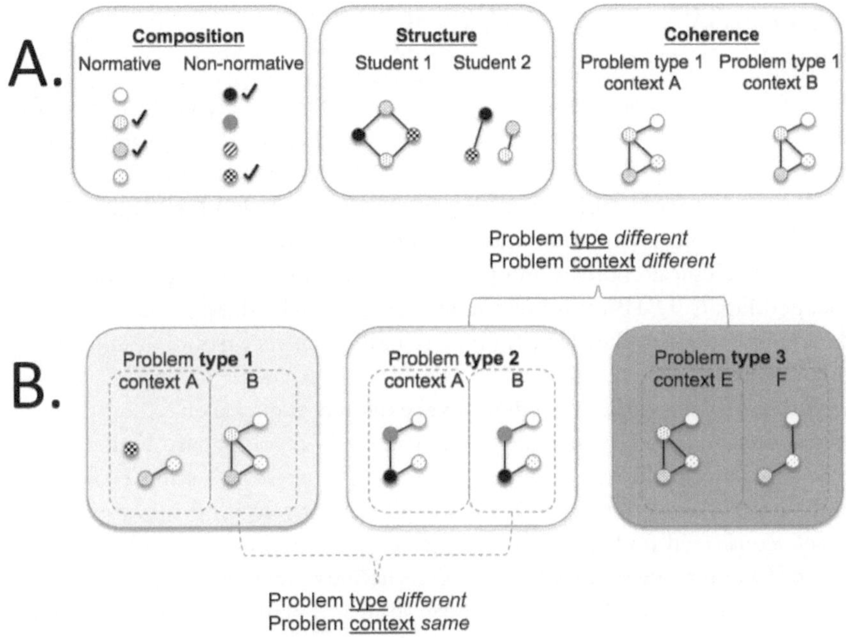

FIGURE 14.1 The composition-structure-coherence framework for approaching the complex task of assessing growth in student understanding of evolutionary change. A. Composition, structure, and coherence of student thinking. First box (composition): a student's idea is composed of some, but not all, of the normative ideas needed to explain a particular phenomenon in a task. Growth is possible by adding new ideas or choosing not to use particular non-normative ideas. Second box (structure): students 1 and 2 may have the *same* normative and non-normative ideas, but they connect them in different ways. Changes in structure (e.g., unlinking normative ideas from non-normative ideas) can occur without changing composition. Box 3 (coherence): conceptual abstraction is emerging as normative ideas are linked and applied across contexts within a problem type. For example, a student was able to build an explanation using normative ideas in a functional trait gain, between-species evolutionary problem using hawks and palm trees. B. Mixing problem types and problem contexts makes it very difficult to monitor growth in student understanding because too many variables have been introduced (e.g., different *problem types* and *different contexts*) to make meaningful sense of conceptual growth. Thus, careful attention to contexts and problem types is essential when monitoring growth in student understanding. Modified and expanded from Nehm and Kampourakis (2016)

A Framework for Conceptualizing Student Thinking About Evolutionary Change

Thus far, student thinking about evolution has been described in terms of (1) the normative and non-normative ideas that students utilize when attempting to solve an evolutionary problem (e.g., Table 14.1), (2) the ways in which students structure or connect these normative and non-normative ideas (Figure 14.1A), and (3) the similarity of composition and structure in evolutionary explanations between biological contexts and problems (e.g., different taxa, polarities of trait change) (Figure 14.1B). It is important to consider why these patterns occur and the implications that they have for understanding the very nature of student ideas about evolution.

Although hundreds of studies have explored the challenges of evolution education, only one has investigated how experts (evolutionary biologists) and novices (undergraduates) perceive, represent, and solve the same evolutionary problems (Nehm & Ridgway, 2011). Despite the novelty of this investigation, it largely confirmed findings from past research on expert/novice problem-solving in other scientific disciplines (e.g., physics, chemistry, mathematics). In many areas of science, novices and experts have been shown to perceive problems differently. Experts and novices have different sensitivities to particular features of the problems (e.g. plants, animals) which in turn lead to differences in how the problem is envisioned by the problem solver. 'Problem representation' refers to the solver's construction of a mental model of the underlying structure, essential nature, or categorization of a problem (Nehm & Ridgway, 2011). Conceptually equivalent problems (from the expert perspective) vary in difficulty for novices because the surface features of the items (e.g., bird flight vs. insect sound) stimulate different mental representations of the situation (Opfer et al., 2012). Different mental representations or problem framing in turn leads to the recruitment of different ideas and schemas (chunks of connected ideas) from long-term memory (Figure 14.2). Thus, differences in problem representation on the part of the learner constrain the likelihood of problem solving success. All too often, novices are unable to see beneath the question 'cover stories' to the 'deep structure' or concept that the question is trying to test.

Many superficial features of evolutionary problems (e.g., plant thorns, animal fur color) activate different suites of cognitive resources during novice problem solving (Nehm, 2010). A student may utilize non-normative ideas (e.g., evolutionary pressures cause mutations in response to the needs of the species) in one situation and normative ideas in another (e.g., existing variation in a population was sorted and only some individuals survived). Indeed, sensitivity to item features is associated with idiosyncratic knowledge activation and the generation of multiple solutions to what experts consider the same problem (Nehm & Ridgway, 2011). Given that different situations evoke different ideas, how should educators address students' non-normative ideas when they arise?

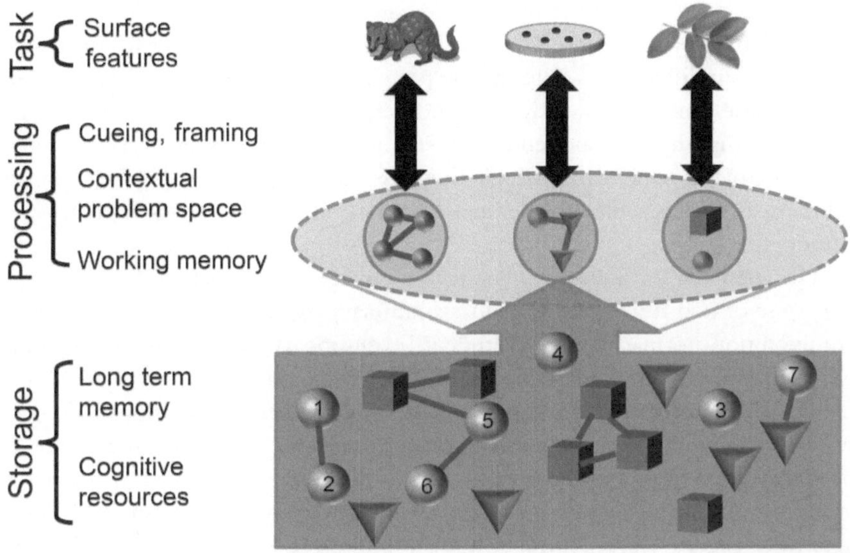

FIGURE 14.2 Cartoon illustrating student thinking about evolutionary change across tasks employing animals, bacteria, and plants. When encountering different tasks (simplified in the cartoon as different organisms), students' sensitivity to surface features (e.g., animal species losing a tail) impact cueing, framing, and problem representation (middle figure), which in turn differentially impacts the recruitment of individual ideas, clusters of ideas, and schemas from long-term memory (storage) into working memory. Different student utterances and artifacts produced in response to these tasks are a result of the differential recruitment and assembly of cognitive resources from long-term memory (and any scaffolding resources within the physical and social context of problem solving). (Nehm & Ridgway, 2011)

Perspectives on Students' Non-Normative Ideas

Prior to discussing how to use our advancing understanding of student thinking to more effectively teach about evolutionary change, it is important to consider alternative perspectives on how teachers should conceptualize the non-normative ideas commonly activated in novice problem solving (Table 14.1). Educators commonly view these ideas as targets for elimination given that they are inappropriate solutions to most evolutionary problems. This view has limitations which may be best illustrated through the use of an analogy. Let's consider a handyman, his tools, and jobs to be done. If the handyman uses his hammer to paint a wall, and his paintbrush to drive in a nail, one response might be to recommend that that hammers and paintbrushes be eliminated from the worker's toolbox in order prevent problematic tool use patterns. An alternative response would be to help the handyman better match his tools to appropriate tasks in order to more efficiently accomplish his jobs. Eliminating particular tools would solve one problem (problematic tool use) but introduce another, namely that the range of tasks he could accomplish would be reduced. Helping the handyman appropriately align his tools to the contexts in which they are most effective would be the best solution, as it allows him to maintain the tools and increase task success (e.g., painting).

The approach to the handyman's tool use patterns has clear parallels to a student's use of non-normative ideas in evolutionary explanations. One of the most common non-normative ideas in students' evolutionary explanations is teleological or goal-based causation (Table 14.1).[1] Teachers could focus on trying to eliminate this non-normative idea from students' minds. Alternatively, they could help students distinguish the contexts or situations in which teleological thinking is productive and those in which it is not. It is important to emphasize that needs and goals explain many actions in the world; teleology is an important, widely applicable cognitive tool central to human thinking. Like the hammer and paintbrush in our analogy, teleology can be used inappropriately in the context of evolutionary explanations, but this fact does not speak to the value of teleology *per se*. Helping students align their cognitive resources to appropriate problems and understand where these ideas fail to work is the most appropriate strategy for teachers.

Recommendations for Teaching About Evolutionary Change

Curricular Considerations

The school curriculum is, in many cases, students' first exposure to normative scientific ideas about evolution (see Rosengren et al., 2012 for discussion of informal educational settings). Nevertheless, the school curriculum often lacks conceptual coherence; evolutionary patterns (e.g., the diversity of plants and animals) are largely isolated in individual chapters or units segregated from biological processes (e.g., mutation, cellular and heredity mechanisms, development; Nehm et al., 2009). This curricular structure is problematic for two major reasons. First, it may foster the development of inaccurate mental models of living systems; specifically, that evolutionary patterns lack meaningful connection to genetics, cell biology, development, or physiology. Second, it may give students the impression that if they reject evolutionary science then they may nevertheless be able to make sense of the rest of biology. For these reasons, interleaving and repeatedly illustrating the cross-cutting centrality of processes to evolutionary patterns would more effectively model conceptual unity across biological contexts and more accurately represent the integrative nature of contemporary life science research.

Evolution textbooks and curricula produced over the past century contain multiple case examples of evolutionary patterns and in some cases their associated evolutionary processes. Common historical examples include horse evolution and peppered moth evolution, and more recent examples include Darwin's finches and antibiotic resistance. Given that a large body of evidence indicates that these common curricular examples have failed to instill robust, abstract models of evolutionary causation in students' minds (see 'Student Thinking About Evolutionary Change', above), we must ask why the presentation of multiple evolutionary examples is not as effective as intended.

Conceptual fragmentation of evolutionary ideas in long-term memory and sensitivity to problem types and contexts (Figure 14.2) explain, in part, why lineage-specific curricular examples (e.g., Darwin's finches, antibiotic resistance) often fail to generate robust, abstract models. From a student's perspective, each case presents unique features worthy of a unique assemblage of ideas (Figure 14.2). On the surface, the functional and ecological consequences of bird beak shapes for survival seem to have little in common with the use of human-produced drugs to kill microscopic bacteria. Yet both are united by the differential survival of variants generated by heritable variation produced through mutation. Teachers who present separate evolutionary examples without highlighting the causal similarities among them should not anticipate significant conceptual progress given that this approach aligns with the situated nature of student reasoning about evolution (Figure 14.2).

Cross-Case Comparisons Must Highlight Surface Feature Dissimilarity and Causal Unity

Studies of student thinking about evolution indicate that curricular examples fostering change in the composition and structure of student ideas relating to animal trait gain may not change the composition and structure of student ideas relating to the evolution of plant trait loss. Indeed, although many interesting and innovative educational interventions focusing on particular evolutionary scenarios have been developed and deployed, efficacy studies rarely test reasoning in scenarios different from the exemplar case (Nehm & Ha, 2011). Given that a goal of evolution education is to foster robust, abstract models of reasoning that transcend individual cases, instructional approaches are needed to effectively promote such abstraction. Evolutionary examples that highlight surface feature dissimilarity while emphasizing causal unity are one approach to this challenge. This approach does not require new curricular examples or new educational materials; rather, it requires a different pedagogical approach when presenting these examples.

The first step in this approach involves reorganizing common curricular examples into contrasting cases (e.g., sickle cell anemia vs. lactase persistence; Darwin's finches vs. antibiotic resistance; loss of thorns in plants vs. loss of eyes in fish). Prior to presenting these cases, it is important for the class to work together to identify the apparent *differences* between them (e.g., one case involves variation in red blood cell shapes and oxygen-carrying abilities and the other variation in the digestion of milk and other dairy products). Highlighting notable differences between the cases helps to draw students' attention to various surface features of the cases, to build on intuition, and to provide a low-stakes entry point to the cases. Summaries of the differences that students identify can be presented in a table.

Once the differences in the case examples have been identified, it is time to begin exploring possible *similarities* between the cases. These similarities can begin at a basic level (e.g., all of the organisms in the cases have cells, use oxygen to metabolize food) and progress to a more advanced level (e.g., heritable mutations constantly occur and can cause differences in the proteins that are made by cells; differences in proteins impact physiological processes like digestion). Teachers can scaffold the activity by providing hints and examples (Do mutations occur in all living things? Do genetic differences among individuals relate to phenotypes? Are there always enough resources and habitats for every living thing to survive?). These similarities should be summarized in a table to parallel the case differences table. In many ways, this pedagogical approach models how novices can think more like experts.

Once the similarities and differences between the cases have been identified and discussed, the more challenging work of connecting process and pattern begins (e.g., the processes causing patterns of lizard limb length change, or processes causing guppy body coloration differences). This step will require scaffolding tools, such as lists of possible ideas (normative and non-normative) for students to discuss and evaluate as potentially relevant to both cases (e.g., lactase persistence and guppy coloration). For example, need-based (teleological) causation is commonly used by students (Table 14.1), and so students could evaluate the degree to which 'needs' can explain the biological patterns in the two cases. Would the lack of food in a human population, and an associated need to consume and digest milk, impact the frequency of individuals with lactase persistence? Would predators differentially eating colorful guppies, and the need to be less colorful, cause individual guppies to become drab? A variety of causes could be evaluated as contributors to the patterns documented in the cases.

Parallel examples with normative ideas should also be used (e.g., Do mutations occur in humans and fish? Do mutations contribute to phenotypic differences in humans and fish? Do phenotypic differences impact survival in humans and fish?). Cross-case comparisons must emphasize *similarity of process* (e.g., mutation and genetic recombination generate large quantities of heritable variation; variations in genomes relate to variation in phenotypes; variation in phenotypes impacts competition for mates and securing resources) and *dissimilarity of pattern* (e.g., lizard limb length, lactase persistence). Evaluating potential causal contributors to different evolutionary scenarios channels thinking about how patterns might relate to processes.

By introducing *multiple* cross-case examples and gradually fading instructional scaffolds (e.g., suggestions of possible processes, summary tables of similarities and differences), teachers can provide multiple opportunities for students to practice thinking about how seemingly dissimilar biological patterns across the tree of life can be understood in terms of small sets of unifying causal processes. Contrasting cases provide an opportunity for students to build cognitive models that transcend exemplar cases, and address the well-documented fragmentation and context specificity of students' evolutionary reasoning (Nehm & Ha, 2011). Current curricula in K–12 and university settings continue to present individual evolutionary cases differing in surface features. Unfortunately, such cases do not challenge students' fragmented and case-specific cognitive models of evolutionary change.

Conclusions

Teaching evolution requires understanding how students make sense of biological change and biological differences. Student thinking about evolution appears to be strongly dependent on *what* is different (or changing) and the *ways* in which it changes. Student explanations of evolutionary change vary significantly across different *lineages*, trait *polarities*, taxon *familiarities*, hierarchical *scales*, and trait *functions*. Students utilize a broad range of both normative and non-normative ideas when approaching these different contexts, and purely correct or purely incorrect explanations are uncommon. Different evolutionary patterns appear to pose predictable challenges for students: explaining biological differences between individuals in different populations tends to be significantly easier for students compared to explaining differences between species; the evolutionary gain of new traits tends to be easier for students compared to the loss of traits; explaining evolutionary changes in familiar animals tends to be easier than explaining change in plants; and useable traits appear to invoke more non-normative ideas compared to static traits. Building abstract schemas of evolutionary change that transcend individual exemplars requires using multiple cross-case curricular examples that highlight surface feature dissimilarity while emphasizing causal unity.

Acknowledgments

Thanks to Robyn Tornabene and Kostas Kampourakis for insightful comments. This work was supported in part by NSF #1322872. Any opinions, findings, conclusions, or recommendations expressed in this publication are those of the author and do not necessarily reflect the views of the National Science Foundation.

Note

1 It is important to emphasize that some scholars view teleological thinking as appropriate insofar as natural selection and not design is invoked. Readers are encouraged to explore these views (see Lennox & Kampourakis, 2013).

References

Bishop, B.A., Anderson, C.W. (1990). Student conceptions of natural selection and its role in evolution. *Journal of Research in Science Teaching*, 27(5), 415–427.

Coyne, J. (2010). *Why Evolution Is True*. New York: Penguin Books.

Gregory, T. R. (2009). Understanding natural selection: essential concepts and common misconceptions. *Evolution: Education and Outreach*, 2, 156–175.

Ha, M., Nehm, R. H. (2014). Darwin's difficulties and students' struggles with trait loss: cognitive-historical parallelisms in evolutionary explanation. *Science & Education*. 23(5), 1051–1074.

Kampourakis, K. (2014). *Understanding Evolution*. Cambridge: Cambridge University Press.

Kampourakis, K. & Nehm, R. H. (2014). History and philosophy of science and the teaching of evolution: students' conceptions and explanations. In M. R. Matthews (Ed.), *International Handbook of Research in History, Philosophy and Science Teaching*, Dordrecht: Springer, 377–399.

Lennox, J. G. & Kampourakis, K. (2013). Biological teleology: the need for history. In K. Kampourakis (Ed.), *The Philosophy of Biology: A Companion for Educators*. Dordrecht: Springer, 421–454.

Mayr, E. (1994). *What Evolution Is*. New York: Basic Books.

Moherrari, K., Ha, M., Nehm, R. H. (2014). EvoGrader: an online formative assessment tool for automatically evaluating written evolutionary explanations. *Evolution: Education and Outreach*, 7(1), 15. doi:10.1186/s12052-014-0015-2

National Research Council. (2001). *Knowing What Students Know*. Washington, DC: The National Academies Press.

National Research Council. (2012). *Thinking Evolutionarily*. Washington, DC: The National Academies Press.

Nehm, R. H. (2010). Understanding undergraduates' problem solving processes. *Journal of Biology and Microbiology Education*, 1(2), 119–121.

Nehm, R. H., Beggrow, E. P., Opfer, J. E. & Ha, M. (2012). Reasoning about natural selection: diagnosing contextual competency using the ACORNS instrument. *The American Biology Teacher*, 74(2), 92–98.

Nehm, R. H. & Budd, A. F. (Eds.) (2008). *Evolutionary Stasis and Change in the Dominican Republic Neogene*. Springer.

Nehm, R. H., et al. (2009). Does the segregation of evolution in Biology textbooks and introductory courses reinforce students' faulty mental models of Biology and evolution? *Evolution Education and Outreach*, 2, 527–532.

Nehm, R. H. & Ha, M. (2011). Item feature effects in evolution assessment. *J Res Sci Teach*, 48(3), 237–256.

Nehm, R. H. & Kampourakis, K. (2014). History and philosophy of science and the teaching of macroevolution. In M. R. Matthews (Ed.), *International Handbook of Research in History, Philosophy and Science Teaching*. Dordrecht: Springer, 401–421.

Nehm, R. H. & Kampourakis, K. (2016). Conceptual change in science and in science education. *Encyclopedia of Educational Philosophy and Theory*. Dordrecht: Springer, 1–5.

Nehm, R. H. & Ridgway, J. (2011). What do experts and novices "see" in evolutionary problems? *Evolution: Education and Outreach*, 4(4), 666–679.

Opfer, J. E., Nehm, R. H. & Ha, M. (2012). Cognitive foundations for science assessment design: knowing what students know about evolution. *Journal of Research in Science Teaching*, 49(6), 744–777.

Rosengren, K. S., Brem, S. K., Evans, E. M., Sinatra, G. M. (Eds.) (2012). *Evolution Challenges: Integrating Research and Practice in Teaching and Learning about Evolution*. Oxford: Oxford University Press.

Simpson, G. G. (1944). *Tempo and Mode in Evolution*. New York: Columbia University Press.

15

MICROBIOLOGY

Katherine Carson, Vaille Dawson and Grady Venville

Introduction

The chapter has four main sections. The first section introduces the discipline of microbiology in the context of recent discoveries in science compared with the standard school science curriculum in most countries. The second section provides a synthesis of the literature on students' misconceptions about microbiology. The third section considers the challenges for teaching the topic of microbiology in school, and the fourth and final section provides a framework of five principles for the effective teaching of microbiology.

The Discipline of Microbiology

The broad field of microbiology, namely the study of microorganisms (also known as microbes) such as bacteria and viruses, is an expanding discipline at the forefront of contemporary scientific discoveries. Since the discovery of 'animalcules' by Van Leeuwenhoek in the 1670s, scientists have continued to investigate and better understand microorganisms and utilise them to advance processes in health, industry, agriculture and the environment. Historically, some of the most significant advances for humanity have been the eradication of microorganisms through pasteurisation, disinfection of drinking water, development of antibiotics and the prevention of a number of important human diseases with vaccines. Today, industrial scientists take advantage of microorganisms and their enzymes in the mass production of chemicals, such as 'biologically active' washing powder. Soil scientists use microorganisms to improve crop production and rehabilitate damaged soil ecosystems. In human health, vaccinations and human medicinal products such as antibiotics are saving millions of lives, and biotechnological processes are advancing faster than the generation time of *Escherichia coli* (which is one generation every 17 minutes).

Microbiology interacts with every aspect of our daily lives, and yet research has shown that many school students perceive all microorganisms as harmful and potentially pathogenic and have little appreciation of the role of microorganisms in the environment or of the beneficial uses of microorganisms (Jones & Rua, 2006). The aim of school science is to produce scientific and health literate citizens capable of making evidence-based decisions for the benefit of themselves, their family and their community. In order to do so, it is important for school students to develop a comprehensive understanding and appreciation of this comparatively new science as well as the interaction of microbiology with society, including the related ethical issues. Microbiology is a complex topic that provides teachers with an opportunity to engage students, to promote the relevance of school science to their daily lives, and to consider scientific and socioscientific issues that span the full spectrum from local to global.

Microorganism Taxonomy

The most commonly accepted classification system used in school science today places all organisms on Earth into five kingdoms based on evolutionary theories: Animalia, Plantae, Fungi, Protista and Monera. The first four kingdoms contain organisms known as eukaryotes. In eukaryotic cells, the DNA is bound within a nuclear membrane to form a nucleus. The organisms within the kingdom Monera are known as prokaryotes. In prokaryotic cells, the DNA is clumped together into a nucleoid and there are no membrane-bound organelles. In this system, the kingdom Monera contains all known bacteria.

Microbiologists today generally have a different view of classification, preferring to recognise a group of microorganisms, the Archaebacteria, first identified in 1977 by Carl Woese and George Fox (Woese et al., 1990). Based on ribosomal RNA analysis and phylogenetic relationships, Woese and his colleagues proposed a new level of taxonomy with three domains that sit above the traditional kingdoms (Figure 15.1). The three domains are Bacteria, Archaea and Eukarya. This system divides prokaryotes into two groups, the Bacteria (sometimes referred to as Eubacteria) and the Archaebacteria. Archaebacteria have unique cell properties, including the arrangement of bases in their ribosomal RNA and the composition of their plasma membranes and cell walls, which indicate they are as different from the Eubacteria as they are from plants and animals. Archaebacteria often are found living in extreme conditions, including very hot, cold, saline, acidic or alkaline environments such as hot spring vents and hyper saline water. Woese et al.'s three-domain system better represents the vast number of species and biomass made up by microorganisms and acknowledges that they comprise the majority of the Earth's genetic, metabolic and ecological diversity.

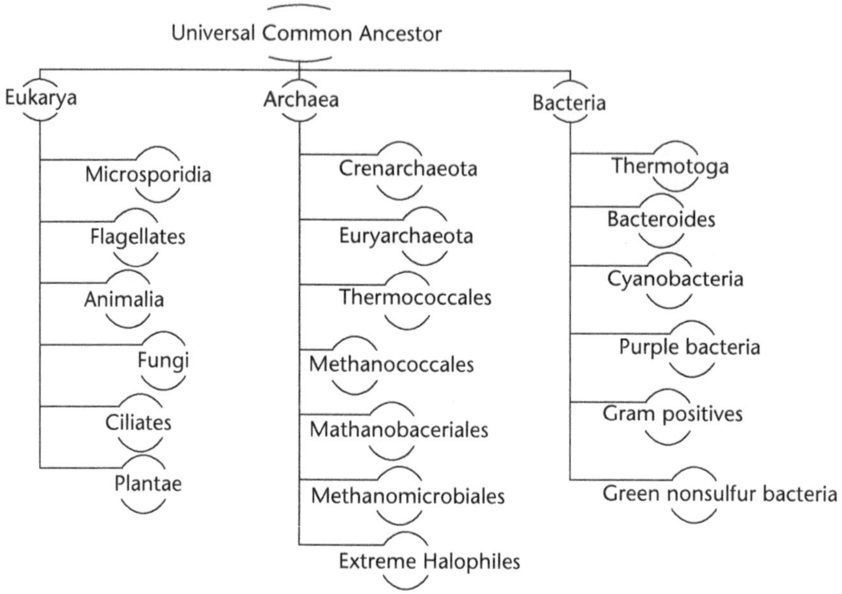

FIGURE 15.1 A diagrammatic representation of the three domains of Archaea, Bacteria and Eukarya proposed by Woese et al. (1990) as a level of taxonomy higher than the traditional kingdoms

It is important to acknowledge that microorganisms are not a recognised taxonomical group of phylogenetically or morphologically related organisms. Microbiology is, however, a discipline that is taught in universities across the globe and a field of research with communities of professionals with accepted norms of practice including conferences, journals and other forums for communication. The Microbiology Society, the largest organisation of its type in Europe and with a global membership, provides the following definition of microbiology on its website (www.microbiologysociety.org/about.html):

> Microbiology is the study of all living organisms that are too small to be visible with the naked eye. This includes bacteria, archaea, viruses, fungi, prions, protozoa and algae, collectively known as 'microbes'.

This definition of microbiology includes bacteria that are pervasive in soil, air and water as well as on and within other organisms (including humans), such as species from the common *Micrococcus*, *Staphylococcus*, *Bacillus* and *Pseudomonas* genera. *Staphylococcus* and *Streptococcus* species are well known for the infections they can cause in human beings, including boils, impetigo and pharyngitis. Examples of archaea include: *Thermococcus literalis*, which can live in the extreme temperatures near deep-sea vents; *Halobacterium salinarium*, which tolerates high salt environments and causes the pink/red tones of the very salty Dead Sea; and *Methanobrevibacter smithii*, which lives in the human gut and has an important role in the digestion of complex sugars and the removal of excess hydrogen. Viruses are much smaller than even bacteria and can only reproduce when they are inside cells of other organisms. Examples include the *Tomato mosaic* virus (species), the Hepatitis B virus and the *Varicella zoster* virus, which causes chicken pox and shingles in human beings.

The above definition of microbiology also includes organisms from the huge spectrum of single-celled species in the Fungi kingdom, including baker's yeast (*Saccharomyces cerevisiae*) and the common and invasive plant pathogen called root rot or dieback (*Phytophthora cinnamomi*). Prions are infectious particles that are made up entirely of unusually folded proteins that, like viruses, reproduce by taking control of another organism's cells. Prions cause diseases like Creutzfeldt-Jacob disease and other wasting diseases in humans and in animals. Protozoa are single-celled, non-photosynthetic eukaryotes, including the intestinal parasite *Giardia muris* and decomposers important in the food chain such as the *Testacea* species. The final group of organisms identified in the definition, microscopic algae, are abundant in aquatic and terrestrial environments and include green algae, diatoms and dinoflagellates. It is generally not well known that microscopic algae in aquatic environments produce a large proportion of the Earth's oxygen.

This very short exploration of microorganism taxonomy provides a tiny insight into the incredible diversity, complexity and, most of all, importance of this field of science to students' general scientific literacy. The next section of this chapter explores the understandings and misconceptions reported in the literature that school students have regarding microbiology.

School Students' Misconceptions About Microbiology

Microbiological learning commences as soon as a child begins to understand spoken language and visual cues. There is parental, social and educational influence as children become trained to wash their hands to get rid of 'bugs' and avoid eating off the floor because it has 'germs'. As children are educated about body and food hygiene, they experience microbiology, and the concepts and terms used influence the construction of their knowledge. Young children assimilate this experiential knowledge along with intuitive knowledge into their microbiological understanding to make sense of new scientific knowledge. This may result in the development of misconceptions.

The formation of many misconceptions in microbiology is conditioned by societal ideas concerning health and disease. Modern society's hygiene culture promotes a human-centred perspective of microorganisms, which is manifested in many documented misconceptions, for example, that all microorganisms cause illness in humans (Jones & Rua, 2006). Other misconceptions can be viewed from an anthropomorphic perspective, giving microorganisms human physical and mental characteristics, for example that microbes 'eat' decaying matter (Byrne, 2011). These perspectives persist in some individuals (Byrne et al., 2009), demonstrating the difficulty in teaching for conceptual change.

The majority of research studies on students' microbiological understanding have been exploratory in nature. Data generation methods have focused on drawings and interviews with school-aged children, and questionnaires and interviews with tertiary students, pre-service and in-service teachers to assess microbiological knowledge.

Misconceptions on Classification and Morphology

'Germ' is the most common term used by primary school students to describe microorganisms (Byrne, 2011; Jones & Rua, 2006). Students from the ages of 10 to 17 have been found to acknowledge bacteria as microorganisms, but very few name viruses or any other groups of microorganisms (Jones & Rua, 2006). Secondary school students have been found to use the terms 'microorganism', 'germ' and 'bacteria' interchangeably (Byrne, 2011), with some students using 'germ' to include both bacteria and viruses (Maxted, 1984).

Although students find classification of microorganisms difficult, they intuitively classify microorganisms into three groups: those which are detrimental and cause illness; those which do not cause illness but are not beneficial; and those of benefit to the human body (Jones & Rua, 2006).

Many primary school students consider microorganisms to be animals and hence animal-like (Byrne, 2011). This perception is reflected in their drawings. If students consider a microorganism to be an animal, they are more likely to draw it as an insect or worm (Byrne, 2011) with anthropomorphic characteristics (Byrne et al., 2009). Secondary students are more likely to draw a recognisable animal cell or bacterium; however, some still draw an inaccurate picture, such as an animal-like image or a piece of dirt (Byrne, 2011; Jones & Rua, 2006).

The majority of students recognise that microorganisms are invisible to the naked eye and are microscopic; however, what this actually means is not well understood (Maxted, 1984; Simonneaux, 2000). When describing the size of microorganisms, standard measures such as micrometres are not used by students who attempt to relate the size of microorganisms to known objects, such as a rice grain or pencil tip (Byrne, 2011). Simonneaux (2000) reported that some high school students compared the size of an unspecified microorganism to that of an atom. Often, students relate the size of a microorganism to its pathogenicity and consider viruses to be larger than bacteria (Byrne, 2011; Jones & Rua, 2006). This is thought to be due to personal experiences with illness and the perception that viruses cause worse illnesses than bacteria.

Misconceptions on Pathogenicity and Disease

There is no question that many of the misconceptions displayed by students are related to an anthropocentric perspective on microorganisms and their perceived strong causal relationship with disease. Jones and Rua (2006) reported that all 63 primary and secondary students in their study related a 'germ' (this term was used to represent a microorganism) to illness and even death. Some primary students consider all microorganisms responsible for illnesses (Byrne, 2011). This conception decreases with age, as with increasing scientific knowledge students become aware that not all microorganisms cause illness (Byrne, 2011) and that different diseases are caused by different microorganisms (Maxted, 1984).

The microbiological disease process and immune system response are poorly understood by high school students (Jones & Rua, 2008). Many see microorganisms as completely exogenous to the human body and as alien invaders attacking the immune system (Dumais & Hasni, 2009; Simonneaux, 2000). Similarly, Jones and Rua (2006) reported that high school students are unsure of how microorganisms make humans symptomatically ill. In terms of pathogenicity, viruses are perceived as more pathogenic than bacteria, with students considering all viruses as aggressive and harmful to humans (Jones & Rua, 2006; Simonneaux, 2000). Some students even consider bacteria to be the cause of an illness and the virus the resultant illness (Jones & Rua, 2006). This is thought to be due to the common usage of the English language term 'virus' to describe an illness.

The purpose and process of vaccination is poorly understood by students (Byrne, 2011; Simonneaux, 2000). Less than half of the 332 high school students in Simon et al.'s (2017) study could name a viral disease prevented by vaccination. Jones and Rua (2008) reported that high school students had little knowledge of the relationship between viruses and vaccination. Some students knew a vaccine contained an altered virus but could not describe how a vaccine worked. Some middle school students believe vaccination is used as a treatment once a person becomes ill (Byrne & Grace, 2010), and some high school students believe it protects people from the flu by giving them antibiotics (Romine et al., 2013). There is also confusion about treatment with antibiotics, with many students believing the treatment kills both bacteria and viruses (Jones & Rua, 2008; Prout, 1985).

Misconceptions on Location

Studies have shown that both primary and secondary students are aware that microorganisms are ubiquitous in the environment (Jones & Rua, 2006; Maxted, 1984). Locations where high numbers of microorganisms can be found are often associated with areas of high human activity, such as door handles, bathrooms and computer keyboards (Byrne, 2011; Jones & Rua, 2006). This anthropocentric view seems to increase with age, possibly due to classroom teaching promoting microorganisms as causal agents of human illness (Byrne, 2011). High school students also associate damp and dirty conditions with large numbers of microorganisms (Byrne, 2011; Simonneaux, 2000). When considering locations where microorganisms could not survive, students of all ages name extreme environments such as inside volcanoes and the frozen Antarctic as environments unsuitable for the survival of microorganisms (Jones & Rua, 2006; Simonneaux, 2000).

Misconceptions on Microorganisms' Role in Decomposition

Many students have a poor understanding of the role of microorganisms in decomposition and nutrient recycling in the environment. Many primary students understand microorganisms are somehow involved in the spoilage of food (Byrne, 2011), most likely due to experiential learning in the home and early childhood education about food hygiene. Secondary school students retain this association; however, many do not understand the decomposition process and believe organic matter can decompose with or without the presence of microorganisms (Byrne, 2011). Leach et al. (1996) reported that more United Kingdom students perceived decomposition as an innate feature of organic matter than those who mentioned the involvement of microorganisms. Students also did not often mention microorganisms as the causal agents of decomposition in ecological systems and had a poor understanding of conservation of matter. Secondary school students more often tended to understand that microorganisms made things disappear and helped to enrich the soil through degradation of organic matter; however, there often was no mention of the importance of microorganisms for the availability of inorganic matter and the cycling of matter, such as nitrogen and carbon.

Misconceptions on Beneficial Use of Microorganisms

Many students possess a limited understanding of the beneficial use of microorganisms in industries such as health and agriculture or in the environment. In the study by Simonneaux (2000), half of the ten French 15-year-olds interviewed could not give a specific example of how humans beneficially use bacteria. Both primary and secondary students considered microorganisms as potential contaminants of food and therefore causes of illness (Byrne, 2011) which is inconsistent with students' association of microorganism use in the food manufacturing process (Simonneaux, 2000).

Few students are aware of the use of microorganisms in the medical industry in the production of vaccines, antibiotics and other medical products (Byrne, 2011; Simonneaux, 2000). When this is understood, it is at only a basic level, such as placing 'good' microorganisms into a vaccine (Byrne, 2011). Furthermore, young people are unaware of the beneficial role of normal microbial flora on the skin and mucous membranes in protecting humans from infection.

The next section of this chapter outlines the challenges in teaching microbiology.

Challenges in Teaching Microbiology

Overcoming misconceptions such as those outlined in the previous section of this chapter is difficult, due in part to the nature of microbiology itself. Microorganisms are abstract, as they can only be seen with magnification devices and are thus detached from everyday student life and experience. It has been suggested that the stage of cognitive development of a student may affect their understanding of microbiological concepts (Carlan et al., 2014). The initial role of teachers, therefore, is to use teaching approaches which make these abstract microscopic organisms as concrete as possible.

Research shows microbiological understanding does not develop in a linear fashion (Byrne, 2011), and taught microbiological concepts may not be remembered from one year to the next (Simonneaux, 2000). Conceptual understanding is complicated by time, as students have experiences with microorganisms in between teaching of microbiology and consequently readjust their microbiological conceptual understanding. If these experiences are not acknowledged and misconceptions remain, new scientific understanding may not be gained from learning (Maxted, 1984).

Microbiology is mostly taught within the broader science discipline of biology; however, it can be included in other disciplines, including chemistry and earth and environmental science (e.g. Australian Curriculum and Assessment Reporting Authority [ACARA], 2016; Department for Education, 2015; Next Generation Science Standards [NGSS] Lead States, 2013). Outside the classroom, microbiological knowledge plays essential roles in a diverse array of fields, including biotechnology, agriculture, environmental science and healthcare (Redfern & Verran, 2015). The non-integrated classroom approach to teaching microbiological concepts and their role in systems such as ecology prevents application of prior knowledge and promotes a siloed knowledge system. For example, without explicit acknowledgement of the role of microorganisms in ecological systems, students may be unaware of the role of bacteria, algae and other microorganisms in the decomposition of matter and ultimate recycling of organic and inorganic matter.

The mass media have been implicated by some researchers in the construction of microbiological misconceptions (Jones & Rua, 2006; Simonneaux, 2000). Headlines such as *Superbugs to kill 'more than cancer' by the year 2050* (Walsh, 2014) may draw readers' attention, but simultaneously promote the anthropocentric view of microorganisms as dangerous to human health. Others like *Darwin fisho attacked by flesh-eating bacteria* (Australian Broadcasting Corporation, 2017) promote an anthropomorphic view of microorganisms, giving them animalistic characteristics. Advertisements of popular household products and over-the-counter medication also promote these views, with animated bacteria and viruses given limbs, facial features and voices, and often are shown engaging in mock battles with 'good' bacteria, healthy cells or cleaning products. People's interaction with media-based information has grown since the studies by Jones and Rua (2006) and Simonneaux (2000), and more recent studies have shown the strong influence of the mass media on students' knowledge in biological fields such as genetics (Donovan & Venville, 2014).

A Framework for the Effective Teaching of Microbiology in School Science

This section of the chapter presents a framework of five principles for the effective teaching of microbiology. The five principles of this framework are: 1. Microbiology should be integrated across the science curriculum; 2. Misconceptions in microbiology should be challenged; 3. The role of microorganisms in human health should be explicitly taught; 4. The abstract nature of microbiology should be addressed; and, 5. Microbiology is a socioscientific topic that can be used to engage students in critical thinking (Figure 15.2).

1. Microbiology Should Be Integrated Across the Science Curriculum

Microorganisms are ubiquitous and should be integrated as examples and specific case studies in all relevant topics in school science, including but not limited to environmental science, human health, biological classification, ecology, animals, plants, soil science, matter and genetics.

Human biological and environmental systems teaching focuses on bacteria within the microorganism realm. Yet fungi, protista and viruses are also microorganisms

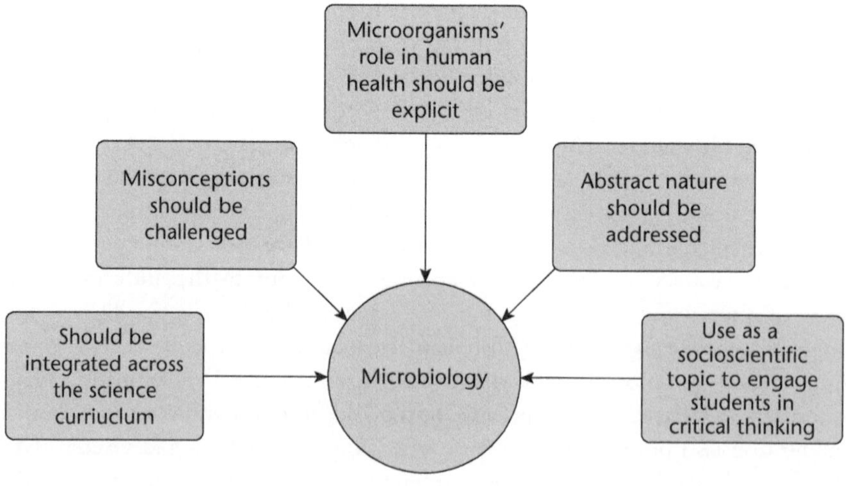

FIGURE 15.2 A diagram of the framework for the effective teaching of microbiology in school science

and should be considered. Fungi and viruses are mentioned in relation to human health and nearly always related to pathogenic conditions such as respiratory and gastrointestinal symptoms. There is a need to promote a healthy awareness of the beneficial role of microorganisms without increasing fears of infection and illness.

2. Misconceptions in Microbiology Should Be Challenged

This chapter has identified a considerable number of misconceptions associated with microbiology; conceptual change teaching approaches that identify, acknowledge and specifically address these misconceptions should be used at all levels of schooling. In particular, misconceptions that may result from social contexts and students' engagement with the mass media, such as an anthropomorphic understanding of microorganisms and the pathogenic nature of microorganisms, should be taken into consideration.

Incorrect terminology forms the basis of many of the misconceptions in microbiology displayed by students. When talking with younger students, it is important that teachers use the organisms' correct names—bacteria, fungi and viruses instead of 'germs' or 'bugs'.

The mass media and advertising play an important role in today's students' scientific understanding of microorganisms (Donovan & Venville, 2014). It would be of benefit to analyse the media's hyperbolic use of terminology and distorted use of facts to support critical thinking skills and understanding of microorganisms.

3. The Role of Microorganisms in Human Health Should Be Explicitly Taught

The importance of microorganisms to human health should be explicitly taught and developed in a developmentally appropriate manner across the early, middle and late childhood levels of schooling.

When considering the role of microorganisms in human health, conceptualising the human body as its own ecosystem can aid students in understanding more about the beneficial microorganisms that inhabit and protect our bodies. Each human carries their own 'micro-biome', which contains more microbial cells than there are human cells in the body. These microorganisms are beneficial to many body systems, aiding digestion of food and absorption of nutrients, producing vitamins and aiding our immune system.

In primary school, it is recommended that teachers use events such as children's illness or vaccination as teachable moments (Jones & Rua, 2006; Byrne et al., 2009), using scientifically accurate terminology and photographic images. For example, during winter when it is more likely that there are rhinoviruses infecting children, the teacher can explain how the virus is transmitted through sneezes and other bodily fluids.

4. The Abstract Nature of Microbiology Should Be Addressed

One of the main reasons microorganisms are abstract concepts for students is because of their very small size. It is difficult for young people to perceive microscopic size and scale in a way that is comprehensible or meaningful. Size and scale are included in the suite of ideas referred to by Duschl (2012, p. 10) as "crosscutting concepts" that are fundamental to science as a way of knowing. Tretter et al. (2006) have pointed out the need for educators to induct students into being able to think like scientists by using sizes and scales very different to those in the human-scale world. Teaching about the sizes and scales commonly used in the microbiological world provides an opportunity to encourage young people to extend their current ways of thinking and understanding.

Practical activities and multimedia graphics, including virtual reality, should be used to teach students about microorganisms, including their classification, morphology and comparative sizes. These types of activities are likely to engage students and, at the same time, address the abstract nature of microbiology. In the United Kingdom it has been reported that only two-thirds of secondary science teachers perform practical microbiological activities in the classroom. The three most common activities utilised are testing antibiotic sensitivity, using agar plates to grow bacterial colonies and aseptic technique (Redfern et al., 2013). However, science teachers may overestimate the cost, technical difficulty and health and safety risks of performing microbiological activities in the classroom.

Teachers may also be bewildered by the use of some microorganisms, such as *Escherichia coli*, when certain strains can be pathogenic to humans. However, it is the non-pathogenic K12 strain of *E. coli* that is commonly used in school laboratory activities due to its rapid generation time and its usefulness in antibiotic activities because of its resistance to penicillin but susceptibility to streptomycin.

Perceived health risks to students is another limitation for teachers when considering growing microbial cultures. There is concern that pathogenic microorganisms may grow at 37°C and at-risk students, such as those who are immuno-compromised, may become ill from improper handling of normally non-pathogenic or spore-producing microorganisms. To support teachers (and science laboratory technicians), most countries have guidelines for best practice when performing microbiological activities in the classroom. For example, Australian guidelines divide microbiological activities into four levels based on risk to students and staff (Science ASSIST, 2017), and the United Kingdom guidelines divide activities into three levels (Society for General Microbiology, 2008). The United States guidelines divide laboratories into two biosafety levels and then offer guidelines as to types of activities that can be undertaken in these areas (American Society for Microbiology, 2012). By using these guides, and performing a risk assessment, teachers and laboratory technicians can assess the potential risk to themselves and students of any microbial activity. Many microbiological activities can be performed with basic equipment in a safe manner for students, provided proper handling and growth techniques are used.

Extended time activities, such as growing bacteria and fungi on agar plates, mould on bread and oranges, observing rotting fruit and composting of organic matter, can be visualised using speed photography and/or computer programs.

5. Microbiology Is a Socioscientific Topic That Can Be Used to Engage Students in Critical Thinking

Microbiology includes many socioscientific issues and thus presents an opportunity to engage students in critical thinking through activities that promote argumentation, ethical thinking and logical problem solving. Studying microorganisms through socioscientific issues provides authentic and relevant contexts to help students understand scientific principles. These activities encourage students to use their microbiological understanding and include web-based research, media analysis, debates and role plays. Examples of topical issues associated with microbiology include access to vaccines and drugs to treat infections (e.g. HIV), global responses to epidemics (e.g. Ebola, SARS), the stockpiling of vaccines (e.g. influenza), the preventable causes of childhood mortality from malaria and cholera in developing nations and germ warfare (e.g. anthrax).

Conclusion

Microbiology is an expanding realm of knowledge, as scientists uncover more about the essential role microorganisms play in the survival of all ecological systems on Earth. This includes the micro-biomes of the human body, in which beneficial microbes interact daily with our external and internal environment to ensure our health and wellbeing. Microbiology is intriguing and fascinating and provides ample opportunities to engage students in activities and discussion, through which they gain a rich understanding of core biology principles. As science educators, we need to ensure our students, our future citizens, both recognise and understand the vital role these tiniest of organisms play in our everyday life.

References

American Society for Microbiology. (2012). *Guidelines for Biosafety in Teaching Laboratories*. Available at: www.asm.org/images/asm_biosafety_guidelines-FINAL.pdf.

Australian Broadcasting Corporation. (2017, May 31). Darwin fisho attacked by flesh-eating bacteria. *Drive*. Available at: www.abc.net.au/radio/darwin/programs/drive/flesh-eating-bug/8577318.

Australian Curriculum and Assessment Reporting Authority [ACARA]. (2016). *Foundation to Year 10 Curriculum: Science*. Available at: www.australiancurriculum.edu.au/science/curriculum/f-10?layout=1.

Byrne, J. (2011). Models of micro-organisms: children's knowledge and understanding of micro-organisms from 7 to 14 years old. *International Journal of Science Education*, 33(14), 1927–1961.

Byrne, J. & Grace, M. (2010). Using a concept mapping tool with a photograph association technique (CoMPAT) to elicit children's ideas about microbial activity. *International Journal of Science Education*, 32(4), 479–500.

Byrne, J., Grace, M. & Hanley, P. (2009). Children's anthropomorphic and anthropocentric ideas about micro-organisms. *Journal of Biological Education*, 44(1), 37–43.

Carlan, F. A., Sepel, L. M. N. & Loreto, E. L. S. (2014). Teaching cell biology in primary schools. *Education Research International*, 1–5.

Department for Education. (2015). *National Curriculum in England: Science Programmes of Study*. Available at: www.gov.uk/government/publications/national-curriculum-in-england-science-programmes-of-study/national-curriculum-in-england-science-programmes-of-study.

Donovan, J. & Venville, G. (2014). Blood and bones: the influence of the mass media on Australian primary school children's understandings of genes and DNA. *Science & Education*, 23(2), 325–360.

Dumais, N. & Hasni, A. (2009). High school intervention for influenza biology and epidemics/pandemics: impact on conceptual understanding among adolescents. *CBE—Life Sciences Education*, 8(1), 62–71.

Duschl, R. A. (2012). The second dimension-crosscutting concepts. *Science and Children*, 49(6), 10–14.

Jones, M. G. & Rua, M. J. (2006). Conceptions of germs: expert to novice understandings of microorganisms. *Electronic Journal of Science Education*, 10(3), 1–40.

Jones, M. G. & Rua, M. J. (2008). Conceptual representations of flu and microbial illness held by students, teachers, and medical professionals. *School Science and Mathematics*, 108(6), 263–278.

Leach, J., Driver, R., Scott, P. & Wood-Robinson, C. (1996). Children's ideas about ecology 2: ideas found in children aged 5–16 about the cycling of matter. *International Journal of Science Education*, 18(1), 19–34.

Maxted, M. A. (1984). *Pupils' Prior Beliefs About Bacteria and Science Processes: Their Interplay in School Science Laboratory Work*. (Unpublished doctoral dissertation), Vancouver, Canada: University of British Columbia.

Next Generation Science Standards [NGSS] Lead States. (2013). *Next Generation Science Standards: For States, by States*. Washington, DC: The National Academies Press.

Prout, A. (1985). Science, health and everyday knowledge: a case study about the common cold. *European Journal of Science Education*, 7(4), 399–406.

Redfern, J., Burdass, D. & Verran, J. (2013). Practical microbiology in schools: a survey of UK teachers. *Trends in Microbiology*, 21(11), 557–559.

Redfern, J. & Verran, J. (2015). What is a microbiologist? a survey exploring the microbiology workforce. *FEMS Microbiology Letters*, 362(24).

Romine, W. L., Barrow, L. H. & Folk, W. R. (2013). Exploring secondary students' knowledge and misconceptions about influenza: development, validation, and implementation of a multiple-choice influenza knowledge scale. *International Journal of Science Education*, 35(11), 1–28.

Science ASSIST. (2017). *Guidelines for Best Practice in Microbiology in Australian Schools*. Available at: https://assist.asta.edu.au/resource/4196/guidelines-best-practice-microbiology-australian-schools.

Simon, U. K., Enzinger, S. M. & Fink, A. (2017). "The evil virus cell": students' knowledge and beliefs about viruses. *PLoS ONE*, 12(3), e0174402.

Simonneaux, L. (2000). A study of pupils' conceptions and reasoning in connection with "microbes", as a contribution to research in biotechnology education. *International Journal of Science Education*, 22(6), 619–644.

Society for General Microbiology. (2008). *Basic Practical Microbiology—A Manual.* Available at: www.microbiologyonline.org.uk/file/ca2189fba3b39d24c5a44c1285d0082c.pdf.

Tretter, T. R., Jones, M. G., & Minogue, J. (2006). Accuracy of scale conceptions in science: Mental maneuverings across many orders of spatial magnitude. *Journal of Research in Science Teaching*, 43(10), 1061–1085.

Walsh, F. (2014, December 11). Superbugs to kill 'more than cancer' by 2050. *BBC News: Health*. Available at: www.bbc.com/news/health-30416844.

Woese, C. R., Kandler, O., & Wheelis, M. L. (1990). Towards a natural system of organisms: proposal for the domains Archaea, Bacteria, and Eucarya. *Proceedings of the National Academy of Sciences*, 87(12), 4576–4579.

16

BIOTECHNOLOGY

Marcus Hammann

Introduction

Biotechnology is defined as the "application of science and technology to living organisms as well as parts, products and models thereof, to alter living or non-living materials for the production of knowledge, goods and services" (OECD, 2005). Some rather traditional biotechnological practices have existed for a long time, like the production of wine, vinegar, bread and yogurt. Recent advances in genomics, proteomics and metabolomics, however, have revolutionized biotechnology, turning it into the key technology of the twenty-first century (Arber & Brauchbar, 1998). Its impact on society is tremendous, mainly through practical applications in medicine, agriculture, nutrition, industry and environment. Biotechnology is used, for example, for the remediation of contaminated environments, production of enzymes, development of improved medications and vaccines and for the development of disease-resistant crops and better foods. To cite an example from agriculture: Golden Rice, a genetically modified organism, was produced in 2000 in response to the world's growing population, significant parts of which rely on rice as their major or only source of food (Al-Babili & Beyer, 2005). Genetic engineering was used to introduce three beta-carotene-producing genes, thus lending Golden Rice an added nutritional value and compensating for people's vitamin-A-deficiency problems. Another important technique used in biotechnology is cloning, which is defined as the process used to create identical copies of DNA fragments, genetically identical cells or organisms. The genetic engineering tool of the future, however, is CRISPR CAS-9. By injecting a cell with a specific CAS-9 complex, it is possible to introduce or disable a specific gene at a certain place. Acting as a pair of DNA scissors, CRISPR CAS-9 is more precise than other gene editing tools and can cut DNA in locations that have previously been inaccessible. Some biotechnological techniques and products are controversially discussed in the public with arguments focusing on risks for human health and the environment as well as other negative effects. Ethical issues are involved as well. Development and application of modern biotechnological products is thus preceded by comprehensive risk/safety assessments, ethical analyses and stakeholder analyses.

Biotechnology Education for High School Students

As biotechnology involves complex socio-scientific issues surrounded by controversy and public debate, biotechnology education, in its most general sense, aims at enabling high school students, as future citizens, to participate in public debate about the risks and benefits of modern biotechnology as well as the ethical issues involved (Bayrhuber et al., 2000; Dawson & Schibeci, 2003; Dawson, 2007; Harms, 2002; Lock, 1996; Steele & Aubusson, 2004). As such, biotechnology education is considered crucial for scientific literacy. The task is demanding, however, as making informed decisions about biotechnology involves the assessment of case-specific risks and benefits, which often requires specialized knowledge, typically far removed from standard high school biology. For example, the potential of genetically modified crops to transfer foreign genes through pollen to related plant species is considered an environmental concern with respect to the emergence of pesticide-resistant "super-weeds" or the introduction of undesirable traits into related crops (Daniell, 2002). Assessment of these risks depends upon understanding the range of available gene containment strategies and their effectiveness, for example transgenic mitigation (transgenes that reduce fitness of the hybrid).

Thus, biotechnology education strives to enable high school students to understand biotechnology as a socio-scientific issue, which involves understanding:

- what biotechnology is and what techniques and processes are used;
- what values are involved and what benefits and risks are associated with specific biotechnological techniques and applications;
- what multiple (often competing) political, social, legal, economic, and ethical claims are made for and against biotechnology.

Investigating the challenges of biotechnology education at high school level, educational researchers have focused on awareness and understanding of biotechnology, for example by asking high school students whether or not they have heard about biotechnology before, giving examples of biotechnology and describing its use in society. Additionally, students were asked to express their attitudes towards biotechnology, make decisions and engage in argumentation about related socio-scientific issues. This article, thus, summarizes research findings about students' *awareness* and *understanding* of biotechnology, as well as *misconceptions* that emerged from these studies. Research has also focused on students' *attitudes*, *argumentation* and *ethical reasoning* in the context of biotechnology. These aspects are central as students are expected to be informed about, willing to engage and able to make informed decisions about biotechnology and the use of biotechnological products in society.

High School Students' Awareness of Biotechnology

Studies from the 1990s, a time when modern biotechnology was a new and rapidly developing field, revealed that high school students lacked awareness of biotechnological techniques and applications. In a study of 188 students (15–16 years old) in the UK, one third of the sample did not know what biotechnology or genetic engineering was (Lock & Miles, 1993). Furthermore, nearly half of the sample could not give examples of either biotechnology or genetic engineering. Likewise, a study with 138 students (16–19 years old) from the UK revealed that 42% of high school students had never heard the term *biotechnology* (Gunter et al., 1998). Giving examples of biotechnology proved difficult for Australian high school students as well, as one-third of respondents in a sample of 1,116 student (15 years old) were unable to provide a single example for biotechnology (Dawson & Schibeci, 2003). Similar findings have been reported by researchers comparing high school students in the UK and in Taiwan, where 50% of the participants (17–18 years old) were unable to give examples of biotechnology (Chen & Raffan, 1999).

The University of Leeds project 'Young people's understanding of, and attitudes to, "the new genetics"' (1994–1997) provided additional information about high school students' awareness of biotechnology (Lewis et al., 1997a). Though mainly interested in exploring participants' understanding of genetics, researchers also included techniques central to biotechnology, for example cloning and genetic engineering. In this case, 482 high school students (14–16 years old) were asked to describe what these technologies involved, give an example of their applications and state their source of information. The same questions were used in a follow-up study in 2011, with 154 high school students of the same age group in the UK (14–16 years old) (Lewis, 2014). Compared to the older study, there was a statistically significant increase in the number of high school students who had heard of cloning. The percentage increased from 53% in 1995 to 96% in 2011. Slight increases were also observed for the percentages of high school students who had heard of genetic engineering, from 80% in 1995 to 84% in 2011. Surprisingly, there was a statistically significant decrease in the number of high school students who had heard of DNA testing, from 83% in 1995 to 67% in 2011. Thus, although this study found a positive trend towards greater awareness—at least for some aspects—of biotechnology, further research is necessary to investigate if there are similar positive trends for high school students in other countries.

High School Students' Knowledge and Understanding of Biotechnology

Findings from several studies suggest that high school students lack knowledge and understanding of biotechnology. In focus-group interviews, high school students (16–19 years old), for example, correctly associated biotechnology with cloning and genetic engineering, but did not show a clear understanding of how these techniques are applied and what their intended uses are (Gunter et al.,1998). In the same study, when encouraged to talk about food derived from biotechnology processes, high school students commented on food production in general ways and did not mention the role of genetic engineering. Elaborating on the role of biotechnology in food production, one high school student from the same study was quoted as saying: "You picture sort of laboratories and things like that you know where scientists are making food you know" (Gunter et al., 1998, p. 86). Other researchers, likewise, report on high school students' limited understanding of the applications and purposes of biotechnology. Researchers in Australia found that high school students struggle with understanding biotechnology, genetic engineering, cloning and genetically modified foods (Dawson, 2007; Dawson & Schibeci, 2003). In a sample of 175 high school students (14–15 years old), 18% of students provided a scientifically correct definition of cloning (Dawson, 2007). In a different Australian sample of high school students (15 years old), 29% of students were reported to confuse cloning and genetic engineering, believing that Dolly, the sheep, was genetically engineered and not cloned (Dawson & Schibeci, 2003). Also, high school students had misconceptions about cloning insofar as they believed that organs, limbs and humans are currently cloned (Dawson & Schibeci, 2003). Additionally, the study found high school students confusing current uses of biotechnology with possible future applications.

Two studies describe a positive trend regarding high school students' understanding of biotechnology. In Taiwan, students' knowledge of biotechnology significantly increased from 1995 to 2014 (Chen et al., 2016). Likewise, researchers from England reported a positive trend from 1995 to 2011. More high school students (14–16 years old) were able to give correct scientific information about selected biotechnological processes in 2011 than in 1995 (Lewis, 2014). For cloning, for example, correct responses (copying genes, genetic information, living things) increased from 9% in 1995 to 59% in 2011. For genetic engineering, correct responses (change, manipulate or edit genes) increased from 19% in 1995 to 58% in 2011. As another main finding from this study, however, many high school students also gave vague or incorrect descriptions. In particular, 23% of high school students (1995) and 17% of high school students (2011) described cloning as "copying something" without further specification or reference to anything biological. Other high school students defined cloning as "making something" without indication of copying.

As many biotechnological practices make use of microorganisms (microbial biotechnology), high school students' understanding of biotechnology is limited by student conceptions about microorganisms. As the predominant conception, high school students perceive microorganisms from an anthropocentric perspective and consider microorganisms harmful to human beings (Byrne et al., 2009; Byrne, 2011). In a brainstorming activity, for example, 42% of high school students (14 years old) associated microorganisms with disease/illness, but only 4% with food decay and 8% with food production. The authors of the study discuss these findings as limiting for biotechnology education and argue for addressing the role of microorganisms in "food technology beyond bread and yogurt production" (Byrne et al., 2009, p. 43). Investigating students' conceptions about microorganisms in the context of the EIBE project (European Initiative for Biotechnology Education), other researchers also found that high school students are unaware of the industrial use of microorganisms, whether genetically modified or not (Simonneaux, 2000). In interview sequences on yogurt making and silage production, students saw microorganisms solely as decomposing agents so that the production of yogurt and silage "is mostly seen as passive, a matter of 'letting it happen'" (Simonneaux, 2000, p. 632).

High School Students' Acceptance of Biotechnology

High school students' acceptance of biotechnology is often studied in conjunction with attitudes. A sample item for assessing students' attitudes towards biotechnology is "we should not alter the genes in potatoes to improve their taste" (Chen et al., 2016, p. 482). Attitudes towards biotechnology have been shown to be context-sensitive, with differences depending on the organisms involved, the techniques used and the aims pursued (e.g., Dawson, 2007; Gebhard, 2000; Gunter et al., 1998; Hill et al., 1999; Lewis et al., 1997b; Lock & Miles, 1993). Studies found higher acceptance among high school students (14–16 years old) for biotechnology involving plants and microorganisms than for animals and human beings (e.g., Dawson, 2007; Lewis et al., 1997b; Lock & Miles, 1993). Additionally, high school students (16–19 years old) perceived higher risks for the cloning of animals than for genetic engineering of animals (Gunter et al., 1998). Students' attitudes, furthermore, depended upon the aims and purposes of biotechnology. One study, for example, found that high school students in the UK and Taiwan (17–18 years old) judged genetic engineering of tomatoes and cattle for disease resistance more positively than for enhanced growth (Chen & Raffan, 1999; Chen et al., 2016). Also, students (mostly 17–19 years old) were more inclined to accept biotechnology and genetic engineering if there was hope that this would ultimately help people (Gebhard, 2000). Similarly, students from Taiwan (17–18 years old) found the making of transgenic mice to study cancer acceptable (Chen et al., 2016).

Further evidence for the context-sensitivity of attitudes towards biotechnology comes from a longitudinal study comparing high school students' attitudes towards biotechnology in Taiwan (Chen et al., 2016). The study involved 183 students (17–18 years old) in 1993 and 199 students (17–18 years old) in 2014. As a main finding, high school students' attitudes towards medical biotechnology remained unchanged, whereas attitudes towards agricultural biotechnology proved significantly more unfavourable in 2014 than in 1993. For assessing attitudes towards agricultural biotechnology, the study used items focusing on improving disease resistance, growth enhancement and improvement in crop quality. According to Norman E. Borlaug (2000), Nobel Prize Laureate for Peace in 1970, these biotechnological applications have the potential to end world hunger. The worsening of attitudes towards agricultural biotechnology described by Chen et al. (2016) for students in Taiwan, thus, is problematic as students need to be able to make fine-grained distinctions, even more so as an earlier study found unconditional acceptance among English high school students (15–16 years old) of genetic engineering for the development of high-yield crops (Lewis et al., 1997b).

Important negative attitudes towards biotechnology are reflected in statements such as "messing with genes is wrong (acting god, messing with nature)" and "things are as they are for a reason (so we shouldn't be trying to change them)" (Lewis et al., 1997b, p. 22). Juxtaposing biotechnology and 'untouched' nature, 80% of a sample of 586 German high school students (mostly 17–19 years old) were reported to believe that biotechnology constitutes a threat to nature and creates ecological crises (Gebhard, 2000). Thus, these high school students tended to believe that biotechnology is "unnatural", takes things too far, and destroys the balance in nature. To illustrate the belief that gene technology interferes with the natural order, one student was quoted as saying: "Genetic engineering is positive, just as long as nature is not destroyed by it" (Gebhard, 2000, p. 144).

To explain the lack of acceptance of biotechnology, researchers investigated the relationship between attitudes and knowledge. Findings across studies, however, are contradictory so that the role of knowledge is presently unclear. While some studies found positive correlations between knowledge and more positive biotechnology attitudes (Prokop et al., 2007; Fonseca et al., 2012), others found that knowledge about biotechnology is not always a predictor for positive attitudes (Klop & Severiens, 2007; Usak et al., 2009). Differences concerning "sample construction, data collection methods and the way that understanding is determined" limit comparability of findings across studies (Dawson, 2007, p. 61). Additionally, due to the context-specificity of attitudes, quantitative investigations of the relationship between knowledge and attitudes suffer from using too general measures.

High School Students' Argumentation and Ethical Reasoning in Biotechnology Contexts

Biotechnology poses contentious socio-scientific problems involving public debate (Sadler & Zeidler, 2004; Sadler & Donelly, 2006). Typically, conflicting ethical issues are involved and high school students need to learn how to negotiate them. Sadler and Zeidler (2004) observed inconsistencies in high school students' argumentation across different gene-therapy and cloning scenarios. An inconsistency consists, for example, in opposing genetic engineering in one biotechnology context because "people should not alter the natural order" and supporting it in another context because "it helps people" (Sadler & Zeidler, 2004, p. 78). In the first case, students hold the view that genetic engineering is wrong in itself, irrespective of its consequences. The second argument, in contrast, focuses on the consequences of genetic engineering, namely the improvement of human well-being through medical advances. In a nutshell, these are the two major types of ethical reasoning fundamental to Western moral philosophy, deontological ethics and consequentialism (Reiss, 2010). On the one hand, there is the deontological position (i.e., rights discourse) focusing on intrinsic rights and obligations. Reference is made to the natural order, which should not be altered, as a moral absolute. Changing the natural order is thus intrinsically wrong. On the other hand, there is consequentialism, focusing on the consequences of an action to decide whether an action is morally right or wrong. Reference is made to human well-being, which is a consequence of applying biotechnology to medicine. Genetic engineering is considered morally correct because it can help people.

Biotechnology needs to be taught in the context of a socio-scientific issues. But this entails discussing ethical issues that biology teachers find difficult to address. One approach, developed in EIBE (European Initiative for Biotechnology Education), acknowledges ethical issues and helps students construct and understand coherent arguments (see Table 16.1). The EIBE approach organizes ethical analysis into six steps, makes students aware of the fact that scientific actions involve values and helps students classify arguments as either deontological or consequentialist. As steps 3–6 are repeated for deontological ethics and consequentialism, students become aware that different types of ethical reasoning exist and that different people may arrive at different judgements. Thus, students are ultimately taught tolerance, as controversy cannot be resolved.

Teaching Approaches to Biotechnology Education

What does it mean to teach biotechnology in the context of socio-scientific issues? According to Lock (1996), teaching biotechnology in the context of socio-scientific issues means addressing controversial biotechnology issues

> as an integral part of workschemes; not as an 'add on' at the end of a unit, nor as an extra for homework or as the final element of extension materials for fast workers, but as a central theme covered by *all* students.
>
> *(p. 237)*

TABLE 16.1 Ethical analysis in teaching biotechnology (according to Bayrhuber, 2000)

Step	*Questions Asked*	*Description*
1	Which action is under debate?	The action is described (descriptive statements). Value judgements are avoided.
2	Which values does the action affect?	Students need to be made aware of the fact that scientific procedures involve values.
3	How does the action affect a particular value?	A particular value is discussed in relation to the action. When well-being is the value, benefits and risks are analysed.
4	What does this mean for the evaluation of the action itself?	Drawing on step 3, the action is labelled morally right or morally wrong.
5	What socio-political or personal consequences are to be drawn?	The aim is to discuss how politicians should decide and what individuals should do as socially responsible beings.
6	Discussion and criticism	Possible inconsistencies in the analysis are discussed.

Equally important, teaching materials should provide students with balanced information so that they can learn to distinguish between facts and opinions and make claims justified by the available evidence. Thus, although biotechnology often triggers intuitive reactions and emotions, the goal, ultimately, is to enable students to analyse biotechnology issues rationally, weigh risks and benefits and give reasons for their decisions. Furthermore, students should develop scientific attitudes, such as curiosity, open-mindedness, respect for evidence and willingness to tolerate uncertainty (Lock, 1996).

A large variety of different teaching approaches to biotechnology education is available. They range from practical activities with microbes, which scale down industrial applications for school laboratory use, to out-of-school visits to biotechnological production sites and commercial biotechnology laboratories (Wray et al., 2001), as well as activities related to adapted primary literature to role-play and debate. The EIBE materials, for example, developed for 16–19-year-olds and updated until 2000, are a valuable online resource (www.archiv.ipn.uni-kiel.de/eibe/HOME.htm). When biotechnology is beyond the scope of a traditional course, computer-based learning is available, for example biotechnology-themed video-games that allow students to fight an unknown viral disease by exploring symptoms and characteristics for various viruses, extract nucleic acids from virtual patient samples, conduct and analyse PCR results and set up reverse transcriptase reactions (Barko & Sadler, 2013). High school students who prepared their own biotechnology projects and presented them at a student biotechnology expo, a new model for a science fair, were found to engage with issues that are of interest and relevance to them (Chowning, 2002). Writing to a younger audience has been found to be an effective way to promote conceptual understanding of biotechnology (Hohenshell et al., 2004). Questions like "Why are restriction enzymes used in gene cloning?", "Why would a person have a DNA fingerprint that is different from that of another person?" and "Why is electrophoresis used in DNA fingerprinting?" have been proposed to promote active learning in a think-pair-share activity (Prahl, 2017). As biotechnology has been surrounded by controversy and public anxieties, engaging students in debates is crucial. An adequate way of doing so is fostering students' argumentation skills through dilemmas (Zohar & Nemet, 2002).

Another important principle of biotechnology education is applying knowledge about biotechnology acquired in the classroom to the real world. Nguyen and Siegel (2015) described a project-based attempt to help students to do so. Three steps are involved: 1) choosing a topic, 2) assessing the needs of a community and planning an action related to biotechnology, and 3) taking action and documenting it. Students were given a list of possible topics to choose from—for example, stem cell research, cloning and genetically modified plants and animals. The second step of the project focused on assessing the needs of the group that have a stake in the topic. Students, for example, used questionnaires, email surveys and interviews to collect information from the group of people they intended to inform. Next, students planned an action related to biotechnology and put it into practice, for example informing farmers in the area about gene-editing in agriculture.

Conclusions

Biotechnology, a scientifically complex and interdisciplinary enterprise, is rapidly developing. Advances in biotechnology require new teaching materials and teaching approaches. Nevertheless, the gap between traditional high school biology and cutting-edge biotechnology is constantly widening. Biology educators around the world strive to prepare students, who, as future citizens, need to be able to participate in public discourse about the risks and benefits of biotechnological applications in society. What renders this endeavour difficult is the fact that decision-making in biotechnology is case-specific—with applications ranging from uncontroversial to tremendously problematic, and knowledge about risks and side-effects ranging from very well-researched to completely unknown. Thus, dealing with biotechnology issues typically requires highly specialized knowledge as well as willingness to tolerate uncertainty. What makes biotechnology education even more challenging is the fact that cases differ and transfer of knowledge from one case to another is rendered difficult. Biotechnology education, on the other hand, profits from the increasing recognition of the importance of teaching socio-scientific issues in biology education, and significant advances have been made in terms of engaging students in argumentation and ethical reasoning in biotechnology contexts (Jones et al., 2010). Generally, in-depth case-based teaching and learning can be recommended for biotechnology education. Although the cases students encounter in high school biology may be few in number, they need to be different enough—and surely we don't know what the cases in 20 years from today will be—so that one can hope that students who learn to ask the right questions in biotechnological contexts in high school biology today will also ask the right questions in everyday life in the future.

References

Al-Babili, S. & Beyer, P. (2005). Golden rice—five years on the road—five years to go? *Trends in Plant Science*, 10, 565–573.

Arber, W. & Brauchbar, M. (1998). Biotechnology for the 21st century. In OECD (Eds.), *21 Century Technologies: Promises and Perils of a Dynamic Future*. Paris Cedex: OECD Publications, 77–95.

Barko, T. & Sadler, T. D. (2013). Learning outcomes associated with classroom implementation of a biotechnology-themed video game. *The American Biology Teacher*, 75(1), 29–33.

Bayrhuber, H. (2000). Ethical analysis in teaching biotechnology. In H. Bayrhuber, W. Garvin & J. Graiger (Eds.), *Teaching Biotechnology at School: A European Perspective*. Kiel: EIBE, IPN, 101–110.

Bayrhuber, H., Garvin, W. & Graiger, J. (Eds.) (2000). *Teaching Biotechnology at School: A European Perspective*. EIBE, Kiel: IPN.

Borlaug, N. E. (2000). Ending world hunger: the promise of biotechnology and the threat of antiscience zealotry. *Plant Physiology*, 124, 487–490. doi:10.1104/pp.124.2.487

Byrne, J. (2011). Models of micro-organisms: children's knowledge and understanding of micro-organisms from 7 to 14 years old. *International Journal of Science Education*, 33(14), 1927. doi:10.1080/09500693.2010.536999

Byrne, J., Grace, M. & Hanley, P. (2009). Children's anthropomorphic and anthropocentric ideas about micro-organisms. *Journal of Biological Education*, 44(1), 37–43. doi:10.1080/00219266.2009.9656190

Chen, S.-Y., Chu, Y.-R., Lin, C.-Y. & Chiang, T.-Y. (2016). Students' knowledge of, and attitudes towards biotechnology revisited, 1995–2014: changes in agriculture biotechnology but not in medical biotechnology. *Biochemistry and Molecular Biology Education*, 44(5), 475–491.

Chen, S.-Y. & Raffan, J. (1999). Biotechnology: student's knowledge and attitudes in the UK and Taiwan. *Journal of Biological Education*, 34(1), 17–23.

Chowning, J. T. (2002). The student biotechnology expo: a new model for a science fair. *The American Biology Teacher*, 64(5), 331–339.

Daniell, H. (2002). Molecular strategies for gene containment in transgenic crops. *Nature Biotechnology*, 20(6), 581–586.

Dawson, V. (2007). An exploration of high school (12–17 year old) students' understanding and attitudes towards biotechnology processes. *Research in Science Education*, 37, 59–73.

Dawson, V. & Schibeci, R. (2003). Western Australian school students' understanding of biotechnology. *International Journal of Science Education*, 25(1), 57–69. doi:10.1080/09500690210126720

Fonseca, M. J., Costa, P., Lencastre, L. & Tavares, F. (2012). Multidimensional analysis of high-school students' perceptions about biotechnology. *Journal of Biology Education*, 46, 129–139.

Gebhard, U. (2000). The role of nature in adolescents' conceptions of gene technology. In H. Bayrhuber, W. Garvin, J. Graiger (Eds.), *Teaching Biotechnology at School: A European Perspective*. Kiel: EIBE, IPN, 137–145.

Gunter, B., Kinderlerer, J. & Beyleveld, D. (1998). Teenagers and biotechnology: a survey of understanding and opinion in Britain. *Studies in Science Education*, 32, 81–112.

Gusky, S. (2014). Students as technicians: screening newborns for Cystic Fibrosis. *The American Biology Teacher*, 76(4), 254–258. doi:10.1525/abt.2014.76.4.7

Harms, U. (2002). Biotechnology education in schools. *Electronic Journal of Biotechnology* 5(3), 205–211.

Hill, R., Stanisstreet, M., O'Sullivan, H. & Boyes, E. (1999). Genetic engineering of animals for medical research: students' views. *School Science Review*, 80(293), 23–30.

Hohenshell, L., Hand, B. & Staker, J. (2004). Promoting conceptual understanding of biotechnology: writing to a younger audience. *The American Biology Teacher*, 66(5), 333–338.

Jones, A., McKim, A. & Reiss, M. (Eds.) (2010). *Ethics in the Science and Technology Classroom: A New Approach to Teaching and Learning*. Rotterdam, Boston, Taipei: Sense Publishers.

Klop, T. & Severiens, S. E. (2007). An exploration of attitudes towards modern biotechnology: a study among Dutch secondary school students. *International Journal of Science Education*, 29(5), 663–679.

Lewis, J. (2014). From Flavr Savr tomatoes to stem cell therapy: young people's understandings of gene technology, 15 years on. *Science and Education*, 23, 361–379. doi:10.1007/s11191-012-9523-z

Lewis, J., Driver, R., Leach, J. & Wood-Robinson, C. (1997a) Understanding of basic genetics and DNA technology. Young people's understanding of, and attitudes to 'The New Genetics'. Working Paper 2. Available at: www.education.leeds.ac.uk/assets/files/staff/lewis/understanding-basic-genetics-dna-technology.pdf.

Lewis, J., Driver, R., Leach, J. & Wood-Robinson, C. (1997b). Opinions on and attitudes towards genetic engineering: acceptable limits. A: the discussion task. Young people's understanding of, and attitudes to 'The New Genetics'. Working Paper 7. Available at: www.education.leeds.ac.uk/assets/files/staff/lewis/understanding-basic-genetics-dna-technology.pdf. www.education.leeds.ac.uk/assets/files/staff/lewis/understanding-basic-genetics-dna-technology.pdf.

Lock, R. (1996). Biotechnology and genetic engineering: student knowledge and attitudes: implications for teaching controversial issues and the public understanding of science. In G. Welford, J. Osbourne & P. Scott (Eds.), *Research in Science Education in Europe: Current Issues and Themes*. London: Routledge, 229–242.

Lock, R. & Miles, C. (1993). Biotechnology and genetic engineering: students' knowledge and attitudes. *Journal of Biological Education*, 27(4), 101–106.

Nguyen, P. D. & Siegel, M. A. (2015). Community action projects: applying biotechnology in the real world. *The American Biology Teacher*, 77(4), 241–247. doi:10.1525/abt.2015.77.4.3

OECD. (2005). Statistical definition of biotechnology. Available at: www.oecd.org/sti/biotech/statisticaldefinitionofbiotechnology.htm.

Prahl, K. (2017). Best practices for the think-pair-share active-learning technique. *The American Biology Teacher*, 79(1), 3–8. doi:10.1525/abt.2017.79.1.3

Prokop, P., Lešková, A., Kubiatko, M. & Diran, C. (2007). Slovakian students' knowledge of and attitudes toward biotechnology. *International Journal of Science Education* 29, 895–907.
Reiss, M. (2010). Ethical thinking. In A. Jones, A. McKim & M. Reiss (Eds.), *Ethics in the Science and Technology Classroom: A New Approach to Teaching and Learning*. Rotterdam, Boston, Taipei: Sense Publishers, 7–17.
Sadler, T. D. & Donnelly, L. A. (2006). Socioscientific argumentation: the effects of content knowledge and morality. *International Journal of Science Education*, 28(12), 1463–1488. doi: 10.1080/09500690600708717
Sadler, T. D. & Zeidler, D. L. (2004). The significance of content knowledge for informal reasoning regarding socioscientific issues: applying genetics knowledge to genetic engineering issues. *Science Education*, 85, 71–93.
Simonneaux, L. (2000). A study of pupils' conceptions and reasoning in connection with "microbes", as a contribution to research in biotechnology education. *International Journal of Science Education*, 22(6), 619–644. doi:10.1080/095006900289705
Steele, F. & Aubusson, P. (2004). The challenge in teaching biotechnology. *Research in Science Education*, 34, 365–387.
Usak, M., Erdogan, M., Prokop, P. & Ozel, M. (2009). High school and university students' knowledge and attitudes regarding biotechnology. *Biochemistry and Molecular Biology Education*, 37(2), 123–130.
Wray, F. P., Fox, M. C., Huether, C. A. & Schurdak, E. R. (2001). Biotechnology for non-biology majors: an activity using a commercial biotechnology laboratory. *The American Biology Teacher*, 63(5), 363–367.
Zohar, A. & Nemet, F. (2002). Fostering students' knowledge and argumentation skill through dilemmas in human genetics. *Journal of Research in Science Teaching*, 39(1), 35–62.

17

USING AUTHENTIC TEXTS TO PROMOTE DISCIPLINARY LITERACY IN BIOLOGY

Moriah Ariely and Anat Yarden

Introduction

"What is biology other than words? If all the words that biologists use were subtracted from the language . . . there would be no biology" (Postman & Weingartner, 1971, p. 86).

When asked about the language of science, most people automatically think of technical terms. In biology, terms like cells, photosynthesis, meiosis, plasma membrane, enzymes, genes and chromosomes are just a small fraction of 'the biological jargon' that students must learn as they enter into the discipline. However, the distinctive quality of scientific language lies in 'the wording' as a whole, and although technical terms are an essential part of the scientific language, they are not the whole story (Halliday & Martin, 1993).

Different purposes of different disciplines are reflected in the way that individuals in each discipline structure their discourses, invent vocabulary and make grammatical choices (Hynd-Shanahan, 2013). Biology, as a discipline, has its own language, its own ways of thinking and explaining the natural world and supporting its knowledge claims. This use of language is mostly apparent in the texts that scientists produce to communicate their findings to other members in their community—the research articles. As such, scientific texts are not merely tools for storage and transmission of scientific knowledge, but are also "constitutive parts of science" and essential vehicles for the expression of scientific thought (Norris & Phillips, 2003, p. 226).

Recent discussions emphasize both the derived sense and the fundamental sense of scientific literacy. The two meanings of scientific literacy complement each other, since the fundamental sense of scientific literacy is critical for supporting the derived sense (Norris & Phillips, 2003). Nonetheless, it is the fundamental sense of scientific literacy that is the focus of this chapter.

Quite often it happens that students who read accurately and fluently have trouble comprehending a scientific text. Learning to read scientific texts requires learning about the unique strategies of reading and writing in a specific discipline. Thus, readers need specific skills and knowledge to support their reading in biology in order to read and understand biology texts. In this view, becoming scientifically literate is an essential part of enculturation and socialization into a specific discipline through its specialized discourse. This kind of advanced literacy is termed 'disciplinary literacy' (Shanahan & Shanahan, 2008), and it is concerned with reading and writing in a specific discipline.

Scientific language is recognized by the combined effect of several linguistic features, such as reliance on technical vocabulary, use of passive voice, nominalizations (Fang, 2005; Halliday, 1993) and hedging (Hyland, 1998). These features serve certain communicative needs; scientists are required to describe observations precisely and to justify their arguments. The language in scientific texts is, therefore, functional in producing and organizing scientific knowledge, and it embodies a unique worldview and reasoning (Halliday & Martin, 1993). Thus, an important aspect of disciplinary literacy is understanding the unique linguistic forms and structures that construct and communicate scientific principles, knowledge and beliefs (Fang, 2005; Martin, 1993).

Students' Difficulties With Reading Biology Texts

Text comprehension is a complex task that involves different cognitive skills and processes. To truly comprehend a text, one should go beyond the words and understand the ideas, and the relationships between ideas, conveyed in the text (McNamara, 2007). A deep comprehension of any kind of text requires readers to link ideas coherently, to scrutinize the validity of claims and to understand the motives of the author. A deep comprehension of scientific texts is even more difficult, especially when readers have minimal knowledge of technical terms and other forms of background knowledge to support their reading (Graesser, 2007).

Why Do Students Find Biology Texts Difficult to Read and Comprehend?

Difficulty in reading scientific texts is often interpreted by readers as a difficulty with the text's vocabulary when, in fact, the difficulty is more likely to lie in the grammar (Halliday, 1993). Indeed, several scholars have claimed that one of the major difficulties in students' comprehension is the language of science (Fang, 2005; Lemke, 1990; Wellington & Osborne, 2001). Research articles were found to share several key aspects of the grammar of scientific language that makes reading them a challenging task. They contain a high density of information-bearing words and grammatical processes, such as extended noun phrases, embedded clauses and nominalizations, that compress complex ideas into a few words

(Halliday, 1993; Osborne, 2014). Some of the key difficulties of the grammar of scientific language are:

- The high informational density poses a great challenge for students' reading comprehension, since they often feel overwhelmed and confused by the density of information during text processing (Fang, 2005; Halliday, 1993).
- Abstraction, which is achieved mainly through nominalization, is effective in creating a flow of discourse, since it allows the author(s) continuously to move from what has been previously established to what follows from it next. However, nominalizations are also challenging for students, since much of the semantic information and meaning is hidden in the nominalized words (Fang, 2005; Halliday, 1993).
- Technicality in scientific texts was also found to be challenging (Fang, 2006). Technical terms allow scientists to establish a relationship among entities and construct classes or categories of entities in their field of research (Fang, 2005, 2006; Wignell et al., 1993). However, everyday words in scientific texts often have specialized meanings, and the texts contain many highly specialized words that are specific to the discipline. Moreover, in scientific texts terms are often organized into taxonomies with several layers of organization. These taxonomies can become very complicated, and they are not made explicit. Therefore, students are left to work them out by themselves from reading the text (Halliday, 1993).
- In scientific texts the information is presented in an objective manner, which is accomplished by making the text as impersonal as possible by removing people from the text (Hyland, 2002). As a result, the language of the article may seem alienating to students, who are used to more informal and interactive language from their everyday lives.

In addition to grammatical difficulties, the distinctive structure of biology texts can also be challenging for novice readers. It was found that research articles have a common organizational structure and a variety of speech acts through which authors create an argumentative structure (Suppe, 1998). Novice readers who have little or no understanding of the structure of a scientific article may be at a total loss about the author's goals and, consequently, the meaning of the paper (Yarden et al., 2015).

Teachers' Difficulties With Using Biology Texts

Some of the major difficulties of teaching using texts lie in the reading difficulties as outlined in the previous section. Indeed, perhaps the most significant difficulty for teachers who would like to use texts in their teaching is their students' poor reading skills. Several studies have shown that a high proportion of students at secondary and high school levels do not read academic texts with proficiency or high interest because of low literacy skills, motivation and engagement (Moje & Speyer, 2008). Others just don't want to read. Unfortunately, some teachers 'solve' the problem by removing texts from their teaching and telling their students what they need to know. There is no easy solution to this problem. Reading complex texts in general and reading scientific texts in particular requires persistence and focus (Hynd-Shanahan, 2013).

Another problem lies in the fact that biology, as a scientific discipline, is often described as a 'hands-on' activity. However, biology, as is the case for any scientific discipline, is more concerned with ideas than with data. In fact, science cannot advance without the 'minds-on' activities of communication, i.e. reading, writing and speaking (Norris et al., 2009). Despite this, many teachers do not pay much attention to text, since they do not perceive the use of text as part of 'doing science'. These teachers lack the knowledge about the vital role that literacy plays in enhancing science learning, and they fail to mentor their students in the necessary literate practices that would help them read science (Osborne, 2014). Thus, language is pushed into the background or ignored, while thinking or doing are brought into the foreground as if they have little to do with language (Gee, 2004).

When teachers do use texts in their classroom, many of them teach content area reading, and wrongly assume that they are teaching disciplinary literacy. Content area reading views literacy as "the ability to use reading and writing to learn subject matter in a given discipline" (Shanahan & Shanahan, 2012, p. 7). Content area reading teachers view reading tasks as being similar across the disciplines. These teachers emphasize a generalizable set of skills or strategies, which they believe can be used by students when reading any text in any field to help them with comprehension (Shanahan & Shanahan, 2012). To change this view and shift to more disciplinary reading skills, teachers must understand the role of literacy in the understanding of science concepts and believe in its importance. But even if teachers believe in the importance of literacy for learning science, many of them do not have a full understanding of the literacy of their own discipline. Biology teachers are content experts, but they often lack the necessary language awareness and literacy strategies to help their students cope with the specific language and literacy demands of the discipline. This is mostly because they were never explicitly taught the ways in which biologists create, communicate and critique knowledge. Teachers should have a better understanding of the central role that language plays in their discipline, so they can unpack the implicit understanding of the ways that experts in biology engage in literacy practices (Fang & Coatoam, 2013; Hynd-Shanahan, 2013).

Last, but not least, texts available for instruction are usually of poor quality, and most biology textbooks lack explicit encouragement for reading instruction and provide little assistance in teaching the specialized language of science (Fang, 2005; Pearson et al., 2010). Moreover, students are usually not exposed to the language of science in its normal social function (Wellington & Osborne, 2001), since most texts that populate the science class do not reflect the core attribute of authentic scientific reasoning and are antithetical to the epistemology of authentic science (Chinn & Malhotra, 2002).

Adapted Primary Literacy as Authentic Scientific Text

Adapted Primary Literature (APL, Falk et al., 2008) refers to an educational genre specifically designed to enable the use of scientific research articles in high school. In the adaptation process, the original research articles are adapted to match students' knowledge, reading ability and cognitive skills, while retaining the authentic characteristics of the original articles and taking into account the practical reasoning involved in producing scientific knowledge when adapting the article (Yarden, 2009; Yarden et al., 2001).

APL articles may have great benefits for developing disciplinary literacy and for teaching the ways of communication that are employed in the scientific community. It has been claimed that reading APL articles can help students to improve their understanding of inquiry, active learning and integration of knowledge (Yarden et al., 2009); their understanding of the nature of science and their ability to criticize (Baram-Tsabari & Yarden, 2005); and their level of inquiry thinking and uniqueness (Brill & Yarden, 2003).

It has been claimed that the APL articles represent authentically the structure, claims, content and uncertainty that are found in the original research articles (Yarden et al., 2015). In recent research, in which we analyzed the language of an APL article compared to a research article and to a popular article, we found that, indeed, the APL article construed specific realms of scientific knowledge and beliefs in a form that was more readable and interpretable to students (Ariely et al., submitted). Thus, APL articles may serve as models of scientific reasoning and communication and may also serve to promote students' disciplinary literacy. Accordingly, instruction should focus on the way knowledge is created and communicated in the specific discipline while taking the language into account (Hynd-Shanahan, 2013; Moje & Speyer, 2008).

In the following section, we suggest three different activities for using APL articles in the biology that which are aimed at promoting students' disciplinary literacy.

Using Contradictory Texts for Promoting Critical Thinking

This teaching and learning activity is aimed at promoting secondary school students' critical thinking when reading scientific texts. In this activity, students read two contradictory APL articles and then participate in a debate. Prior to the reading and the debate, students participate in three to four lessons in which they are taught about common fallacies of scientific arguments. The way the abovementioned activity was enacted in a 10th grade biology class is detailed below.

Two contradictory APL articles that describe the effects of the herbicide Atrazine on frogs' reproductive system and sexual development were used. One APL article (Ariely & Yarden, unpublished) was adapted from a PSL article that provides evidence that Atrazine is safe for use (Kloas et al., 2009), and the other APL article (Ariely & Yarden, 2013) was adapted from a PSL article that provides evidence that Atrazine is not environmentally safe for use (Hayes et al., 2002). For the debate, the class was divided into two groups: one group was instructed to speak in favor of using Atrazine, and the other group was instructed to speak against the use of Atrazine. Each group (for and against) was given both articles and was asked to prepare itself for a debate about the use of Atrazine in agriculture. The groups were given time for discussion, and then each group chose one or more presenters to represent the group in the debate. It should be noted that the goal for the debate was to promote the students' critical thinking by having them look deeply into the two APL articles, elucidating their strengths and weaknesses. Therefore, it was not a 'classic', 'by the book' debate, but rather a discussion with the following rules:

- Claims should be presented respectfully.
- Groups present their claims in turns.
- At each turn, the presenter has exactly one minute to present a claim.
- When the presenters speak, no one is allowed to interrupt.
- Groups are allowed to ask for time-off for consulting, questioning and reorganizing their thoughts.

We found this activity useful in promoting the critical reading of scientific text. After the activity, students' ability to recognize fallacies in scientific arguments presented to them and their ability to criticize a scientific text significantly improved compared to their ability to do so prior to the activity. In addition, this activity had a positive effect on students' views of the argumentative nature of science (Ariely & Yarden, unpublished).

Using Texts for Learning About the Unique Features of Different Genres of Scientific Text

This activity is aimed at promoting students' understanding of the different genres of scientific text and their unique linguistic features.

Let us demonstrate with a concrete example that was handed to 11th grade students learning biology. Below are two paragraphs taken from two different scientific texts genres. One is taken from an APL article (Zer-Kavod & Yarden, 2013), and the other is taken from a popular article (Guynup, 2000). Can you tell which is which?

a. Why edible vaccines? The World Health Organization (WHO) has called for new strategies to deliver vaccines. WHO estimates that 10 million children die in developing countries each year from infectious diseases that could be prevented with vaccines. Existing vaccines are expensive and require a semi-skilled person to give the injection, with needles that are hard to come by in developing countries. Reused needles can transmit viruses such as Hepatitis B and C and HIV. Injectable vaccines also require refrigeration. "In some countries, you never know when the electricity is going to go off," says Arntzen.

b. Recently, a few attempts have been made at creating an edible vaccine with plants, by inserting genes into the plant cell and integrating them into the cell nuclear genome. The main problem with these attempts was the low level of expression of engineered proteins in the cell (0.02–0.5% of the total soluble proteins in the cell). Another disadvantage of integrating genes into the cell nuclear genome is the ecological risk inherent in the transition of genes from the transgenic plants to the non-genetically modified plants. Inserting the genes and integrating them into the plant chloroplast genome may serve as a solution to both of these problems, since each plant cell contains hundreds of chloroplasts and each chloroplast has a large number of chloroplast DNA molecules, which increases the expression level.

The first paragraph is taken from the popular article, and the second is taken from the APL article. All the high school students who read these two paragraphs identified correctly the origin of each text, but more interesting were the students' explanations for how they could distinguish between the two genres. Students were able to identify differences in the language of the paragraphs, in the rhetorical structure of the sentences, in the level of technicality and in the texts' authoritativeness. This

TABLE 17.1 A comparison between a research article and a popular article (adapted from Ariely & Yarden, 2013; Yarden, 2009)

	Research Articles	*Popular Articles*
Source of Information	Scientists	Journalists
Target Audience	Scientists	Non-scientists
Structure	Canonical	Non-canonical
Presentation of Science	Uncertain, using evidence to support conclusions	Certain, with limited evidence to support conclusion
Language	'Scientific': technical, distant, 'dry', abstract, a lot of information (dense), etc.	Everyday language, rhetorical questions, not so technical, more fluent, more attractive, etc.
Writer's Purpose	To get claims accepted by the scientific community	To explain and summarize scientific discoveries to the general public

is interesting because this was done intuitively by the students, before learning anything about the unique features of the scientific language.

In the next stage of the activity, the full texts were presented to the students, and a comparison between research articles and popular articles was made by the students (with the help of the teacher) using Table 17.1.

The differences between the research article and the popular article are apparent to everyone who reads the articles, even though they are implicit. This activity can make the implicit knowledge explicit. By doing so, teachers can help their students to understand and reflect on the different communicative goals of each text genre.

Using APL Articles for Learning About the Linguistic Features of Scientific Texts

This teaching and learning activity is aimed at promoting secondary school students' understanding of the language of science as a system for construing meaning in a specific context. In this activity, students are given two paragraphs from an APL article (Zer-Kavod & Yarden, 2013); one is taken directly from the article, and the other one is modified. The paragraphs are essentially identical, except for one grammatical feature which is purposely modified. Students are asked to compare the paragraphs and decide which paragraph is the original one, and which is a 'fake' one. Here we present two different examples, where each is designed to present a different feature of scientific language. In the following paragraphs, taken from the methods section of the article, can you tell which is the original and which is the fake one?

a. In order to examine the development of immunity against cholera and malaria, we fed the mice with the genetically modified tobacco plants. We crushed and melted transgenic tobacco leaves expressing the two cholera and malaria proteins, and fed them to the mice. An additional group of mice that we fed with unmodified tobacco leaves served as a control group. There were ten mice in each trial group. We fed the mice with the genetically modified plants ten times for approximately 300 days of the experiment. After administering the last dose of the vaccine, we exposed the mice to the disease agents (the cholera toxin or the *Plasmodium* parasite) and examined the immune response to the disease.
b. In order to examine the development of immunity against cholera and malaria, the mice were fed with the genetically modified tobacco plants. Transgenic tobacco leaves expressing the two cholera and malaria proteins, were crushed and melted, and were fed to the mice. An additional group of mice fed with unmodified tobacco leaves served as a control group. There were ten mice in each trial group. The mice were fed with the genetically

modified plants ten times for approximately 300 days of the experiment. After administering the last dose of the vaccine, the mice were exposed to the disease agents (the cholera toxin or the *Plasmodium* parasite) and the immune response to the disease was examined.

The second paragraph is taken directly from the APL article, and the first paragraph is the modified one. The difference between the paragraphs lies in the way verbs are used in them. In the first paragraph the verbs are active, and in the second the verbs are passive (e.g. "*we fed* the mice with the genetically modified tobacco plants" vs "the mice *were fed* with the genetically modified tobacco plants"). The use of verbs also makes a difference to the presentation of human participation. In the first paragraph the actors (i.e. the scientists) are present, but in the second they are absent. Absence of human actors is an important feature of scientific writing, and it has a function in producing objectivity in the text (Fang, 2005; Schleppegrell, 2002).

Another example is taken from the discussion section of the same APL article.

a. We suspect that the new spatial structure created by fusing the two proteins also contributed to enhancing the immune response against them. It is also possible that the cholera antigen served as an immunological carrier for the malaria antigen and so contributed to enhancing the immune response against it. An elevation in the immune response against the malaria antigen can also explain the effective response of the malaria antigen antibodies in inhibiting the penetration of the parasite into red blood cells.
b. We are sure that the new spatial structure created by fusing the two proteins also contributed to enhancing the immune response against them. It is certain that the cholera antigen served as an immunological carrier for the malaria antigen and so contributed to enhancing the immune response against it. An elevation in the immune response against the malaria antigen definitely explains the effective response of the malaria antigen antibodies in inhibiting the penetration of the parasite into red blood cells.

In the above example, the difference between the paragraphs is in the hedging. The first paragraph is taken directly from the APL article, and the second paragraph is the modified one. The difference between the paragraphs lies in the level of certainty concerning the factuality of statements. In the first paragraph the statements are presented as uncertain by using modal verbs and other types of hedging, while in the second paragraph the statements are presented as certain (e.g. "We *suspect* that . . ." vs "We are *sure* that . . .", "It is also *possible* that . . ." vs It is *certain* that . . .", "An elevation in the immune response against the malaria antigen *can* also explain the . . ." vs. "An elevation in the immune response against the malaria antigen *definitely* explains the . . ."). Hedging is a significant communicative resource and a common feature of scientific texts. It allows writers to anticipate possible opposition to claims by expressing statements with precision and caution (Hyland, 1998).

We found that students who are not familiar with the role of hedging wrongly assume that hedged statements are hypotheses, when in fact these statements are the conclusions in the article.

Thus, by comparing the paragraphs as presented above, teachers can help their students to understand grammatical features in scientific language and, more importantly, to appreciate why they are important, what their role is, and how the changes in the grammar change the reasoning in the article and the author's message.

Conclusions

Readers need specific skills for understanding science texts other than dealing with the specialized vocabulary. Knowing the vocabulary of science without understanding how it is used, or why, has little value for understanding or becoming literate in science (Osborne, 2002). Accordingly, education programs should support students in acquiring the facility of science language and the ability to use and comprehend the full range of science text and representations (Gee, 2004).

Students should be taught about the specialized meaning-making grammatical resources of science, since without this knowledge they are likely to experience difficulties when reading scientific texts.

Is it the science teacher's job to teach her students how to read scientific texts? Our answer to this question is undoubtedly—Yes! Students will not learn to read scientific texts by reading other forms of literature. Students need to be explicitly taught strategies for reading biology texts, and for understanding the tradition of biology investigation, including the way knowledge is created, shared and evaluated by the biology research community (Hynd-Shanahan, 2013). Unfortunately, despite the importance of language in learning science, biology teachers typically view their job as teachers of content and prioritize content coverage rather than reading or language instruction in their content area (Fang, 2005). However, arming students with specific and more efficient strategies for learning scientific information can result in increased knowledge and more intrinsic interest in science (Shanahan, 2004).

It is our hope that the information and activities outlined in this chapter provide an insight into some important aspects of disciplinary literacy and suggest strategies for promoting disciplinary literacy of secondary school students who learn biology.

References

Ariely, M., Livnat, Z. & Yarden, A. *Authentic Scientific Texts in the Science Classroom: Taking the Language into Account*. Submitted for future publication.

Ariely, M. & Yarden, A. (2013). Exposure to low doses of atrazine results in sexual developmental problems in frogs (An Adapted Primary Literature Article). In B. Eylon, A.

Yarden & Z. Scherz (Eds.), *Exploring Life Systems* (in Hebrew). Rehovot: Department of Science Teaching, Weizmann Institute of Science, 211–213.

Baram-Tsabari, A. & Yarden, A. (2005). Text genre as a factor in the formation of scientific literacy. *Journal of Research in Science Teaching*, 42(4), 403–428.

Brill, G. & Yarden, A. (2003). Learning biology through research papers: a stimulus for question-asking by high-school students. *Cell Biology Education*, 2(4), 266–274.

Chinn, C. A. & Malhotra, B. A. (2002). Epistemologically authentic inquiry in schools: a theoretical framework for evaluating inquiry tasks. *Science Education*, 86(2), 175–218.

Falk, H., Brill, G. & Yarden, A. (2008). Teaching a biotechnology curriculum based on adapted primary literature. *International Journal of Science Education*, 30(14), 1841–1866.

Fang, Z. (2005). Scientific literacy: a systemic functional linguistics perspective. *Science Education*, 89(2), 335–347.

Fang, Z. (2006). The language demands of science reading in middle school. *International Journal of Science Education*, 28(5), 491–520.

Fang, Z. & Coatoam, S. (2013). Disciplinary literacy: what you want to know about it. *Journal of Adolescent & Adult Literacy*, 56(8), 627–632.

Gee, J. P. (2004). Language in the science classroom: academic social languages as the heart of school-based literacy. In E.W. Soul (Ed.), *Crossing Borders in Literacy and Science Instruction: Perspectives on Theory and Practice*. Arlington, VA: NSTA Press, 13–32.

Graesser, A. C. (2007). An introduction to strategic reading comprehension. In D. S. McNamara (Ed.), *Reading Comprehension Strategies: Theories, Interventions, and Technologies*. New Jersey: Lawrence Erlbaum Associates, Inc, 3–26.

Guynup, S. (2000). Seeds of a new medicine. genes, plants, and edible vaccines, J. Craig Venter Institute GNN. Genome News Network. Available at: www.genomenewsnetwork.org/articles/07_00/vaccines_trees.shtml.

Halliday, M. A. K. (1993). Some grammatical problems in scientific English. In M. A. K. Halliday & J. R. Martin (Eds.), *Writing Science*. Pittsburgh: University of Pittsburgh Press, 69–85.

Halliday, M. A. K. & Martin, J. R. (1993). *Writing Science: Literacy and Discursive Power* (Vol. 8). Pittsburgh: CRC Press.

Hayes, T. B., et al. (2002). Hermaphroditic demasculinized frogs after exposure to the herbicide atrazine at low ecologically relevant doses. *Proceedings of the National Academy of Sciences*, 99(8), 5476–5480.

Hyland, K. (1998). *Hedging in Scientific Research Articles* (Vol. 54). Amsterdam and Philadelphia: John Benjamins Publishing.

Hyland, K. (2002). Authority and invisibility: authorial identity in academic writing. *Journal of Pragmatics*, 34(8), 1091–1112.

Hynd-Shanahan, C. (2013). What does it take? the challenge of disciplinary literacy. *Journal of Adolescent & Adult Literacy*, 57(2), 93–98.

Kloas, W., et al. (2009). Does atrazine influence larval development and sexual differentiation in Xenopus laevis? *Toxicological Sciences*, 107(2), 376–384.

Lemke, J. L. (1990). *Talking Science: Language, Learning, and Values*. Norwood, New Jersey: Albex Publishing.

Martin, J. R. (1993). Literacy in science: learning to handle texts as technology. In M. A. K. Halliday & J. R. Martin (Eds.), *Writing Science: Literacy and Discursive Power*. Pittsburgh: University of Pittsburgh Press, 166–202.

McNamara, D. S. (2007). Preface. In D. S. McNamara (Ed.), *Reading Comprehension Strategies: Theories, Interventions, and Technologies*. New Jersey: Lawrence Erlbaum Associates, Inc, xi–xv.

Moje, E. B. & Speyer, J. (2008). The reality of challenging texts in high school science and social studies: how teachers can mediate comprehension. In K. A. Hinchman & H. K. Sherian-Thomas (Eds.), *Best Practices in Adolescent Literacy Instruction*. New York: Guilford Press, 185–211.

Norris, S. P., Falk, H., Federico-Agraso, M., Jiménez-Aleixandre, M. P., Phillips, L. M. & Yarden, A. (2009). Reading science texts—epistemology, inquiry, authenticity—a rejoinder to Jonathan Osborne. *Research in Science Education*, 39(3), 405–410.

Norris, S. P. & Phillips, L. M. (2003). How literacy in its fundamental sense is central to scientific literacy. *Science Education*, 87(2), 224–240.

Osborne, J. (2002). Science without literacy: a ship without a sail? *Cambridge Journal of Education*, 32(2), 203–218.

Osborne, J. (2014). Teaching scientific practices: meeting the challenge of change. *Journal of Science Teacher Education*, 25(2), 177–196.

Pearson, P. D., Moje, E. & Greenleaf, C. (2010). Literacy and science: each in the service of the other. *Science*, 328(5977), 459–463.

Postman, N. & Weingartner, C. (1971). *Teaching as a Subversive Activity*. New York: Delta.

Schleppegrell, M. J. (2002). Linguistic features of the language of schooling. *Linguistics and Education*, 12(4), 431–459.

Shanahan, C. (2004). Better textbooks, better readers and writers. In E. W. Soul (Ed.), *Crossing Borders in Literacy and Science Instruction: Perspectives on Theory and Practice*. Arlington, VA: NSTA Press, 370–382.

Shanahan, T. & Shanahan, C. (2008). Teaching disciplinary literacy to adolescents: rethinking content-area literacy. *Harvard Educational Review*, 78(1), 40–59.

Shanahan, T. & Shanahan, C. (2012). What is disciplinary literacy and why does it matter? *Topics in Language Disorders*, 32(1), 7–18.

Suppe, F. (1998). The structure of a scientific paper. *Philosophy of Science*, 381–405.

Wellington, J. J. & Osborne, J. (2001). *Language and Literacy in Science Education*. Buckingham: Open University Press.

Wignell, P., Martin, J. R. & Eggins, S. (1993). The discourse of geography: ordering and explaining the experiential world. In M. A. K. Halliday & J. R. Martin (Eds.), *Writing Science: Literacy and Discursive Power*. London: Falmer, 136–165.

Yarden, A. (2009). Reading scientific texts: adapting primary literature for promoting scientific literacy. *Research in Science Education*, 39(3), 307–311.

Yarden, A., Brill, G. & Falk, H. (2001). Primary literature as a basis for a high-school biology curriculum. *Journal of Biological Education*, 35(4), 190–195.

Yarden, A., Falk, H., Federico-Agraso, M., Jiménez-Aleixandre, M. P., Norris, S. P. & Phillips, L. M. (2009). Supporting teaching and learning using authentic scientific texts: a rejoinder to Danielle J. Ford. *Research in Science Education*, 39(3), 391–395.

Yarden, A., Norris, S. P. & Phillips, L. M. (2015). *Adapted Primary Literature*. Dordrecht, The Netherlands: Springer.

Zer-Kavod, G. & Yarden, A. (2013). Engineered bacteria produce biofuel from switchgrass (an Adapted Primary Literature article) In Hebrew, *Gene Tamers—Studying Biotechnology Through Research*, 2nd ed. Rebovot: Department of Science Teaching, Weizmann Institute of Science, 1–13.

18

NATURE OF SCIENTIFIC KNOWLEDGE AND SCIENTIFIC INQUIRY IN BIOLOGY TEACHING

Norman G. Lederman

Introduction

The phrase 'scientific literacy' has been around for well over half a century, and its connection to an understanding of the nature of scientific knowledge and the nature of scientific inquiry was, perhaps, most formalized by the work of Showalter (1974) and by a National Science Teachers Association position statement on science-technology-society (NSTA, 1982). Although the aforementioned citations are from the U.S., scientific literacy is a primary goal of science education throughout the world. Scientific literacy has always been at least partially associated with an individual's ability to make informed decisions about scientifically based personal and societal issues. However, meeting the stipulations of what it means to be scientifically literate requires that an individual understands subject matter, nature of scientific knowledge (NOSK), and nature of scientific inquiry (NOSI). In short, making informed decisions about scientifically based issues requires that one understands the limits of knowledge and how it was developed.

The meanings of NOSK and NOSI are not always familiar to teachers, and often they are contentious within the science and science education communities. Consequently, this chapter begins with discussion of the meaning of these constructs. It is important to note that when understandings of NOSI and NOSK are viewed as instructional outcomes, the audience of interest is K–12 students. Hence, one must consider their developmental appropriateness, the empirical evidence that students can learn the constructs, and a justification that knowledge of such constructs is necessary for students' scientific literacy (and general citizenry). Using this lens, the often contentious debates about what constitutes NOSK and NOSI disappear (Lederman, 1998, Lederman & Lederman, 2014).

Conceptualizing Nature of Scientific Knowledge

Given the manner in which scientists develop scientific knowledge (i.e., NOSI), the knowledge is engendered with certain characteristics. These characteristics are what typically constitute NOSK (Lederman, 2007). As mentioned earlier, there is a lack of consensus among scientists, historians of science, philosophers of science, and science educators about the particular aspects of NOSK. This lack of consensus, however, should be neither disconcerting nor surprising given the multifaceted nature and complexity of the scientific endeavor. Conceptions of NOSK have changed throughout the development of and systematic thinking about science and are reflected in the ways the scientific and science education communities have defined the phrase 'nature of science' during the past 100 years (e.g., AAAS, 1990, 1993; Lederman, 2007; NSTA, 1982).

However, many of the disagreements about the definition or meaning of NOSK that continue to exist among philosophers, historians, and science educators are irrelevant to K–12 instruction. The issue of the existence of an objective reality as compared to phenomenal realities is a case in point. There is an acceptable level of generality regarding NOSK that is accessible and relevant to K–12 students and their daily lives. At this level, little disagreement exists among philosophers, historians, and science educators. What follows is a brief consideration of the characteristics of the scientific enterprise corresponding to this level of generality. These aspects of NOSK are not an exhaustive list, or 'THE' list, but are filtered with respect to the K–12 audience.

First, students should be aware of the crucial distinction between observation and inference, and that all scientific knowledge is derived from observation and at least partially from inference. Observations are descriptive statements about natural phenomena that are 'directly' accessible through the senses (or extensions of the senses) and about which several observers can reach consensus with relative ease. By contrast, inferences are statements about phenomena that are not 'directly' accessible to the senses. In short, it is one thing to observe a phenomenon, but another to make inferences about the explanation for what is observed. For example, scientists have noticed over long periods of time that frequencies of genes and various phenotypes have changed, and that old species have become extinct and new species have emerged. The concept of biological evolution has been observed, but the process behind evolution is an inference. Some of these inferences have been Lamarckism and natural selection.

Second is the distinction between scientific laws and theories. Individuals often hold a simplistic, hierarchical view of the relationship between theories and laws, whereby theories become laws depending on the availability of supporting evidence. It follows from this notion that scientific laws have a higher status than scientific theories. Both notions, however, are inappropriate because, among other things, theories and laws are different kinds of knowledge and one cannot develop or be transformed into the other. Laws are *statements or descriptions of the relationships* among observable phenomena. Mendel's Laws of Segregation and Independent Assortment relate phenotype reproduction pairs to phenotypes to offspring (Lederman, 1998). Theories, by contrast, *are inferred explanations* for observable phenomena. Mendel's Laws of Segregation and Independent Assortment did not offer an explanation for the law-like observations. Scientists did not know about genes for another 40 years. After the development of gene/chromosome theory, an explanation for the law was provided. Scientists do not usually formulate theories in the hope that one day they will acquire the status of 'law.' Scientific theories serve important roles in their own right, such as guiding investigations and generating new research problems, in addition to explaining relatively huge sets of seemingly unrelated observations in more than one field of investigation. Although many would debate whether there are any 'real' laws in the life sciences, the message is the same.

Third, even though scientific knowledge is at least partially based on and/or derived from observations of the natural world (i.e., empirical), it nevertheless involves human imagination and creativity. Science, contrary to common belief, is not a totally lifeless, rational, and orderly activity. Science involves the *invention* of concepts and explanations, and this requires a great deal of creativity by scientists. Linus Pauling had access to more or less the same data as Watson and Crick, yet the Pauling model of base pairing for DNA was far different than the Watson and Crick model. Watson and Crick's creativity developed the model of DNA that has existed with little modification for over 50 years. This aspect of science, coupled with its inferential nature, entails that scientific concepts, such as atoms, black holes, and species, are functional theoretical models rather than faithful copies of reality.

Fourth, scientific knowledge is subjective or theory-laden. Scientists' theoretical commitments, beliefs, previous knowledge, training, experiences, and expectations actually influence their work. All these background factors form a *mind-set* that *affects* the problems scientists investigate and how they conduct their investigations, what they observe (and do not observe), and how they make sense of or interpret their observations. It is noteworthy that, contrary to common belief, science never starts with neutral observations (Chalmers, 1982). Observations (and investigations) are always motivated and guided by and acquire meaning in reference to questions or problems. These questions or problems, in turn, are derived from within certain theoretical perspectives.

Fifth, science as a human enterprise is practiced in the context of a larger culture, and its practitioners (scientists) are the product of that culture. Science, it follows, affects and is affected by the various elements and intellectual spheres of the culture in which it is embedded. These elements include, but are not limited to, social fabric, power structures, politics, socioeconomic factors, philosophy, and religion. An example may help to illustrate how social and cultural factors impact scientific knowledge. Telling the story of the evolution of humans (*Homo sapiens*) over the course of the past seven million years is central to the biosocial sciences. Scientists have formulated several elaborate and differing story lines about this evolution. Until recently, the dominant story was centered about "the man-hunter" and *his* crucial role in the evolution of humans to the form we now know (Lovejoy, 1981). This scenario was consistent with the white-male culture that dominated scientific circles up to the 1960s and early 1970s. As the feminist movement grew stronger and women were able to claim recognition in the various scientific disciplines, the story about hominid evolution started to change. One story that is more consistent with a feminist approach is centered about "the female-gatherer" and *her* central role in the evolution of humans (Hrdy, 1986).

Sixth, it follows from the previous discussions that scientific knowledge is never absolute or certain. This knowledge, including 'facts,' theories, and laws, is tentative and subject to change. Scientific claims change as new evidence, made possible through advances in *theory* and technology, is brought to bear on existing theories or laws, or as old evidence is reinterpreted in the light of new theoretical advances or shifts in the directions of established research programs. A prime example of this is the evolution of ideas concerning the structure of cell membranes. Originally, the structure was proposed to be a simple bilayer model. However, with the advancement of technology and the reinterpretation of existing data, the model was changed to a mosaic bilayer model with proteins embedded within the lipid bilayer.

Conceptualizing Nature of Scientific Inquiry

Although closely related to science processes, scientific inquiry/processes extend beyond the mere development of process skills such as observing, inferring, classifying, predicting, measuring, questioning, interpreting, and analyzing data. Scientific inquiry includes the traditional science processes, but also refers to the combining of these processes with scientific knowledge, scientific reasoning, and critical thinking to develop scientific knowledge. From the perspective of the *National Science Education Standards* (NRC, 1996), students are expected to be able to develop scientific questions and then design and conduct investigations that will yield the data necessary for arriving at answers for the stated questions.

Benchmarks for Science Literacy (AAAS, 1993) is a bit less ambitious as it does not advocate that all students be able to design and conduct investigations in total. Rather, it is expected that all students at least be able to understand the rationale of an investigation and critically analyze the claims made from the data collected. Scientific inquiry, in short, refers to the systematic approaches used by scientists in an effort to answer their questions of interest. Pre-college students, and the general public for that matter, believe in a distorted view of scientific inquiry that has resulted from schooling, the media, and the format of most scientific reports. This distorted view is called THE SCIENTIFIC METHOD. That is, a fixed set and sequence of steps that all scientists follow when attempting to answer scientific questions. A more critical description would characterize THE METHOD as an algorithm that students are expected to memorize, recite, and follow as a recipe for success. The visions of reform, as well as any study of how science is done, are quick to indicate that there is no single fixed set or sequence of steps that all scientific investigations follow. The contemporary view of scientific inquiry advocated is that the research questions guide the approach and the approaches vary widely within and across scientific disciplines and fields (Lederman, 1998).

The perception that a single scientific method exists owes much to the status of classical experimental design. The examples of scientific investigations presented in science textbooks most often are experimental in nature. The problem, of course, is not that investigations consistent with 'the scientific method' do exist. The problem is that experimental research is not representative of scientific investigations as a whole. Consequently, a very narrow and distorted view of scientific inquiry is promoted in the K–12 science curriculum.

At a general level, scientific inquiry can be seen to take several forms (i.e., descriptive, correlational, and experimental). The purpose of research in many areas is often simply to describe. On the other hand, there are numerous examples in the history of science that have led to more than a description. The initial research concerning the cardiovascular system by William Harvey was descriptive in nature. However, once the anatomy of blood vessels had been described, questions arose concerning the circulation of blood through the vessels. Such questions led to research that correlated anatomical structures with blood flow and experiments based on models of the cardiovascular system (Lederman, 1998).

Scientific inquiry has always been ambiguous in its presentation within science education reforms. It can be viewed as a set of skills to be learned by students and combined in the performance of a scientific investigation. It can also be viewed as a cognitive outcome that students are to achieve. In particular, the current visions of reform (e.g., NGSS, 2013; NRC, 1996) are very clear in distinguishing between the performance of inquiry (i.e., what students will be able to do) and what students know about inquiry (i.e., what students should know). For example, it is one thing to have students set up a control group for an experiment, while it is another to expect students to understand the logical necessity for a control within an experimental design. The third use of 'inquiry' in reform documents relates strictly to pedagogy and further muddies the water. In this sense, 'scientific inquiry' is viewed as a teaching approach used to communicate scientific knowledge to students (or allow students to construct their own knowledge) as opposed to an educational outcome that students are expected to learn about and learn how to do. In summary, the knowledge about inquiry included in the NGSS includes the following (NGSS Lead States, 2013):

- Scientific investigations all begin with a question, but do not necessarily test a hypothesis.
- There is no single set and sequence of steps followed in all scientific investigations (i.e., there is no single scientific method).

- Inquiry procedures are guided by the question asked.
- All scientists performing the same procedures may not get the same results.
- Inquiry procedures can influence the results.
- Research conclusions must be consistent with the data collected.
- Scientific data are not the same as scientific evidence.
- Explanations are developed from a combination of collected data and what is already known.

Research on Teaching and Learning of the Nature of Scientific Knowledge and Scientific Inquiry: A Brief Review

Research on teaching and learning of the nature of scientific knowledge (NOSK) has a long history, dating back over 50 years. Early research in this area conflated inquiry/practices and the nature of scientific knowledge, but more importantly it assumed that students learned NOSK implicitly and that teachers' conceptions of NOSK necessarily influenced their teaching. With respect to the NGSS (NGSS Lead States, 2013), understandings of inquiry/practices are subsumed under the rubric of NOSK. To be historically correct, research that focused solely on understandings of inquiry/practices are discussed in the latter part of this section. Space does not permit a detailed review of all the research, but rather a summary of the general findings and perspectives of research in both areas, referring readers to some of the primary reviews of the empirical research for details of individual investigations.

The research on students' understandings has taken a different perspective during the past 15–20 years. In particular, much of the research on teachers indicated that understandings of NOSK were best promoted through an 'explicit' and reflective approach. In this approach, NOSK is treated as any other cognitive outcome and is not assumed to be a natural outcome of simply having students 'do' science or discussing historical case studies. Few studies have directly studied the effectiveness of explicit, reflective approaches to teaching NOSK relative to implicit approaches with K–12 students within the same study. However, in these few studies, the explicit, reflective approach is clearly more effective.

The presumed relationships between teachers' conceptions of science and those of their students, as well as that between teachers' conceptions and instructional behaviors, have been directly tested and demonstrated to be too simplistic relative to the realities of the classroom. It is safe to assume that teachers cannot possibly teach what they do not understand. However, research on the translation of teachers' conceptions into classroom practice indicates that even though teachers' conceptions of NOSK can be thought of as a *necessary* condition, these conceptions, nevertheless, should not be considered *sufficient* (Lederman, 1992; Lederman & Lederman, 2014). Contemporary research efforts "extend well beyond simply studying teachers' understandings of nature of science, as the translation of these understandings into classroom practice is mediated by a complex set of situational variables" (Lederman, 1992, p. 351). In addition to investigations that assessed the relationship between teachers' conceptions and classroom practice, efforts to identify those factors that do influence students' conceptions have also been pursued.

Inclusion of history of science in the science curriculum has often been advocated as a way to improve students' understandings of NOSK. As part of their review of literature, Abd-El-Khalick & Lederman (2000) noted that the use of history of science to improve understandings of NOSK has had ambivalent findings, with success primarily seen when aspects of NOSK were made explicit with historical case studies. The use of 'hands-on' activities has also been recommended as a way to improve students' understandings of NOSK. Again, the results have indicated that 'hands-on' activities are effective if explicit attention is given to NOSK. Simply doing science activities does not reliably lead to improved students' understandings of NOSK.

Science apprenticeship programs have been a popular approach to engaging high ability students in science in order to promote their interest in future science careers. Interestingly, a common goal of apprenticeship programs is that students develop improved conceptions of NOSK. Although scientists typically thought that their students had learned a lot about inquiry and NOSK, student data indicated that changes only occurred in students' abilities to do inquiry.

Shifting to research on NOSI/Science Practices, the reader is reminded that scientific inquiry, as an educational objective, has been viewed as a set of abilities and process skills that students develop, and as a set of cognitive understandings. Unfortunately, there has been little research on students' understandings about inquiry, and what exists is primarily at the elementary school level. The reader is again reminded that the NGSS has replaced the phrase 'scientific inquiry' with 'science practices.' The term 'science practices' did not formally exist before the unveiling of the NGSS. Therefore, the phrase 'scientific inquiry' will be used in this review of the research. The primary reason for the relative lack of research on students' understandings of scientific inquiry (as opposed to NOSK) is because, until recently, valid and reliable assessments of understandings of inquiry did not exist.

Relevant research studies on scientific inquiry have been conducted within the research on NOSK or epistemological beliefs. Early researchers in the 1960s and 1970s considered understandings about scientific inquiry as part of the nature of science (similar to the NGSS). In addition, some science educators also combined understandings about inquiry and epistemology of science. In general, these studies of inquiry (using either specified activities or various curricula) tended to show that students did not possess adequate understandings of NOSK and NOSI, nor did the doing of science result in increased understandings of NOSK or NOSI.

Recently, assessments for knowledge solely about scientific inquiry have emerged, and research in this area has gained some traction. In this regard, authentic scientific research does not appear to influence views of NOSK or NOSI. Research has also indicated that scientists in different fields of science do not have differing understandings of NOSK or NOSI. Recent research has investigated elementary students' (grades 1–5) understandings of science, scientists, and how scientists do their work. Results indicated that there were very few gains in NOSK and NOSI understandings between grades 1 and 5. Finally, and most recently, Lederman and Lederman (2017) just completed an ambitious international collaborative investigation of seventh grade students' understandings about scientific inquiry. Using the Views About Scientific Inquiry (VASI) (Lederman et al., 2014), seventh grade students' understandings of scientific inquiry, from 19 countries (spanning six continents), were assessed. A total of 2,960 students were surveyed. The understandings of scientific inquiry assessed were the same as those listed in the section on "conceptualizing" nature of scientific inquiry. Overall, with no exceptions, informed views of any of these aspects were well below 50 percent (Lederman & Lederman, 2017).

Summary of Conclusions of Research on NOSK/NOSI

After over 50 years of empirical research on NOSK and the relatively recent focus on scientific inquiry/practices, a few generalizations can be made:

- K–12 students and teachers do not typically possess 'adequate' conceptions of NOSK and NOSI.
- Conceptions of NOSK and NOSI are best learned through explicit, reflective instruction as opposed to implicitly through experiences with simply 'doing' science.
- Teachers' conceptions of NOSK and NOSI are not automatically and necessarily translated into classroom practice.
- Teachers do not regard NOSK and NOSI as an instructional outcome of equal status with that of 'traditional' subject matter outcomes (Lederman, 2007; Lederman & Lederman, 2014).

These generalizations should not be construed to mean that teachers and students are incapable of understanding NOSK or NOSI. Rather, they are a consequence of little pedagogical attention being given to these two constructs beyond simply having students, preservice teachers, and practicing teachers 'do' science. That is, it has always been assumed that teachers and students will necessarily induce understandings of NOSI and NOSK by doing authentic scientific investigations. For detailed reviews of empirical research, the reader is referred to Lederman (1992), Lederman (2014), Lederman and Lederman (2014), and Abd-El-Khalick and Lederman (2000) for research on NOSK, and Crawford (2014) for research on NOSI.

Some Concrete Instructional Examples for NOSK/NOSI

What follows are classroom-tested activities in the life sciences at the upper elementary/middle and secondary school levels. These activities are directly tied to 'traditional' topics in the life science/biology curriculum, while at the same time facilitating students' understandings of both NOSK and NOSI.

Finger Print Activity/Activities (Upper Elementary/Middle School)

Grade Level: 6–8

Targeted NOSK Aspects: tentativeness, subjectivity, observation, inference, empirically based

Targeted NOSI Aspects: investigations begin with a question, there is no single scientific method, procedures are guided by the question

Discipline: Biology (introduction to genetics)

Prerequisite knowledge: none

Instructional Scenario:

1. Ask students to answer the question "Do you think people have the same finger prints?" "If they have different finger prints, is there any pattern?"

   The discussion that follows will likely lead to clarifying whether each finger has the same print.
2. Give students a finger print chart (Figure 18.1) that shows different types of finger prints. Ask students to describe the different types of finger prints. The language they use will differ, but focus the discussion so that the class has a general understanding of the differences among the finger print types.
3. Ask the students what type of finger print they personally have. After a few answers are given, ask students if they have any questions. Are they curious about anything? If the question is not raised, ask the students which finger print they believe is most common in their class.
4. As the students if there is a difference between what they said about their own finger print and what they think is the most common finger print in the class. The students will eventually say that they can see their own finger print, but they haven't seen everyone's finger print for sure. At this point, the students understand the difference between an observation and an inference, although they may not know the words.
5. Tell students that what they said about their own finger print is an observation and the prediction about the most common finger print in the class is an inference.
6. Ask the students if scientists make observations and inferences. Ask for examples from several students.

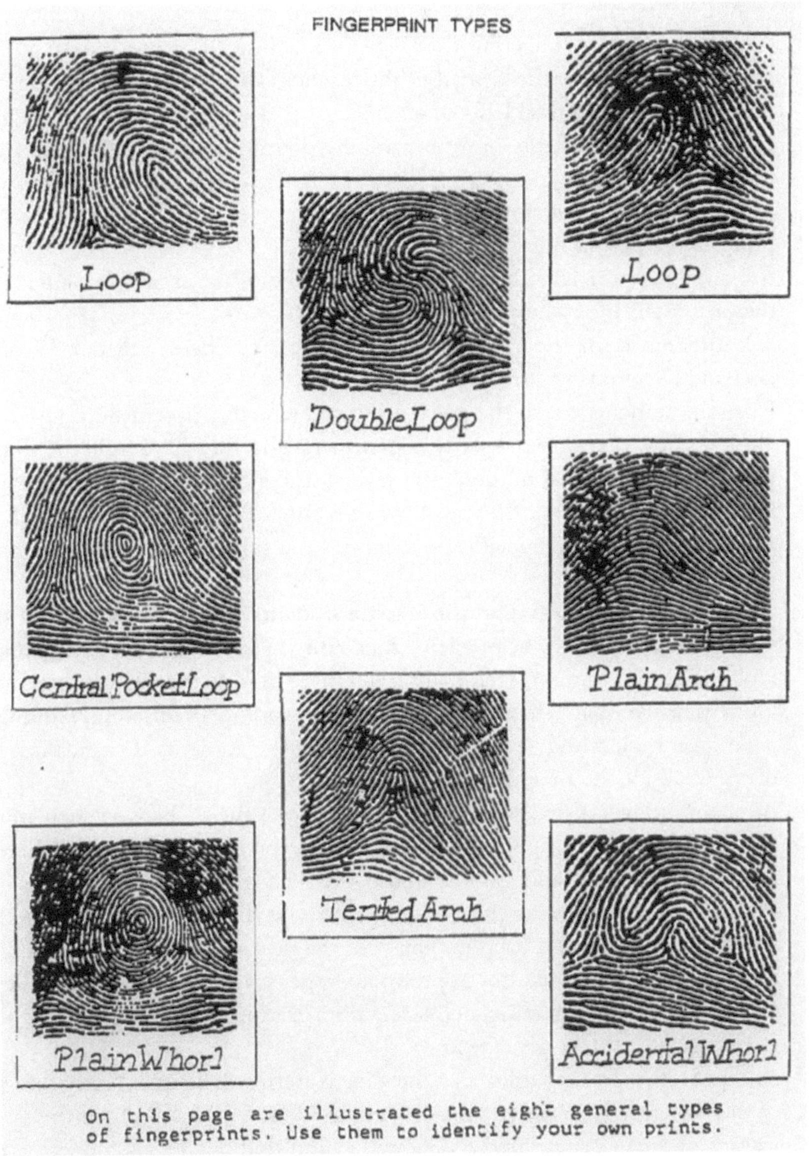

FIGURE 18.1 Major types of fingerprints

7. Engage the students in a conversation of how they can determine what is the most common finger print in the class. The students will quickly decide that they need to have everyone determine their type of finger print and then they can add up the number of each type. Ask students why this is better

than just inferring. They will quickly arrive at the conclusion that they can decide by actually collecting the finger prints instead of just guessing. The students won't know the word, but the teacher can label what they have just decided as the empirical basis of science.

8. Illustrate to the students how to carefully determine their personal finger print.
9. Provide each student with a 2B pencil, a piece of scotch tape, a piece of paper, and a post-it note.
10. Ask students to make a black spot on the piece of paper with a pencil and then rub their index finger on it.
11. Ask students to roll the inked finger onto the sticky side of a piece of Scotch tape and to remove it slowly.
12. Have the students attach the piece of Scotch tape to a post-it note.
13. Students now determine what type of finger prints they have from the finger print chart. There will no doubt be several students with prints that are difficult to categorize. Ask the class how they should decide. This will result in a class discussion that specifically identifies the critical aspects of each type of print.
14. Throughout the above discussion, some students will likely categorize the same print differently from each other. Ask the students why two people looking at the same finger print categorize it differently. Students will quickly realize that different people may interpret data differently and this is because their minds 'work differently.' Ask the students if scientists ever disagree about the meaning of the same data.
15. After this conversation, ask the students if any of them changed their mind about what type of finger print they have. Several minds will have changed. Asked students why and ask if scientists ever change their minds.
16. Put an enlarged copy of each finger print type (from Figure 18.2) on the white/black board and have students put the post-it notes, on which finger prints are attached, under the appropriate finger print type. This will result in a class histogram, providing a concrete visualization of the class data on the distribution of finger print types.
17. Ask the students if they think that the same pattern will be true in other classes of the same grade. What about other grades? If girls and boys have the same pattern of finger print types? Ask how they can find out. Have students work in groups of 3–4 and develop a question they want to ask. Have each group present their question and data collection design.
18. If two or more groups have the same question, but different designs, ask if this is okay. If all groups with the same question have the same design, ask why they have the same design. This will make it clear that the question determines the research design. Also, ask the students if any of their designs follow the steps of the scientific method in their textbook.

FIGURE 18.2 A typical owl pellet

Follow this with a discussion of whether their proposed investigation is scientific.

19. Students are asked to conduct their designed study and write up the results within two days.
20. It is important to note that this activity is situated within a life science class, and the students are expected to learn life science subject matter in addition to NOSK and NOSI.
21. After the students present the results of their investigations, ask them where their finger prints come from. Return their attention the fact that they all did not have the same finger print types and neither did the subjects in their investigation. Ask the students why. What determines their fingerprint types?
22. Students will quickly offer their parents as the source, and this is the introduction to heritable traits. Again, have students develop a question and research design to determine what determines their fingerprint type.
23. After the completion of this series of activities, the class can move to other traits that are inherited, and those that are not. The mechanism for inheritance naturally will follow.

Mystery Bones (High School)

This activity/experience actually takes 3–4 days (depending on students' grade level) and is ideal for a biology or life science course when skeletal systems are introduced. The overall focus is on structure and function, as opposed to the less useful memorization of bones and their locations. As students learn about the skeletal system, they also learn about various aspects of NOSK and NOSI. Here students have the particular opportunity to learn that scientific knowledge is tentative; scientific conclusions are a result of human creativity and subjectivity, derived from observations and inferences; and that scientific investigations can take a variety of forms as opposed to a single step-wise scientific method.

Grade Level: 10–12
Targeted NOSK Aspects: tentativeness, subjectivity, creativity, observation, inference, empirically based
Targeted NOSI Aspects: no single scientific method
Discipline: Biology
Prerequisite knowledge: minimal knowledge of skeletal systems
Instructional Scenario:

1. On the first day that students begin their study of skeletal systems, they are given an owl pellet (see Figures 18.2 and 18.3). Students are asked to work in pairs as they dissect the bones found within the pellet. Owl pellets are indigestible food regurgitated by barn owls several times a day. Since bones are not digestible by the owls, they are embedded within the pellets. These can be collected from barns or purchased from any biological supply company.
2. After all the bones are removed from the pellets, students are asked to match the bones on the vole diagram and tape the dissected bones in the appropriate position on the diagram (see Figure 18.4).
3. The next day, hold a class discussion about the structure/form of the various bones and their locations in the vole skeleton. Students are asked where the largest and thickest bones are found and where the smallest and thinnest bones are found. The goal of the discussion is to have students realize that the structure/form of the various bones is related to their function and location (e.g., supporting weight, protection, etc.).
4. Next, students are given a disarticulated skeleton of an unidentified animal; their task, in groups of 4–5, is to put together the skeleton. Students are expected to use the knowledge of skeletons that they learned from the owl pellet activity to infer the structure of this new and unidentified animal (see Figure 18.5). We have used disarticulated skeletons from rabbits, cats, and minks that have been purchased from a biological supply company or borrowed from a local university.

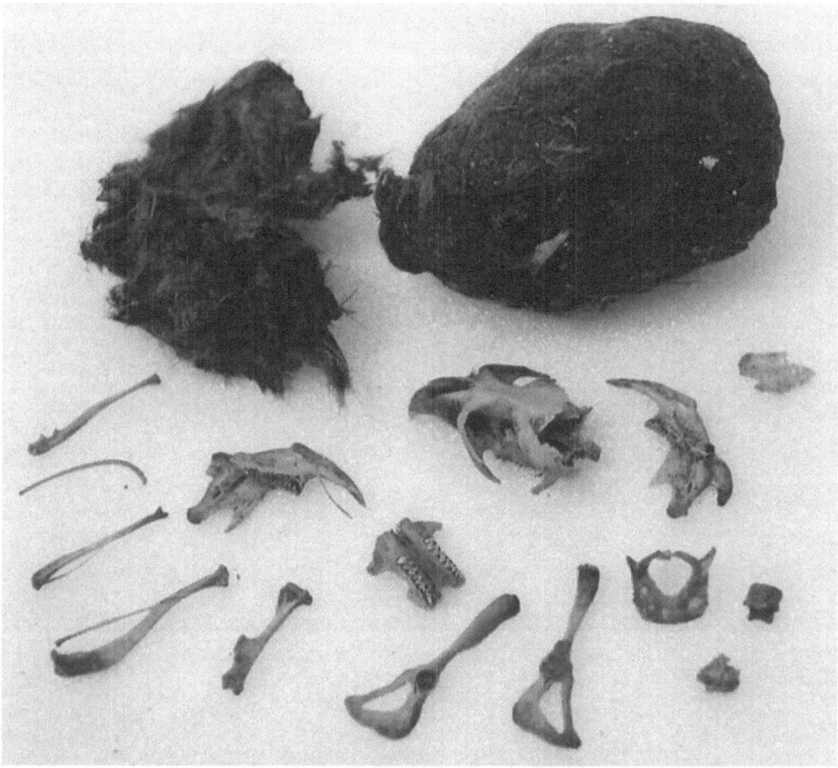

FIGURE 18.3 Common materials found within an owl pellet

5. Following this, engage students in another discussion about the structure and functions of the bones in the skeleton. Students are not expected to identify from which animal the bones come from, but rather mainly solidify the relationship between the structure and functions of bones in the skeletal system. Students are also asked if what they have been doing is related to the work of scientists. That is, do some scientists spend their time piecing together fossilized bones? They are also asked if this type of scientific investigation follows the 'scientific method' as it is portrayed in their textbooks.
6. To equate the work of the students to that of scientists, the next day groups of students are given envelopes that have a set of bones, from an extinct animal, that are represented on laminated paper (see Figure 18.6). They are asked to use their knowledge of skeletal systems to construct the animal's skeletal system, just as paleobiologists do.
7. Students should be encouraged to circulate and view the constructions of other groups out of curiosity or to help them with their own constructions.

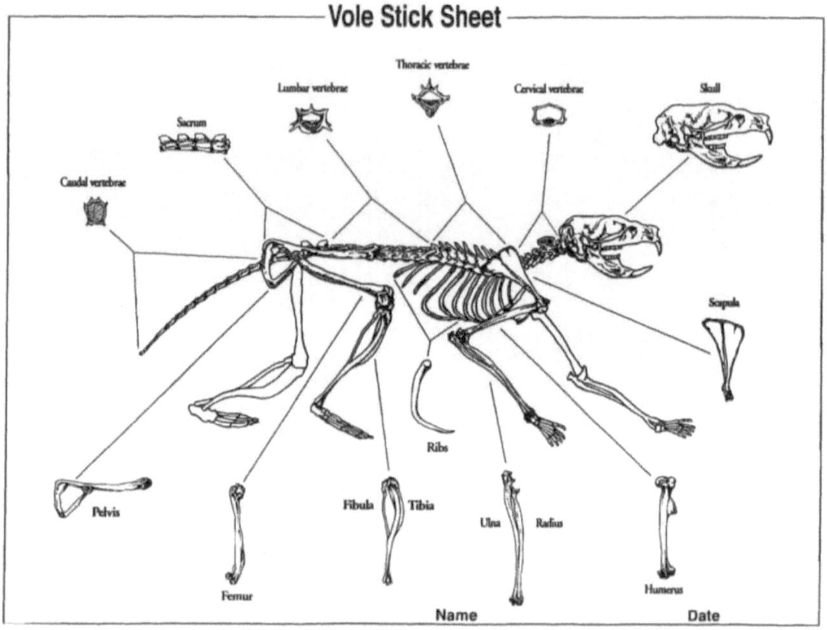

FIGURE 18.4 Vole bones

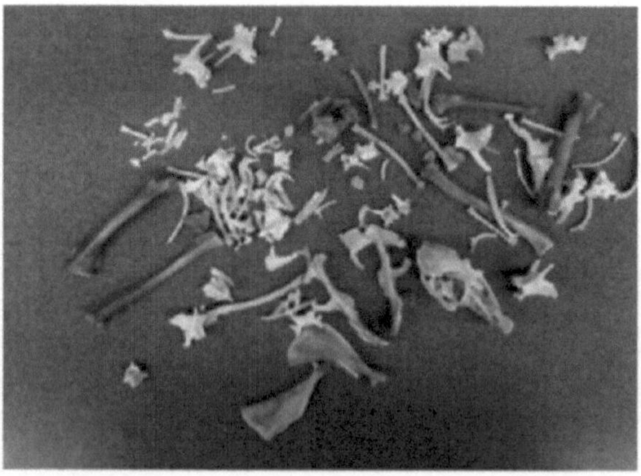

FIGURE 18.5 Disarticulated skeleton (cat, rabbit, or mink)

As constructions near completion, the teacher should take pictures of the various constructions with available technology for later projection in the classroom.

8. The constructions of each group are projected for class discussion. Each group explains the reasons for the placement of the bones in the constructed

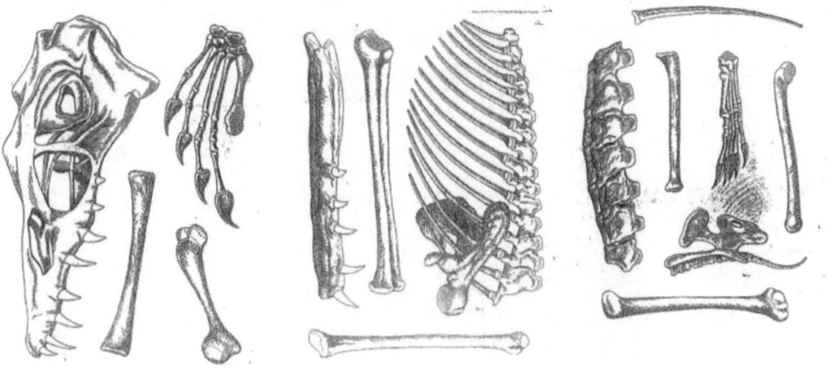

FIGURE 18.6 Mystery bones

skeleton. As this is done, the teacher questions students about the process and the various aspects of NOS that are evident in students' construction of the skeletal (e.g., creativity, tentativeness, subjectivity, etc.).

9. The teacher should now reveal scientists' construction of the skeleton and the inferred appearance of the animal with its skin on (see Figures 18.7 and 18.8). Students are often surprised to see the placement of the bones extending from the forearm digit of the animal because they have previously only seen the skeletons of terrestrial animals.
10. The teacher explains that the organism was believed to be one of the first dinosaurs or reptiles that could fly (more accurately, glide). Its resemblance to a reptile is emphasized because of the lingering debates between the relationship between dinosaurs and reptiles. As an aside, this creature lived in Europe during the late Jurassic period. From its first discovery in 1831, there has always been some confusion concerning whether this creature was an early dinosaur or extinct reptile.
11. The students are then informed that within the past decades scientists have decided that the bones supporting the wing in Figure 18.9 should be moved to the second forearm digit to better support the wing. Students are asked how this relates to NOS and they easily discuss that scientific knowledge is subject to change.
12. Show students Figures 18.9 and 18.10.
13. The teacher asks students about the appearance of Figure 18.8 as opposed to Figure 18.10. The students quickly notice that Figure 18.10 is more birdlike as opposed to resembling a reptile. The teacher explains that we now currently believe that dinosaurs are related to birds and this has led to a different interpretation of what the animal looks like with its skin on. If it is a high school class that has studied, or are studying, evolution, the teacher could ask students why the inferred appearance has changed.

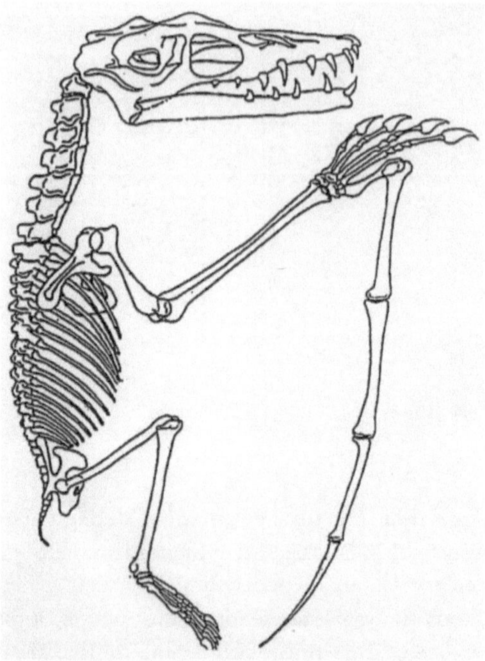

FIGURE 18.7 Reconstructed skeleton of *Scaphognathus crassirostris*

FIGURE 18.8 One paleontologist's imagination of *Scaphognathus crassirostris*

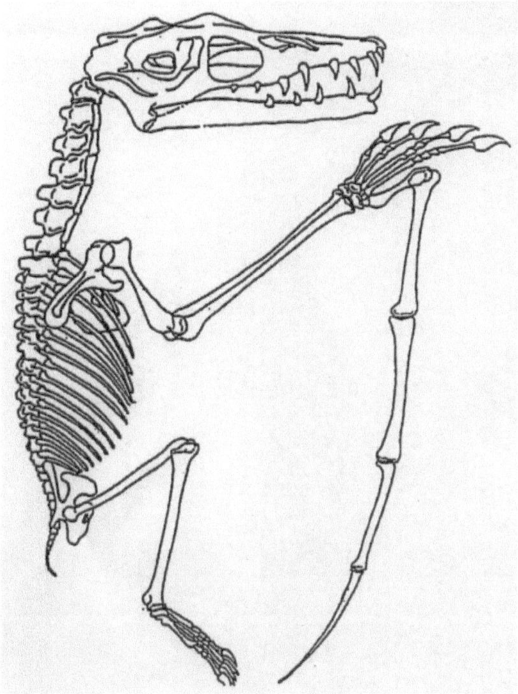

FIGURE 18.9 Reconstructed skeleton of *Scaphognathus crassirostris*

This activity fits quite nicely in a biology class during the study of skeletal systems and the form and function of bones. At the same time, aspects of NOS are easily integrated into the accomplishment of the 'typical' subject matter objectives.

Summary of Activities

The two activities, which span elementary/middle and high school life science, have all been classroom-tested with students. They all illustrate how NOSK and NOSI aspects can be seamlessly incorporated into life science topics typically included in the school curriculum. The integration of NOSK and NOSI does not disrupt the flow of any of the activities, and the investment of additional instructional time is minimal. Research has shown that teachers can be taught to revise existing instructional activities (as opposed to being expected to create totally new activities) in a reasonably short period of time. Most importantly, with the integration of NOSK and NOSI into science instruction, the perennial goal of achieving a scientifically literate citizenry is within reach.

FIGURE 18.10 Another paleontologist's reconstruction of *Scaphognathus crassirostris*

References

Abd-El-Khalick, F. & Lederman, N. G. (2000). Improving science teachers' conceptions of the nature of science: a critical review of the literature. *International Journal of Science Education*, 22(7), 665–701.

American Association for the Advancement of Science. (1990). *Science for all Americans*. New York: Oxford University Press.

American Association for the Advancement of Science. (1993). *Benchmarks for Science Literacy*. New York: Oxford University Press.

Chalmers, A. F. (1982). *What Is This Thing Called Science?* (2nd ed.). Queensland, Australia: University of Queensland Press.

Crawford, B. A. (2014). From inquiry to scientific practices in the science classroom. In N. G. Lederman & S.K. Abell (Eds.), *Handbook of Research on Science Education*. New York, NY: Routledge, Vol. 2, 515–541.

Hrdy, S. B. (1986). Empathy, polyandry, and the myth of the coy female. In R. Bleier (Ed.), *Feminist Approaches to Science*. Perganon Publishers, 119–146.

Lederman, J. S. & Lederman, N. G. (2017, April). *International collaborative investigation of beginning seventh grade students' understandings of scientific inquiry*. An invited symposium at the Annual Conference of the National Association for Research in Science Teaching, San Antonio, TX.

Lederman, J. S., Lederman, N. G., Bartos, S. A., Bartels, S. L., Antink Meyer, A. & Schwartz, R. S. (2014). Meaningful assessment of learners' understandings about scientific inquiry—the views about scientific inquiry (VASI) questionnaire. *Journal of Research in Science Teaching*, 51(1), 65–83.

Lederman, N. G. (1992). Students' and teachers' conceptions of the nature of science: a review of the research. *Journal of Research in Science Teaching*, 29(4), 331–359.

Lederman, N. G. (1998). The state of science education: subject matter without context. *Electronic Journal of Science Education* [On-Line], 3(2), December. Available: http://unr.edu/homepage/jcannon/ejse/ejse.html

Lederman, N. G. (2007). Nature of science: past, present, and future. In S. K. Abell & N. G. Lederman (Eds.), *Handbook of Research on Science Education*. Mahwah, NJ: Lawrence Erlbaum, 831–879.

Lederman, N. G. & Lederman, J. S. (2014). Research on teaching and learning of nature of science. In N. G. Lederman & S. K. Abell (Eds.), *Handbook of Research on Science Education*. New York, NY: Routledge, Vol. 2, 600–620.

Lovejoy, C. O. (1981). The origin of man. *Science*, 211, 341–350.

National Research Council. (1996). *National Science Education Standards*. Washington, DC: National Academy Press.

National Science Teachers Association. (1982). *Science-Technology-Society: Science Education for the 1980s. (An NSTA Position Statement)*. Washington, DC: Author.

NGSS Lead States. (2013). *Next Generation Science Standards: For States, by States*. Washington, DC: National Academies Press.

Showalter, V. (1974). What is unified science education? program objectives and scientific literacy. *Prism II*, 2, 3–4.

19

EXPLANATION IN BIOLOGY EDUCATION

Kostas Kampourakis and Kai Niebert

Introduction

Science education mostly focuses on the teaching of concepts and the related scientific theories. Concepts are, of course, important; however, simply learning and understanding them is of no great value if students do not also learn how to use these concepts in explanations for natural phenomena. For instance, knowing what a gene or an adaptation is does not on its own constitute sufficient scientific knowledge. But knowing what a gene is and being able to provide explanations of gene function in the development of traits, or knowing what an adaptation is and being able to provide explanations for the origin of particular adaptive traits in a population, is a lot more useful. Acquiring knowledge makes science education pointless if students do not also learn how to use this knowledge; explaining is one of the most important endeavors in science, as it is structured around explanatory aims.

Now, even when secondary students learn to use their scientific knowledge to construct explanations, do they always have the opportunity to reflect upon their structure? For instance, students may be able to explain the origin of adaptations based on natural selection, or the role of genes in the development of traits. However, how many students would be able to describe the structure of the explanations that they provide? In this chapter, we argue that teaching and learning about the structure of scientific explanations is important in order for students to acquire a better understanding of science. We focus on explanations for the origin of biological traits, because it is relevant to many fundamental questions in biology at the species level (e.g. why does a species exhibit a trait not found in other species?), at the population level (e.g. why are some traits more frequent in particular populations of a species than in others?), and at the individual level (e.g. why does an individual exhibit a trait that runs through the family?). We begin with a brief review of the related science education literature.

Explanation Versus Description

Understanding the structure of scientific explanations is important in order for students to also understand which are the criteria for considering a statement as an explanation. In order for a statement to qualify as an explanation, it has to have the appropriate form ('because' response to a 'why?' question), be logically coherent, and be scientifically correct (see Taber, 2014, p. 120). Explanation is thus different from mere description. While a description aims to answer *what* has happened, explanations try to give an answer to *why* it took place. For example, explaining why the individuals of a white bear population living in the Arctic have white color requires not only a description of the process of natural selection that took place but also an account of the antecedent conditions that caused the process of natural selection in the first place. Therefore, for valid explanations, secondary students should be able to distinguish between mere descriptions and causal explanations.

However, a closer look at students' explanations of biological phenomena reveals that conceptually clarifying what explanations are and discerning them from descriptions is not self-evident. For instance, Riemeier (2009) found that students—even at the secondary level—cannot make a systematic distinction in the generation of descriptions and explanations of biological phenomena. This finding is consistent with those of Treagust and Harrison (1999), who found in their investigations on the explanations of physical relationships that students cannot distinguish between descriptions and explanations, using instead these terms synonymously.

A closer look at different types of explanations and descriptions in biology may help to produce a typology (see Table 19.1). Explanations and descriptions in biology can be delineated as following:

- *What* is there? In the diversity of life, biological science aims at recognizing the features of and differences between traits. This is traditionally done by morphological description and complemented by molecular-genetic analysis.
- *How* does something work? Functional explanations are at the center of biology, both in classical morphological and physiological disciplines as well as in molecular biology. For example, the white color of a polar bear can be explained by a reduction to physical and chemical properties: the hair-cells do not contain pigments; therefore, a lot of light is reflected and the fur seems white. These explanations are sometimes described as proximate (e.g. Mayr, 1961).
- *Why* did a trait originate? When we ask why something exists in nature, students regularly think about the purpose of biological phenomena. For

TABLE 19.1 A typology of explanations and descriptions in biology education (adapted from Gropengiesser et al., 2016)

	Epistemic Interest	*Question*	*Example*	*Discipline (Example)*
Description				
constative	structures, phenomena	What?	I see a swarm of birds flying in a formation.	Morphology, classification
functional	teleonomic structures	What for?	The muscle is connected to the bone to move it.	Physiology
Explanation				
actual causal	proximate causes (chemical, physical, physiological causes)	How?	The color of the fur may be genetically determined.	Physiology
historical causal	ultimate causes (historic causes)	Why?	Bacteria with a resistance gene are better adapted in penicillin environments than bacteria without that gene.	Evolutionary biology
teleological	conscious actions, reasons, motives	Which aim?	Sperm want to fertilize ova.	Psychology, behavioral biology

example, if we asked "Why do polar bears have white fur?", we would ask about the adaptive significance of organisms, properties, and processes. This is a kind of question that we do not find in physics and chemistry. In biology, we think of such traits as adaptations to the respective living conditions. From this, it is then inferred that these features have become prevalent through a selection process. Such answers are somehow speculative because they assess the value of a feature in the past from its current function, even though the environmental conditions and ways of life may have changed.

- Answers to questions *what for* are also called functional explanations. However, as they do actually make causal attributions, they are better described as functional descriptions. Functional descriptions are to be applied to structures and processes and have been called teleonomic (teleonomy, from telos:

goal, nomos: law) (e.g. Mayr, 1961). For instance, developmental processes can be described in a teleonomic manner. However, functional descriptions can therefore not replace historical explanations.

- *Which aim* does something have? In addition to causal explanations, some teleological explanations are based on the intentions of organisms. It is usually stated that teleological explanations are not valid for all organisms, but only for those whose neural structures allow conscious actions to the extent to which intentions can be attributed. However, this refers to design teleological explanations because there also exist teleological explanations based on natural selection that are legitimate in biology (Lennox & Kampourakis, 2013).

In an interview study, Riemeier (2009) found that students can refer to causal relations when asked to explain biological phenomena such as the phototropism of plants, birds' flight formations, or sweating. When they were asked to describe these phenomena, 94% of the students did indeed provide descriptions. However, when they were asked to give explanations of these phenomena, only 41% of the students did so—the others gave descriptions. In addition, and independently from the phenomenon, students strive to give teleological explanations—even about objects for which a conscious decision cannot be assumed (Abrams & Southerland, 2001; Southerland et al., 2001; Tamir & Zohar, 1991). These findings show a discrepancy between the performance of students to explain biological phenomena and the curricular demands for descriptions and explanations in science classrooms.

Scientific Explanation in Biology Education

Braaten and Windschitl (2011) examined how the science education community has conceptualized scientific explanation. They found that whereas several studies engaged teachers and students in constructing, analyzing, and evaluating scientific explanations, researchers were not always clear about how they conceptualized explanation. Overall, Braaten and Windschitl (2011) documented the following uses of explanation in science education:

- *Explanation as explication*: In these cases, an 'explanation' was considered as a clarification for the meaning of a term or explication of reasoning about a problem. Therefore, the discussion was often focused on explications of terminology, meaning, and reasoning, and not actual explanations of natural phenomena.
- *Explanation as simple causation*: In these cases, emphasis was put on simple cause-effect relations accounting for an effect, thus oversimplifying the nature of causation that is better represented by multiple rather than single relations.

- *Explanation as justification*: In these cases, emphasis was given to the construction of arguments, and so requests for explanations were actually requests for evidence and justification for believing in a claim. Therefore, statements provided as explanations were often assertions or descriptions.

Brigandt (2016) has provided an excellent account of why the distinction between explanation and argument matters for science education. Whereas it is possible to develop an argument about a natural phenomenon, i.e. justify why one has drawn a specific conclusion on the basis of the available data, or in other words develop an argument for why something should be considered as evidence supporting a particular conclusion, this is different from explaining the phenomenon. The existence of causal factors that are explanatorily relevant is at the heart of this distinction; if one argued that factor F is responsible for effect E based on the available evidence, one might still be unable to explain why and how factor F actually brought about effect E. Only the latter could count as a genuine explanation, and it would be distinct from the former. We therefore suggest that teaching about causal explanation could help address all the issues related to the uses in explanation in science education identified above by Braaten and Windschitl (2011). An important point raised by Brigandt (2016) is that one aim of science is often to intervene in nature, which in turn requires explanatory knowledge. In other words, in order to make experimental or other interventions, we need explanatory knowledge about the relevant causes, which in turn informs us about possible points of successful intervention. Therefore, identifying causes and proposing causal accounts for natural phenomena should be core goals in teaching about scientific explanation in science education.

However, there are two kinds of problems with students' understanding of causality. One problem is the narrow range of types of causal models with which students are usually familiar, such as simple causal chains (e.g. of the kind A affects B affects C), whereas many explanations in science require significantly more complex thinking. This is usually something that science teaching typically does not prepare students for. At the same time, there is evidence that teaching students about complex causal structures improves their understanding, and they thus become capable of analyzing problems at a more systemic level by making reference to causal models (Perkins & Grotzer, 2005). This means that students need to be guided explicitly to reflect upon complex causal structures in order for them to reject the simple, linear causal models they usually have in mind. In other words, this means that students need to be guided to analyze complex phenomena and then learn to develop complex causal explanations.

Another problem is that secondary students often misunderstand the causes that are relevant to the explanation of a phenomenon or base their explanations on their misconceptions about natural phenomena. For instance, in the context of evolution, students may explain the existence of traits as the outcome of design or purpose, even after explicit evolution teaching. In particular, whereas they are able to explain some traits as the outcome of selection processes, when they are not given adequate information and they are asked to explain the existence of a trait in a population, they may tend to explain it as being there for a purpose. The problem here is that in such cases students fail to identify causes that precede the effects that they are asked to explain, and rather explain these effects in terms of final ends or causes (e.g. Kampourakis & Zogza, 2009). Similarly, in the context of genetics, secondary students have been found to explain the existence of traits as the result of the direct action of genes, thus overlooking the important role of proteins (or of other molecules such as RNAs) in development of traits. In such cases, students prioritize some of the explanatorily relevant causes and overlook the complex processes that bring about the effects (e.g. Duncan & Reiser, 2007).

A recent study has aimed not only at suggesting which competencies secondary students should have in developing scientific explanations, but also which are the respective stages of a learning progression (Yao & Guo, 2018). The researchers proposed a two-facet model of the learning progression of scientific explanation. The first facet describes the completeness of the language structure in an explanation, whereas the second facet is about delineating levels (basic level and in-depth level) for each component. After conducting an empirical study, the researchers concluded that a learning progression of scientific explanation could start by simply identifying the phenomenon that needs explaining and end up being able to identify the phenomenon that needs explaining; select the scientific concepts, laws, theories, and principles through a systematic analysis of the context; search data in a large data set; and develop a causal chain or clarify the mechanism that connects the phenomenon, evidence, and theory through different kinds of scientific reasoning.

So, we should teach about causes. But what should the starting point be? Philosophy of science provides some useful guidance.

Scientific Explanation in Philosophy of Science

An explanation consists of whatever is being explained, the *explanandum*, and whatever is doing the explaining, the *explanans*. For example, if one asks 'why X?' and the answer is 'because Y', then X is the explanandum and Y is the explanans.

The classic account of scientific explanation used to be the deductive-nomological (D-N) account (Hempel & Oppenheim, 1948). According to that, in order to explain a natural phenomenon, it should be shown how it could be deduced from a law of nature and particular antecedent conditions. In this way, an explanation would be a deductive argument that showed that the phenomenon was something to be expected because of particular antecedent conditions and regularities described by natural laws (Hempel & Oppenheim, 1948, p. 137). Hempel and Oppenheim considered this as a kind of causal explanation and argued that the same analysis would also apply to prediction. They actually noted that the explanation of a particular event would be inadequate if its explanans could not have served as the basis for the prediction of that event (Hempel & Oppenheim, 1948, p. 138).

One of the first critiques of the D-N account came from Michael Scriven, who criticized the relation between explanation and prediction described in the D-N account, arguing that one might sufficiently explain why some individuals and not others survived and reproduced, although one could not have predicted this particular outcome in advance. Extending his reasoning from individuals to species, Scriven argued that evolutionary change could be explained with reference to the features of organisms and to the nature of environmental changes. He argued that a cause could be reliably identified as long as one could show that it was in fact present, and that independent evidence suggested that it could indeed have produced the observed outcome (Scriven, 1959, 1969).

Wesley Salmon (1984) later proposed the causal mechanical model of explanation, according to which an explanation of an event should identify the causal processes and interactions that led to that event. He also suggested that causal explanations were crucial in biology in order to explain functions, whereas evolutionary explanations were based on another account, that of explanatory unification (Salmon, 1989). This account was developed by Philip Kitcher (1981, 1989). Explanatory unification yields understanding when few patterns of argument are needed to derive many beliefs. Explanatory unification relied on regularities or patterns that exist in nature, without making any reference to specific causes. In this way, the same general explanatory pattern could be applied to multiple cases with different causes, as emphasis was not given to the causes themselves but to the processes that produced each effect.

More recently, Michael Strevens (2009) has proposed the kairetic account of explanation, which is an account that borrows elements from both the causal and the unification accounts of explanation. This account is not a hybrid account of explanation but a causal one in which the rationale of the unification account of explanation is applied in order to consider a crucial concept: the difference-making factors. In the causal account, explanation is based on the causal process that produced some effect. However, not all causal factors have an explanatory role. Therefore, a criterion to assess explanatory relevance is necessary. According to Strevens, this can be determined by identifying the causes that made the difference for the occurrence of the effect (Strevens, 2009, pp. 69–141).

The idea of difference-makers is a common feature among the various theories of causation (Hitchcock, 2008). Generally speaking, there can be several potential difference-making causes of a particular effect. To take a classic example, a forest fire can be the effect of several different causes, such as a lighted match, oxygen, and perhaps more, e.g. flammable materials such as dried grass and strong winds. There will be no forest fire if there is no oxygen; flammable materials and strong winds can also be considered as causes of a forest fire because these are conditions under which an initial spark by a lighted match can develop into a large fire. However, in the case of the actual destruction of a forest because of a fire, it is a lighted match that is considered to have caused the fire. Nobody usually considers the presence of oxygen, the flammable materials, or the strong winds as the main causes of the fire, although there is no question that these are causes too—the fire would have not existed without them. Nevertheless, none of this was differentiated; it was the difference in the condition of the match (unlighted→lighted) that brought about the fire, and it is in this sense that the match was the difference-maker. This example shows that whereas there can be several different causes for a particular effect, there is typically one or a few that are causally more important than the others because it is their differentiation that brings about the effect.

Kenneth Waters (2007) has suggested that these causes are more important because they can account for the actual difference, in the case of biological traits, for the actual variation in a population. Therefore, in causal explanations it is difference-makers that are doing the explaining. It should be noted that identifying difference-makers depends on the population considered. For instance, in the case of lactose intolerance, different individuals in a population could have exhibited symptoms due to an inability to digest lactose for different reasons. In populations in which consumption of milk products is common and the allele related to lactase deficiency is rare, it is the latter that makes the difference and thus considered as the cause: people who consume milk products have the disease because they have the allele. In other populations in which consumption of milk products is rare and the allele related to lactase deficiency is common, it is the former that makes the difference and thus considered as the cause: people who have the allele have the disease because they consume milk (Gannett, 1999).

Philosophers suggest that it is not possible to come up with a general account of biological explanation, because biologists have a variety of explanatory aims and practices. As a result, different explanations are used in different contexts, but can also be provided for the same phenomenon depending on the research questions and the explanatory aims of scientists (Brigandt, 2013; Potochnik, 2013). The above very brief consideration of the literature in philosophy of science suggests that it is important to conceive of causes as difference-makers for their effects. With this in mind, we propose in the next section an account of causal explanation for the origin of biological traits. This is not the only kind of causal explanation in biology, but it is an important one because students often have misconceptions about the origin of traits, on the basis of which they tend to provide scientifically illegitimate explanations.

The Nature and Structure of Causal Explanations for the Origin of Biological Traits

The explanations for the origin of traits are causal explanations, as they aim to provide an account of what caused the emergence of a trait. In order to explain the origin of a biological trait, one therefore needs to provide an account of it as being the effect of a particular cause. An influential proposal was the distinction between two different types of causes—proximate causes and ultimate causes—that form the basis for explanations that are answers to 'How?' and 'Why?' questions, respectively (Mayr, 1961). Therefore, one could further distinguish between two kinds of explanations (based on Ariew, 2003):

a) proximate explanations, which are explanations for individual-level events, citing proximate causes,
b) evolutionary explanations, which are explanations for population-level events, citing ultimate causes.

If we focus on the origin of traits, then we can derive two main types of explanations: evolutionary explanations, which are population-level explanations for the origin of traits across generations (evolutionary time), and developmental explanations, which are individual-level explanations for the origin of traits within generations (developmental time).[1]

Mayr (1961) also described final causes and teleological explanations, considering them as illegitimate. However, teleological explanations in biology are not necessarily wrong. In fact, evolutionary explanations based on natural selection are inherently teleological in the sense that the advantage that a trait confers to its bearers explains its presence in the particular population. Teleological explanations that are applied to non-intentional processes such as natural selection are causal explanations of a special kind, suggesting that whatever a trait allows an organism to do in a given environment is causally relevant to the possession of this trait by the organism. Thus, organisms may have certain functional advantages that produce differences in survival and reproduction that result in natural selection. This kind of explanation is causal because the explanation of the existence of an organ or structure is based on its function: an organ exists because what it does provides an advantage to its bearers who survive and reproduce. Therefore, in order to distinguish between legitimate and illegitimate teleological explanations, it is necessary to identify the causes on which the explanations are based. As a result, there are two types of teleological explanations: (a) those based on design, which suggest that a feature exists for some purpose because it was intentionally designed to fulfil it, and which are not legitimate in biology, and (b) those based on a natural process, which explains a feature's presence in a population by appealing to that feature's beneficial consequences for an organism, and which are legitimate in biology (Lennox & Kampourakis, 2013).

Now, an important distinction is that between *causes that make a difference to the production of effects* and the *processes that bring about these effects*. This distinction applies to both developmental and evolutionary explanations, and it is crucial because both the cause of an effect and the process that produces this effect are part of an explanation. To give a simple example, in the case of a forest fire, oxygen and a lighted cigarette are both causes, and they are both distinct from the process of combustion through which the wood of trees burns. The process of combustion alone is not sufficient to explain why there was a forest fire. To provide a complete explanation, we also need to cite a difference-maker, such as a lighted cigarette. At the same time, simply citing this difference-maker is not enough, as one also needs to be aware of the process of combustion in order to fully explain the effect. Therefore, students should be able to distinguish between antecedent conditions in a habitat (e.g. mutations or conditions that make a particular trait advantageous or disadvantageous in the particular environment etc.) that cause a population to evolve through natural selection, and the overall process of natural selection itself. Students should also be able to distinguish between the conditions (e.g. the expression of a particular DNA sequence or the molecules existing and interacting in a particular cellular context) that cause the production of a particular RNA or protein molecule, and the overall physiological process through which this molecule is implicated in the development of a particular trait.

We might thus state that causal explanations in biology are those that identify at least one actual difference-maker in the past that somehow has a causal connection to the phenomenon under discussion. If one is interested in explaining the origin of biological traits, one should look for an actual difference-maker in the development or the evolution of a particular trait. In the case of developmental explanations, causes are to be found in the recent past, e.g. some individuals in a population have a particular DNA sequence that was expressed in particular cellular contexts and produced a particular molecule that under the particular environmental conditions caused the trait. It is important to note the difference between the cause (the particular DNA sequence being expressed in a particular cellular context) and the process that produced the effect (e.g. a particular physiological/developmental process that resulted in the emergence of this trait). In the case of evolutionary explanations, causes are to be found in the distant past, e.g. individuals in a population of some species had an advantage in surviving and reproducing in a particular environment due to their color, which caused differential survival among them through selection for this color. Again, it is important in this case to distinguish between the cause (the advantage the particular color conferred in survival and reproduction in the particular environment) and the process that took place (e.g. natural selection) that brought about the particular effect (existence of the trait). In the account of developmental and evolutionary explanations we present here, both the cause and the process are considered as parts of an explanation (Figure 19.1).

Implications for Biology Education

Whereas scientific explanations are complex and rich, students' intuitive explanations tend to focus on some aspects of an explanation. In the case of evolution, students cite final ends for explaining the origin of traits across generations

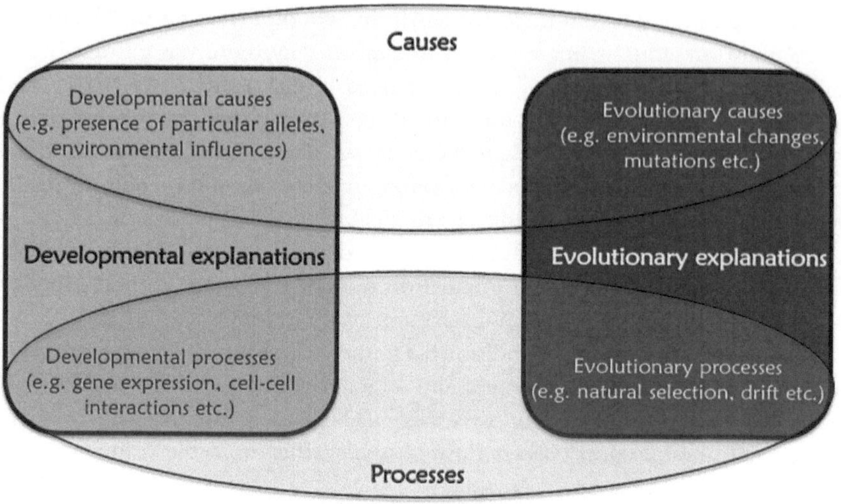

FIGURE 19.1 Explanations (evolutionary or developmental) for the origin of traits should include reference to both causes and processes. Referring to causes only (e.g. genes only in developmental explanations) or processes only (e.g. natural selection in evolutionary explanations) produces incomplete explanations for the origin of biological traits

(e.g. Kampourakis & Zogza, 2009). In the case of genetics, students consider genes as causes that directly determine outcomes (e.g. Duncan & Reiser, 2007). This suggests that students' conceptions should be taken into account when planning teaching interventions about scientific explanations. In the case of the origin of traits across generations, the aim could be to explain to students that causes are to be found in the past, and that natural processes bring about particular outcomes. In the case of the origin of traits within generations, the aim could be to show to students that genes do not determine anything but are implicated, through complex interactions, in the developmental processes that produce the traits in individuals.

If a secondary student is asked why eagles have wings, research suggests that s/he will most likely reply that eagles have wings in order to fly. If a student is asked to explain how eagles have come to have wings, research suggests that s/he most likely reply that or that eagles have genes that determine the production of their wings. These responses can be problematic. Students need to understand that traits evolve in populations because of particular antecedent conditions with causal influence and through natural processes such as natural selection, and not in order to fulfill some purpose. Students also need to understand that genes can be causes of traits, yet the latter are brought about by developmental processes, and thus genes do not produce traits but are implicated in their development. In other words, and as already argued, the identification of the relevant causes of a trait and the description of the process through which this trait came to exist are important in any explanation for the origin of a trait.

Note

1 Recent advances in evolutionary developmental biology have prompted some discussion on whether a distinction between 'developmental' causes and 'evolutionary' causes makes any sense (see e.g. Laland et al., 2011). However, this distinction is important for pedagogical purposes. Before considering the interaction of developmental and evolutionary causes and processes, students need to understand their own characteristics.

References

Abrams, E. & Southerland, S. (2001). The how's and why's of biological change: how learners neglect physical mechanisms in their search for meaning. *International Journal of Science Education*, 23(12), 1271–1281.

Ariew, A. (2003). Ernst Mayr's 'ultimate/proximate' distinction reconsidered and reconstructed. *Biology and Philosophy*, 18(4), 553–565.

Braaten, M. & Windschitl, M. (2011). Working towards a stronger conceptualization of scientific explanation for science education. *Science Education*, 95, 639–669.

Brigandt, I. (2013). Explanation in biology: reduction, pluralism, and explanatory aims. *Science & Education* 22, 69–91.

Brigandt, I. (2016). Why the difference between explanation and argument matters to science education. *Science & Education*, 25(3–4), 251–275.

Duncan, R. G. & Reiser, B. J. (2007). Reasoning across ontologically distinct levels: students' understandings of molecular genetics. *Journal of Research in Science Teaching*, 938–959.

Gannett, L. (1999). What's in a cause? The pragmatic dimensions of genetic explanations. *Biology and Philosophy*, 14, 349–373.

Gropengiesser, H., Harms, U. & Kattmann, U. (2016). *Fachdidaktik Biologie* [Biology Education]. Velber: Friedrich Verlag.

Hempel, C. & Oppenheim, P. (1948). Studies in the logic of explanation. *Philosophy of Science*, 15, 135–175.

Hitchcock, C. (2008). Causation. In S. Psillos & M. Curd (Eds.), *The Routledge Companion to Philosophy of Science*. New York: Routledge, 317–326.

Kampourakis, K. & Zogza, V. (2009). Preliminary evolutionary explanations: a basic framework for conceptual change and explanatory coherence in evolution. *Science & Education*, 18(10), 1313–1340.

Kitcher, P. (1981). Explanatory unification. *Philosophy of Science*, 48(4), 507–531.

Kitcher, P. (1989). Explanatory unification and the causal structure of the world. In P. Kitcher & W. C. Salmon (Eds.), *Minnesota Studies in the Philosophy of Science Vol. 13: Scientific Explanation*. Minneapolis: University of Minnesota Press, 410–505.

Laland, K. N., Sterelny, K., Odling-Smee, F. J., Hoppitt, W. & Uller, T. (2011). Cause and effect in biology revisited: is Mayr's proximate—ultimate dichotomy still useful? *Science*, 334, 1512–1516.

Lennox, J. G. & Kampourakis, K. (2013). Biological teleology: the need for history. In K. Kampourakis (Ed.), *The Philosophy of Biology: A Companion for Educators*. Dordrecht: Springer, 421–454.

Mayr, E. (1961). Cause and effect in biology. *Science*, 134, 1501–1506.
Perkins, D. N. & Grotzer, T. A. (2005). Dimensions of causal understanding: the role of complex causal models in students' understanding of science. *Studies in Science Education*, 41(1), 117–165.
Potochnik, A. (2013). Biological explanation. In K. Kampourakis (Ed.), *The Philosophy of Biology: A Companion for Educators*. Dordrecht: Springer, 49–65.
Riemeier, T. (2009). Wie erklären Schüler biologische Phänomene?—Eine Analyse von Schülerbeschreibungen und—erklärungen. [How do students explain biological phenomena— An analysis of students' descriptions and explanations.] In U. Harms & I. Mackensen-Friedrichs (Eds.), *Heterogenität erfassen—individuell fördern im Biologieunterricht*. Studienverlag: Innsbruck, 71–83.
Salmon, W. (1984). *Scientific Explanation and the Causal Structure of the World*. Princeton: Princeton University Press.
Salmon, W. C. (1989). Four decades of scientific explanation. In P. Kitcher & W. C. Salmon (Eds.), *Minnesota Studies in the Philosophy of Science Vol. 13: Scientific Explanation*. Minneapolis: University of Minnesota Press, 3–219.
Scriven, M. (1959). Explanation and prediction in evolutionary theory. *Science*, 130, 477–482.
Scriven, M. (1969). Explanation in biological sciences. *Journal of the History of Biology*, 2(1), 187–198.
Southerland, S. A., Abrams, E., Cummins, C. L. & Anzelmo, J. (2001). Understanding students' explanation of biological phenomena: conceptual framework or p-prims? *Science Education*, 85, 328–348.
Strevens, M. (2009). *Depth: An Account of Scientific Explanation*. Cambridge, MA: Harvard University Press.
Taber, K. S. (2014). *Student Thinking and Learning in Science: Perspectives on the Nature and Development of Learners' Ideas*. New York: Routledge.
Tamir, P. & Zohar, A. (1991). Anthropomorphism and teleology in reasoning about biological phenomena. *Science Education*, 75(1), 57–67.
Treagust, D. F. & Harrison, A. G. (1999). The genesis of effective scientific explanations for the classroom. In J. Loughran (Ed.), *Researching Teaching: Methodologies and Practices for Understanding Pedagogy*. London: Falmer Press, 28–43.
Waters, C. (2007). Causes that make a difference. *The Journal of Philosophy*, CIV, 551–579.
Yao, J. X. & Guo, Y.Y. (2018). Validity evidence for a learning progression of scientific explanation. *Journal of Research in Science Teaching*, 55(2), 299–317.

20

ARGUMENTATION IN BIOLOGY EDUCATION

María Pilar Jiménez-Aleixandre and Maria Evagorou

Introduction: What Is Argumentation, and How Does It Relate to Biological Knowledge?

In this chapter, argumentation is framed as a scientific practice—alongside modelling and inquiry—in an approach seeking to place these practices at the centre of science education (Osborne, 2014). Scientific practices may be described as epistemic practices in the context of specific learning contexts or content areas, such as biology (Jiménez-Aleixandre & Crujeiras, 2017). This approach means going beyond learning biological theories and concepts, towards an engagement in the practices through which biological knowledge gets constructed.

Argumentation can be characterized as a complex practice involving both justification—the evaluation of knowledge claims by combining available evidence and relevant theory—and persuasion of an audience (Jiménez-Aleixandre & Erduran, 2008). These two dimensions are already present in seminal works: justification in Toulmin (1958) and persuasion in Perelman and Olbrechts-Tyteca (1958), although the latter has received less attention in argumentation studies. Argumentation involves building and using discursive tools within social practices. From different argumentative contexts posing particular discursive challenges (Jiménez-Aleixandre et al., 2014), our focus is on two of relevance to biology education: (1) the evaluation of the explanatory power of alternative explanations in the context of the construction of causal explanations in biology; and (2) the evaluation of the acceptability of alternative options in decision making about socioscientific issues (SSIs). We agree with Osborne and Patterson (2011) in distinguishing between argumentation, the weighing of merits of alternative claims, and explanation, seeking the understanding of a causal mechanism, although in educational settings there is overlap between them.

Argumentation skills are generally agreed to be domain-general, while the use of particular pieces of evidence, justifications, and background knowledge are regarded as domain-specific. In Toulmin's model, the three essential components are:

(a) *Claims*: the knowledge statement or conclusion that is being evaluated.
(b) *Data*: the facts or experiments used to evaluate a claim; in science education, the term *evidence* is widely used, although their meanings are slightly different, as evidence is considered as data in the context of a justification or explanation.
(c) *Justification*: a statement that connects the data to the claim, showing that the step from data to claim is legitimate.

While Toulmin considered that the components are field-invariant, he argued that how they are to be judged—'what counts' as evidence or justification—is field-dependent. This is of relevance for the examination of argumentation in biology education. A dimension related to field dependency may be that what counts as evidence in some biology fields, such as evolution, palaeontology, or ecology, is different from data obtained in a physics experiment. It is not possible to carry out experiments in order to know the phylogenetic relationships among reptiles, dinosaurs, and birds. However, the combination of available fossil evidence, cladistics, and historical narrative points to the claims that these three groups belong to the sauropsida (reptilians), and that birds should be thought of as a type of dinosaurs, avian dinosaurs.

A second dimension of the relevance of disciplinary knowledge to argumentation lies in the process of evaluation of claims, which requires not only the synthesis of evidence, but also coordinating it with theoretical models. In their proposal of an initial learning progression for using evidence in decision-making contexts, Bravo-Torija and Jiménez-Aleixandre (2017) describe the highest level in terms of two performances: "Students *compare the consequences of different options* on the basis of available evidence. Students articulate their option by synthesizing evidence from multiple sources, both *supporting their choice and disconfirming others, coordinating it with theoretical models*" (p. 10, original emphasis). The task in their study is about marine resources management and fishing options, so the relevant theory required is, for instance, the energy flow model in ecosystems, an understanding of the higher energetic efficiency of fishing from a lower trophic level. Without this biological knowledge, the dataset about production and biomass of the different trophic levels, or about the diet of marine organisms, cannot be adequately interpreted. We will return below to the first performance.

As noted in Sampson and Clark's (2008) review, the majority of analytic frameworks focus on the components and structure of arguments, while only a few address the scientific content. Authors who do address it are Zohar and Nemet (2002), who in their study about argumentation in genetics assess the accuracy and correctness of disciplinary knowledge as a part of their rubric, and Kelly and Takao (2002), who in their examination of epistemic levels of arguments on oceanography include appeal to discipline-specific theories.

We propose that discursive contexts should be taken into account, both when designing classroom environments in order to foster argumentation and when studying their outcomes. There are many common operations in the practice of argumentation in the contexts of evaluating (1) the explanatory power of alternative explanations when constructing causal explanations (CEX) in biology; and (2) the acceptability of alternative options in decision-making (DEC) about socioscientific issues. But there are also some subtle differences in challenges, as, for instance, in CEX recognizing that claim and evidence are distinct epistemological categories. For CEX, it is necessary to use evidence to support given claims and to challenge others, while for DEC it is necessary to use it rather to predict the consequences of alternative options, as illustrated in the performances from Bravo-Torija and Jiménez-Aleixandre's (2017) study. In their discussion about the interrelations between argumentation and conceptual change, Jiménez-Aleixandre and Brocos (2018) focus on the modification of the epistemic status of competing explanations, in CEX, or options, in DEC. They propose that the process through which explanations are evaluated, and eventually changed, is rather an individual one, while decision-making is rather a process of social interaction. For these authors, some differences emerge, for instance that in CEX the explanations are considered plausible (true, consistent with other notions) or not, and in DEC the options are considered acceptable (feasible to carry out, consistent with other notions and with values) or not. Thus, evidence evaluation is articulated with theory in causal explanations about biology models such as gene expression or evolution, and the focus is rather on justification. In SSIs, such as diet choices about meat consumption, farm animal welfare, or the use of pesticides, the process involves complex weighting of dimensions such as scientific evidence, socioeconomic issues, cultural identities, ethical concerns, and emotions. In DEC contexts justification is relevant, but persuasion also plays a part. The role of emotive resources in argumentation, pointed out by Plantin (2011), should be also acknowledged.

Why and How to Engage Students in Argumentation in Biology Education

Engaging students in argumentation means that our goal is for them to participate in this scientific practice. In other words, learning argumentation involves taking an active role in knowledge evaluation operations (Jiménez-Aleixandre, 2008) such as: generating proposals, claims, options; contrasting them with evidence, either to support or to criticize them; or appropriating and using epistemic criteria, for instance to evaluate or judge the quality of evidence. We purposely begin this section by discussing successful engagement in argumentation, because we

FIGURE 20.1 A kindergartener drawing of an experiment about whether snails have a sense of taste

believe that research has sometimes focused on the difficulties experienced by students rather than on what they are able to perform, given appropriate support. We suggest that a shift is necessary, meaning a focus on how to promote their engagement.

Most studies about argumentation in biology report on secondary or middle school. However, in recent years there have been findings about how even young children, given appropriate learning environments, can participate in this practice. Monteira and Jiménez-Aleixandre (2016) studied kindergarteners (5–6 years old) who generated first-hand data that were used to build evidence, in the context of a five-month project about snails.

For the purpose of identifying entry points for the use of evidence in kindergarten, data were considered evidence (1) when their discursive role in children's talk was to evaluate a claim, and (2) when they were appropriate. This means leaving out the sufficiency condition proposed by McNeill and Krajcik (2012) for other ages. Monteira and Jiménez-Aleixandre identified two levels of epistemic judgment, in other words, in the development of data into evidence: (1) descriptive evidence statements, used to support a claim, closer to data; and (2) evidence statements involving evaluative judgments, such as offering justifications or explicitly evaluating several alternatives. These young children used evidence from purposeful observation to evaluate and revise their emerging models about snails, for instance about their mouthpiece (radula), starting from an observation of deep holes or "tunnels" in the eaten carrots—"they don't eat carrots like we do"—and several months later proposing a mechanism of "spinning and digging" that accounts for the deep holes.

Research with elementary school students arguing about socioscientific issues (Evagorou, 2011) also suggests that younger students are able to improve their argumentation when engaging in discussion and decision making of a socioscientific issue that is relevant to their community, such as environmental problems caused by a nearby pig farm. An implication of socioscientific issues as found in this study, however, relates to how students argue when they identify with the issue they are discussing. Specifically, even though the students were able to appreciate the role of evidence to support their claims and could provide better arguments, they still insisted on their original claims because how they identified with the issue.

The interest of engaging children in long-term projects and teaching sequences is also emphasized in Lehrer and Schauble's (2012) study about evolutionary thinking with elementary pupils from third to sixth grade. For instance, in their project pupils engage in comparative studies, comparing different ponds or the variability of one characteristic of a given organism, such as the height of plants, with the variability of another one, such as the number of seeds. Variability is one of the three interrelated themes that they consider, in their own terms, seeds for evolutionary reasoning. About variability, pupils were required, for instance, to develop a model of process accounting for the distribution observed and evaluate that model against evidence. Lehrer and Schauble consider students' assessment of the qualities of 'convincing evidence' to be participation in the aesthetics of inquiry.

Studies in middle and secondary school have reported students' difficulties in engaging in argumentation. These problems are similar across scientific domains, and here we focus on a few that are frequently cited in biology education research:

- *Recognizing that claim and evidence are distinct epistemological categories*: this distinction is of particular relevance in CEX contexts but is not always drawn.
- *Using evidence to support or challenge claims or to test predictions*: Sandoval and Millwood (2008) found that most 7th grade students did not provide data to back their claims about plant adaptation to the environment; the authors point out that students needed to understand how the photosynthesis-transpiration compromise might affect plant survival in order to cite it as evidence for their claims.
- *Taking a range of evidence into account, providing sufficient evidence*: as opposed to ignoring some observations or evidence, or to attending only to evidence confirming their own hypothesis. In a study with 9th graders (Sandoval & Millwood, 2005), students cited evidence, but often it was not sufficient. In the literature, it is reported that students sometimes use evidence in ways that support their decision and tend to ignore evidence if it contradicts their

decision, for instance, in Evagorou et al.'s (2012) study about whether the UK government should kill grey squirrels in order to protect the native red ones.

- *Coordinating claim with evidence or connecting them through justifications*: students experience difficulties in explaining how a given piece of evidence supports a claim, for instance in the context of natural selection (Sandoval & Millwood, 2005). Engaging in arguments in genetics would need, for some issues, to challenge determinist views that circulate in the media and are widely accepted (Jiménez-Aleixandre, 2014). Causality in genetics, according to contemporary epistemology, should be viewed as sets of interacting causal factors, not in deterministic approaches attributing complex performances only to genes. Thus, as represented in Figure 20.2, there is a set of factors accounting for gene expression, chance, genes, regulation, and environment that can yield a range of potential phenotypes, rather than a single one.

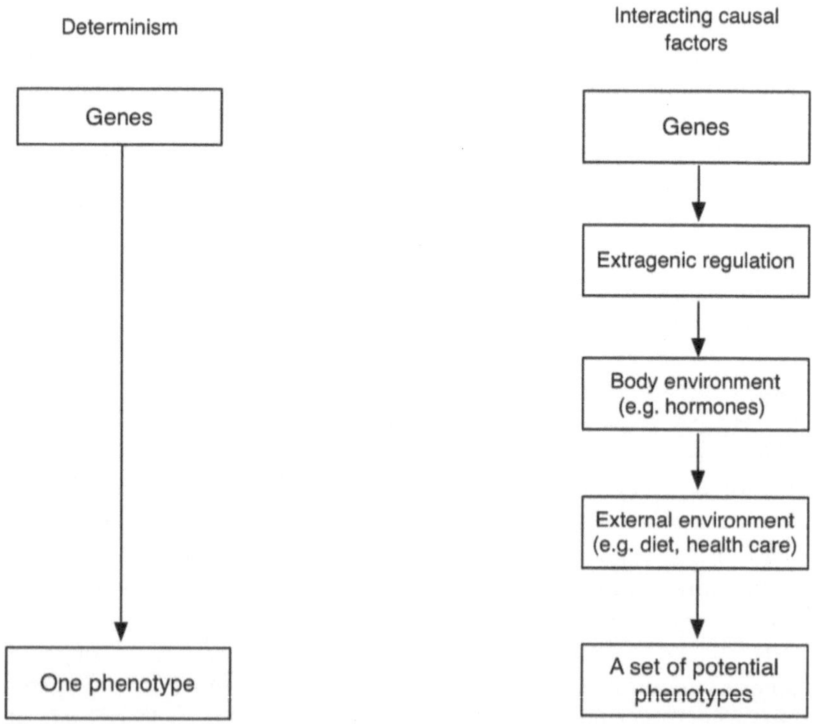

FIGURE 20.2 Two models of causality in gene expression. Determinism (left) and Interacting set of factors resulting in a range of potential phenotypes (right)

- *Metaknowledge about argumentation*: drawing on Schwarz et al.'s (2009) framework for modelling, by metaknowledge we mean understanding the nature and purpose of argumentation, the role of arguments, and the criteria for evaluating them—that is, why arguments are constructed and how they are used. For instance, Evagorou and Osborne (2013) found that one of the dyads they studied engaged in discussion about the structure of their argument and how to communicate it, while the second dyad did not.

This brief overview reveals that while students experience some problems taking part in the scientific practice of argumentation, research shows that given adequate learning environments they engage in sophisticated performances. Two books synthesizing findings from research and recommendations for teaching argumentation are McNeill and Krajcik's (2012) for primary and middle school, and Sampson and Schleigh's (2013) specifically for biology. But why should we integrate argumentation into biology education? Jiménez-Aleixandre and Erduran (2008) proposed contributions from argumentation, such as supporting scientific literacy, critical thinking, and engagement in the reasoning practices of science. For Sampson and Schleigh (2013), argumentation may improve students' understanding of both the nature of science and content knowledge, for instance in genetics; they also point to the improvement of investigative competencies. In the next section, we discuss teaching practices designed to promote argumentation.

Pedagogical Practices and Strategies When Teaching Argumentation

Even though engaging students in argumentation is considered an important aspect of science education (e.g. Achieve, 2013), studies show that students rarely have the opportunity to engage with this scientific practice in the classroom (Roth et al., 1999), because teachers often lack the pedagogical knowledge to engage their students in argumentation (Sampson & Blanchard, 2012; Simon et al., 2006). This is mainly because implementing argumentation in the classroom goes beyond implementing a new curriculum (Evagorou & Dillon, 2011). In order to engage students in the practice of argumentation, a fundamental pedagogical shift may be necessary (Simon et al., 2006; Zohar, 2008): the teacher should shift away from her role as authority towards the role of a facilitator. More specifically, when it comes to argumentation, and particularly in biology, teaching requires a different classroom culture and a different discourse, since argumentation entails shifting teaching from what we know, to how we know what we know, and why we believe what we know (Duschl, 2008).

In order to promote argumentation in the classroom, teachers' subject matter knowledge, their knowledge and understanding of argumentation, and their ability to transform knowledge in teaching activities are important aspects (Dawson & Venville, 2013; Park & Chen, 2012). In the case of biology, teachers need to have established knowledge of biology, but also an understanding of the role and function of argumentation, and specifically of causal explanations, socioscientific issues, and decision-making (Dawson & Venville, 2010; Evagorou & Dillon, 2011). Teachers also need to be aware of the difficulties students have with biology content, and then how to transform these ideas into teaching activities.

Furthermore, as Crawford (2007) concludes in one of her studies, teachers' personal views of teaching science are strong predictors of their actual practice. This is similar to findings from Sampson and Blanchard (2012) in a study exploring science teachers' views and practice about argumentation. In this study, the researchers concluded that teachers' views about the role of argumentation in science, and what a teacher values as important, can influence how argumentation is implemented in the classroom. Hence, an important aspect of preparing biology teachers to teach argumentation will be to provide them with experiences that will change their view of teaching biology into one that includes the understanding of causal explanations and decision making in socioscientific issues, and also an appreciation of the societal and ethical aspects of science (Evagorou & Puig Mauriz, 2017). According to Sampson and Blanchard (2012),

> it is unreasonable to expect science teachers to be able to participate in scientific argumentation or understand how the nature of scientific argumentation and arguments differ from what often takes place in other contexts if they have never had a chance to learn about this complex practice.
>
> *(p. 1143)*

Therefore, what is evident from the aforementioned is that teachers need specific support in teaching argumentation in biology. A number of studies have explored the impact of teacher professional development on teachers' argumentation practices in science. For example, Simon et al. (2006) concluded that a teacher professional development program based on an explicit presentation of an argumentation framework can support the teachers to understand:

> the importance of talking and listening to others, conveying the meaning of argument through modelling and exemplification, positioning oneself within an argument and justifying that position using evidence, constructing and evaluating arguments, exercising counter-argument and debate, and reflecting upon the nature of argumentation.
>
> *(p. 255)*

Additionally, Zembal-Saul (2009) described a pre-service teacher program for elementary school teachers in which the emphasis is on teaching science as argument and supporting students during a methods course to teach argumentation. More specifically, in the science methods course the pre-service teachers experienced what it means to learn science as argument; they observed their mentors teaching using practices consistent with evidence-based explanations and worked with video cases showing argumentation lessons in real classrooms. Through these activities, the pre-service teachers learned about the structure of the argument, which helped them think about the importance of scientific argumentation. Furthermore, findings from McNeill and Krajcik (2008) support the claim that important instructional practices that can help students construct better arguments include discussing the rationale of using evidence-based explanations in science. Along similar lines, other studies suggest that modelling what a good argument is, or providing examples of good arguments, can help students to improve their argumentation (Crawford, 2007). Teachers, however, according to Sampson and Blanchard (2012), are unlikely to incorporate argumentation into their teaching without a great deal of support. The process of incorporating argumentation in teaching becomes more complex when biology teachers are asked to engage their students in argumentation in different discursive contexts, as, for example, in socioscientific issues in which there is a great uncertainty on the underlying knowledge of the topic, and various ethical and societal aspects to discuss (Evagorou, 2011, 2014).

Sampson and Gerbino (2010) and Sampson and Schleigh (2013) describe various techniques to engage students in argumentation in biology:

(a) Engage students in the production and evaluation of scientific arguments.
(b) Design activities or tasks that require students to examine and evaluate alternative theoretical interpretations of a phenomenon.
(c) Ask students to write refutational essays.

Furthermore, Sampson and Gleim (2009) describe argumentation activities in biology that follow the steps: (1) identification of the task, (2) generation of data, (3) production of a tentative argument, (4) interactive argumentation session, (5) creation of a written investigation report, (6) double blind review, (7) revision process, and (8) reflective round table discussion. Sampson and colleagues provide examples of biology activities using these steps in their book *Scientific Argumentation in Biology: 30 Classroom Activities* (Sampson & Schleigh, 2013). Furthermore, these authors provide information on how an argumentation activity differs from a non-argumentation activity for each one of the steps and explain what the role of the teachers is in each step.

There are also a number of research projects that provide activities that promote argumentation in biology contexts. The PROFILES project has developed in collaboration with biology teachers a number of inquiry-based argumentation activities. A published example of PROFILES is presented in Hadjichambis et al. (2016). This is an activity on human reproduction and how it was implemented in the context of the Cypriot curricula. More examples of these activities are available on the project's website and have been developed and implemented by biology teachers themselves (www.cut.ac.cy/profiles/learning_environments_2011_12_EN.html).

FIGURE 20.3 Screenshot from the 3 Parents Lesson on the Engage website

Furthermore, the ENGAGE project (www.engagingscience.eu/en) placed an emphasis on argumentation in biology, especially for socioscientific issues. The research team prepared activities that link biology issues to the news, and the teachers are supported through detailed teaching guides on how to implement these activities (Bayram et al., 2017). An example of an ENGAGE activity is presented in Figure 20.3. This is an activity based on news about the possibility of having a baby with DNA from three parents; students are guided by their teachers to explain whether they agree with this technology or not and explain their arguments. In order to accommodate the needs of novice and expert teachers, there are two different types of argumentation activities on the ENGAGE platform: short tasks (40 minutes) that the teachers can use as evaluation or introductory activities, and longer investigation (3–4 sessions) in which the students have to collect evidence to support their arguments. All activities are supported by detailed lesson plans and worksheets.

Finally, the PreSEES project (www.ssieurope.net) provides a detailed framework and teaching materials on how to support teachers to teach socioscientific argumentation in science- and biology-related issues (Evagorou, 2014). Table 20.1 presents the framework, based on four pillars:

(a) *Contextualization*: The problem is set within an authentic context.
(b) *Discursive activity*: Teachers discuss the problem that was set in the previous tasks in their groups and engage in argumentation.
(c) *Elaboration*: Making connections between scientific knowledge and the knowledge of teaching (pedagogical knowledge).
(d) *Further activity*: Additional readings or online readings, discussions, and reflections.

TABLE 20.1 Framework designed to be used as guideline for teacher professional development in SSI argumentation (PreSEES project)

Modules	*Module 1: Nature of SSI*	*Module 2: Teaching SSI*	*Module 3: Assessment of SSI learning*
Contextualisation	Introducing the SSI through global warming	Introducing the teaching SSI argumentation through eating insects	Introducing the SSI assessment through current issues
Discursive Activity	Reflection on nature of SSI and argumentation	Reflection on issues of teaching SSI argumentation	Reflection on assessment
Elaboration	Selecting/Connecting and justifying a curricular topic as SSI argumentation	Planning and designing materials to teach SSI argumentation	Designing materials to assess SSI learning
Further Work	Reading research on SSI argumentation	Sharing SSI designs	Sharing SSI assessment designs

New Trends Relating to Argumentation in Biology and SSIs

Argumentation in science and in biology has been extensively studied over the last 20 years and research studies provide us with examples on how to train teachers and how to support students to engage in this scientific practice. However, recent trends in Europe are moving towards Responsible Research and Innovation (RRI), or responsible citizenship, which is an approach that anticipates and assesses potential implications and societal expectations with regard to research and innovation, with the aim to foster the design of inclusive and sustainable research and innovation (Commission, 2015).

The main emphasis of responsible citizenship is on science for and with society (Owen, et al., 2012), and it therefore highlights the importance of presenting the societal aspects of science to students. This approach is directly associated with argumentation and is especially relevant to biology. Many of the new scientific discoveries for which the students and the public need to be informed are linked to biology (e.g. growing artificial body parts, baby engineering), and in order to present and discuss these SSIs to the class, biology teachers should first familiarize themselves with socioscientific decision making and causal explanations before attempting to teach them. With RRI being a new concept in Europe, numerous researchers link RRI to SSIs and responsible citizenship (Owen et al., 2012). Therefore, the main idea behind RRI and SSIs, both of which are emphasised by new curricula in Europe, is that by including issues of social relevance in science learning and teaching, we could move science classes towards unwrapping and engaging in discussions about the intersections of science and society and promote scientific practices. Therefore, the inclusion of socioscientific issues in the curriculum offers a means of expanding both the curriculum and the range of instructional practices commonly experienced in the school science classroom (Evagorou & Puig, 2017).

Acknowledgements

M. P. Jiménez-Aleixandre's work is supported by the Spanish Ministerio de Economía y Competitividad (MINECO), contract grant number: EDU2015–66643-C2–2-P.

References

Achieve. (2013). *Next Generation Science Standards: For States, by States.* Washington, DC: National Academies Press.

Bayram, D., Henze, I., Evagorou, M., Shwartz, Y., Aschim, E., Alcaraz-Dominguez, S. & Barajas, M. (2017). Exploring the impact of educative materials on teachers' pedagogical content knowledge. In *ESERA 2017.* Dublin: ESERA.

Bravo-Torija, B. & Jiménez-Aleixandre, M. P. (2017). Developing an initial learning progression for the use of evidence in decision-making contexts. *International Journal of Science and Mathematic Education.* doi:10.1007/s10763-017-9803-9

Commission, E. (2015). *Science Education for Responsible Citizenship.* Luxembourg: Publications Office of the European Union.

Crawford, B. A. (2007). Learning to teach science as inquiry in the rough and tumble of practice. *Journal of Research in Science Teaching,* 44(4), 613–642.

Dawson, V. & Venville, G. (2010). Teaching strategies for developing students' argumentation skills about socioscientific issues in high school genetics. *Research in Science Education,* 40(2), 133–148.

Dawson, V. & Venville, G. (2013). Introducing high school biology students to argumentation about socioscientific issues. *Canadian Journal of Science, Mathematics and Technology Education,* 13(4), 356–372.

Duschl, R. (2008). Quality argumentation and epistemic criteria. In S. Erduran & M. Jimenez-Aleixandre (Eds.), *Argumentation in Science Education: Perspectives from Classroom-Based Research.* Dordrecht, the Netherlands: Springer.

Evagorou, M. (2011). Discussing a socioscientific issue in a primary school classroom: the case of using a technology-supported environment in formal and nonformal settings. In S. Troy (Ed.), *Socio-Scientific Issues in the Classroom.* Dordrecht: Springer, 133–159.

Evagorou, M. (2014). Preparing pre-service science teachers to teach socio-scientific (SSI) argumentation. *Science Teacher Education,* 69, 39–49.

Evagorou, M. & Dillon, J. (2011). Argumentation in the teaching of science. In D. Corrigan, J. Dillon & R. Gunstone (Eds.), *The Professional Knowledge Base of Science Teaching* Dordrecht: Springer, 189–203.

Evagorou, M., Jiménez-Aleixandre, M. P. & Osborne, J. (2012). 'Should we kill the grey squirrels?' A study exploring students' justifications and decision-making. *International Journal of Science Education,* 34(3), 401–428.

Evagorou, M. & Puig Mauriz, B. (2017). Engaging elementary school pre-service teachers in modeling a socioscientific issue as a way to help them appreciate the social aspects of science. *International Journal of Education in Mathematics Science and Technology*, 5(2), 113–123.

Evagorou, M. & Osborne, J. (2013). Exploring young students' collaborative argumentation within a socioscientific issue. *Journal of Research in Science Teaching*, 50(2), 209–237.

Hadjichambis, A. C., Georgiou, Y., Paraskeva-Hadjichambi, D., Kyza, E. A. & Mappouras, D. (2016). Investigating the effectiveness of an inquiry-based intervention on human reproduction in relation to students' gender, prior knowledge and motivation for learning in biology. *Journal of Biological Education*, 50(3), 261–274.

Jiménez Aleixandre, M. P. & Brocos, P. (2018). Shifts in epistemic statuses in argumentation and in conceptual change. In T. Amin & O. Levrini (Eds.), *Converging Perspectives on Conceptual Change: Mapping an Emerging Paradigm in the Learning Sciences*. Oxford: Routledge, 171–179.

Jiménez-Aleixandre, M. P. (2014). Determinism and underdetermination in genetics: implications for students' engagement in argumentation and epistemic practices. *Science & Education*, 23(2), 465–484.

Jiménez-Aleixandre, M. P. & Crujeiras, B. (2017). Epistemic practices and scientific practices in science education. In B. Akpan & K. Taber (Eds.), *Science Education: An International Comprehensive Course Companion*. Rotterdam: Sense, 69–80.

Jiménez-Aleixandre, M. P. & Erduran, S. (2008). Argumentation in science education: an overview. In S. Erduran & M. P. Jiménez-Aleixandre (Eds.), *Argumentation in Science Education: Perspectives from Classroom-Based Research*. Dordrecht: Springer, 3–27.

Jiménez-Aleixandre, M. P., Puig, B., Bravo, B. & Crujeiras, B. (2014). The role of discursive contexts in argumentation. Paper presented at the National Association of Research in Science Teaching (NARST) Annual Meeting. Pittsburgh, PA.

Kelly, G. J. & Takao, A. (2002). Epistemic levels in argument: an analysis of university oceanography students' use of evidence in writing. *Science Education*, 86(3), 314–342.

Lehrer, R. & Schauble, L. (2012). Supporting inquiry about the foundations of evolutionary thinking in the elementary grades. In S. M. Carver & J. Shrager (Eds.), *The Journey from Child to Scientist*. Washington: American Psychological Association, 171–205.

McNeill, K. L. & Krajcik, J. (2008). Scientific explanations: characterizing and evaluating the effects of teachers' instructional practices on student learning. *Journal of Research in Science Teaching*, 45(1), 53–78.

McNeill, K. L. & Krajcik, J. (2012). *Supporting Grade 5–8 Students in Constructing Explanations in Science. The Claim, Evidence and Reasoning Framework for Talk and Writing*. Boston: Pearson.

McNeill, K. L. & Pimentel, D. S. (2010). Scientific discourse in three urban classrooms: the role of the teacher in engaging high school students in argumentation. *Science Education*, 94(2), 203–229.

Monteira, S. F. & Jiménez-Aleixandre, M. P. (2016). The practice of using evidence in kindergarten: the role of purposeful observation. *Journal of Research in Science Teaching*, 53(8), 1232–1258.

Osborne, J. (2014). Scientific practices and inquiry in the science classroom. In N. G. Lederman & S. K. Abell (Eds.), *Handbook of Research on Science Education*. New York: Routledge. Vol II, 579–599.

Osborne, J., & Patterson, A. (2012). Authors' response to 'For whom is argument and explanation a necessary distinction? A response to Osborne and Patterson' by Berland and McNeill. *Science Education*, 96(5), 814–817.

Owen, R., Macnaghten, P. & Stilgoe, J. (2012). Responsible research and innovation: from science in society to science for society, with society. *Science and Public Policy*, 39(6), 751–760.

Park, S. & Chen, Y.-C. (2012). Mapping out the integration of the components of pedagogical content knowledge (PCK): examples from high school biology classrooms. *Journal of Research in Science Teaching*, 49(7), 922–941.

Perelman, C. & Olbrechts-Tyteca, L. (1958). *Traité de L'argumentation. La nouvelle rhétorique.* Bruxelles: Éditions de l'Université de Bruxelles. (*The New Rhetoric: A Treatise on Argumentation.* Notre Dame: University of Notre Dame Press, 1969.)

Plantin, C. (2011). Les Bonnes Raisons des Emotions. (*The good reasons of emotions.* Bern: Peter Lang.)

Roth, K. J., Druker, S. L., Garnier, H. E., Lemmens, M., Chen, C., Kawanaka, T., . . . Schneider, M. (1999). *Teaching Science in Five Countries: Results From the TIMSS 1999 Video Study Statistical Analysis Report.* Retrieved from https://nces.ed.gov/pubs2006/2006011.pdf

Sampson, V. & Blanchard, M. R. (2012). Science teachers and scientific argumentation: trends in views and practice. *Journal of Research in Science Teaching*, 49(9), 1122–1148.

Sampson, V. & Clark, D. B. (2008). Assessment of the ways students generate arguments in science education: current perspectives and recommendations for future directions. *Science Education*, 92(3), 447–472.

Sampson, V. & Gerbino, F. (2010). Two instructional models that teachers can use to promote and support scientific argumentation in the biology classroom. *The American Biology Teacher*, 72(7), 427–431.

Sampson, V. & Gleim, L. (2009). Argument-driven inquiry to promote the understanding of important concepts and practices in Biology. *The American Biology Teacher*, 71(8), 465–472.

Sampson, V. & Schleigh, S. (2013). *Scientific Argumentation in Biology: 30 Classroom Activities.* Arlington, Virginia: NSTA Press.

Sandoval, W. A. & Millwood, K. A. (2005). The quality of students' use of evidence in written scientific explanations. *Cognition and Instruction*, 23(1), 23–55.

Sandoval, W. A. & Millwood, K. A. (2008). What can argumentation tell us about epistemology? In S. Erduran & M. P. Jiménez-Aleixandre (Eds.), *Argumentation in Science Education: Perspectives from Classroom-Based Research.* Dordrecht: Springer, 71–88.

Schwarz, C. V., Reiser, B. J., Davis, E. A., Kenyon, L., Acher, A., Fortus, D., Shwartz, Y., Hug, B. & Krajcik, J. (2009). Developing a learning progression for scientific modelling: making scientific modelling accessible and meaningful for learners. *Journal of Research in Science Teaching*, 46(6), 632–654.

Simon, S., Erduran, S. & Osborne, J. (2006). Learning to teach argumentation: research and development in the science classroom. *International Journal of Science Education*, 28(2–3), 235–260.

Toulmin, S. (1958). *The Uses of Argument.* Cambridge: Cambridge University Press.

Zembal-Saul, C. (2009). Learning to teach elementary school science as argument. *Science Education*, 93(4), 687–719.

Zohar, A. (2008). Science teacher education and professional development in argumentation. In S. Erduran & M. Jimenez-Aleixandre (Eds.), *Argumentation in Science Education: Perspectives from Classroom-Based Research.* Dordrecht, the Netherlands: Springer, 245-268.

Zohar, A. & Nemet, F. (2002). Fostering students' knowledge and argumentation skills through dilemmas in human genetics. *Journal of Research in Science Teaching*, 39(1), 35–62.

21

WORLDVIEWS IN BIOLOGY EDUCATION

Michael J. Reiss

What Is A Worldview?

In *World Views: From Fragmentation to Integration*, Diederik Aerts et al. (1994) stated that "a world view is a coherent collection of concepts and theorems that must allow us to construct a global image of the world, and in this way to understand as many elements of our experience as possible" (p. 17). In an oft-cited paper, Bill Cobern (1996) pointed out that science educators generally assume that if students can see that scientifically orthodox conceptions are more intelligible, plausible and fruitful than other conceptions, they will come to accept these scientific conceptions. Cobern described this as a 'rationalistic view'. He went on to point out that the notion of worldviews provides a different way of understanding why people hold the views that they do. The word itself has quite a long history and, as with almost any important term, has shifted in meaning over time and exists in more than one version—in particular, some people, including Cobern himself, are not keen on the term 'scientific worldview'.

In science education, the notion of worldviews is increasingly being employed. For example, in the edited volume *Science, Worldviews and Education* (Matthews, 2009), a number of philosophers, scientists and science educators use the thinking behind worldviews to explore a range of issues, including whether science itself is a worldview and whether science can test supernatural worldviews. The term has recently been explored as a way of helping conceptualise why, despite the best efforts of many science educators, so few students leave their schooling with the sort of scientific understanding and disposition that most science teachers wish they had. The principal conclusion is that school science fails to enable most students to see the world from a scientific perspective (Reiss, 2011).

At its simplest, a worldview is a reasonably coherent way of understanding ('viewing') reality ('the world'). In science education, the term, as I discuss below, has been most extensively used in the context of evolution, where creationists can be said to hold a particular version of a religious worldview (a creationist worldview) that sits in opposition to the standard scientific understanding. I go on to argue that it can be useful to biology educators to consider contexts other than evolution where the notion of worldviews may be employed.

Evolution

The concept of worldviews can be illustrated by the ways in which science and religion attempt to explain the history of life and today's biodiversity.

The scientific understanding of these topics is far from complete, but the narrative is a powerful one (Reiss, 2009). Around 3.8 billion years ago, life evolved on Earth. Rather little is known with any great confidence about this early history, far less than is known, for example, about how stars form, grow and die. By the time of the earliest fossils, life was unicellular and bacteria-like. Over the ages, natural processes such as natural selection, genetic drift, etc. eventually resulted in the ten million or so species, including our own, that we find today.

The key point is that the scientific worldview associated with this standard scientific understanding is materialistic in the sense that it is neither idealistic nor admits of non-physical, and thus supernatural, explanations (here, 'physical' includes such things as energy and the curvature of space). There is much about the history of life on Earth that remains unknown. How did the earliest self-replicating molecules arise? What caused membranes to exist? How key were the earliest physical conditions—temperature, the occurrence of water and so forth? But the scientific presumption is that either these questions will be answered by science or they will remain unknown. Although some scientists might (sometimes grudgingly) admit that science cannot disprove supernatural explanations, scientists do not employ such explanations in their work (the tiny handful of exceptions only attest to the strength of the general rule).

Religious understandings of the history of life and today's biodiversity are more diverse. Many religious believers, including scientists (e.g. Ecklund, 2012), are perfectly comfortable with the scientific understanding, either on its own or accompanied by a belief that evolution in some sense takes place within God's holding, whether or not God is presumed to have intervened or acted providentially at certain key points (e.g. the origin of life or the evolution of humanity). But many other religious believers adopt a more creationist perspective. Creationism exists in a number of different versions but about 40% of adults in the USA, fewer in other Western countries, believe that the Earth came into existence as described in the early parts of the Bible or the Qur'an and that the most that evolution has done is to change species into closely related species (Miller et al., 2006). For a creationist, it is perfectly possible that the various species of deer had a common ancestor, but this is not the case for deer, bears and squirrels—still less for monkeys and humans, for birds and reptiles or for fish and fir trees.

In this sense creationism can usefully be considered an alternative worldview. One of the key points about worldviews is that it can be difficult for someone who has never inhabited a particular worldview to imagine what it is really like to do so. In Christianity, a creationist is also likely to think that the world will soon be coming to an end, very possibly in their lifetime, that there is a world to come and that the eternal fate of each and every one of us in that world depends on whether or not in this world we accept the teachings of scripture. If you are not a creationist and want to get more of an idea as to how many creationists see the world, try some of the 16 best-selling religious novels by Tim LaHaye and Jerry B. Jenkins that constitute the *Left Behind* series (e.g. LaHaye & Jenkins, 1995).

Climate Change

Evolution is not, of course, the only topic about which people hold strong views that lead them to hold a position that runs counter to the standard scientific one. I well remember one Australian biology educator telling me that I may once I have lost my job because of a controversy around evolution education (Baker, 2010) but a controversy around climate change education had caused her to receive death threats.

The standard scientific account around climate change is that while the climate has changed very considerably over the course of the Earth's history for reasons that have nothing to do with humans, many of the recent, geologically very rapid changes (higher temperatures, rising sea levels, more extreme weather events) are, at least in large measure, anthropogenic—i.e. the result of human activity, chief amongst which is the burning of fossil fuels (principally, coal, oil, gas). So great is the impact of humans on the rest of the world (not only changes to the climate but species extinction, habitat loss and the way in which more and more of the Earth's surface is being used for agriculture and other human activities) that a growing number of scientists, including many geologists, have called for a formal recognition that we are living in the 'Anthropocene'—an epoch variously argued to have started anywhere from the beginning of the Agricultural Revolution (about 12,000 years ago) to the mid-twentieth century.

It is important to emphasise that the consensus among scientists over the role of human activity in climate change is not as great as is the consensus among scientists as to the veracity of the standard scientific understanding of evolution. Nevertheless, almost with each passing year, the consensus among climate scientists is growing. However, as is well known, there are many people who passionately disagree with this consensus. The reasons for this are more varied than are the reasons why many people do not accept the standard scientific account of religion. Religion features, but there is more to it than this. In the USA, Democrats are substantially more likely than Republicans to acknowledge anthropogenic global climate change (Mayer et al., 2013). Stevenson et al. (2014) propose a path model in which climate change knowledge and worldview jointly predict acceptance of anthropogenic global warming, which in turn predicts climate change risk perception.

Use of Animals

People vary greatly in how they consider that animals (a shorthand for 'non-human animals') can legitimately be used by humans. Most school biology teachers, if they ever allow the issue to be discussed in their lessons, will be aware that there is likely to be quite a diversity of opinion. At one pole are those students, quite often in a majority, who feel, for reasons they find hard to articulate, that we have a right to use animals, within reason, for human ends. The phrase 'within reason' is typically used to mean 'without causing unnecessary harm', so that there is a presumption that if, for example, we kill animals for food, any associated pain should not be too great and should not last too long. However, when it comes to using animals in medical experiments, while obviously minimisation of suffering is desirable, such students are likely to countenance some animal suffering provided that the benefits, or potential benefits, to humans are considerable.

At the other pole are students who either reject or are far more hesitant about almost any use of animals for human ends. Such students may be vegetarians or vegans, may refuse to wear leather and may even, though this is rarer, reject the use of animals as companion animals (e.g. pets and guide dogs) on the grounds that this is a form of slavery. Perhaps more relevantly for biology educators, they are likely to reject the use of animals for dissection (Reiss, 2017).

The use of 'worldview' seems to be useful in such instances. There is considerable coherence in terms of how those opposed to the use of animals for human ends put their beliefs into practice and, although I am unaware of any systematic research, my experience, and I suspect that of many other readers, is that while such views, as with any views, may change, they often persist for decades and can be articulated and defended coherently and with conviction.

Modern Biotechnology

To the surprise of many industrialists and scientists, there has been considerable resistance to the use of modern biotechnology in agriculture and medicine. In particular, many people are opposed to the use of genetically modified organisms (GMOs) in agriculture, and some people, notably many Roman Catholics, are opposed to any medical procedures that rely on embryo experimentation or separate the act of sexual intercourse from the possibility of parenthood.

Reasons for objections to the use of GMOs are various, but a common complaint is that such organisms are 'unnatural' (Reiss & Straughan, 1996). To a biologist, this objection may not sound very convincing. After all, what is 'natural' in agriculture? All our crops are very different from their ancestral plants. Cereal crops, for example, are typically much shorter, retain their seeds when ripe, are genetically extremely uniform and have far higher yields. Furthermore, the technologies for making GMOs increasingly offer greater precision than is provided by conventional crop breeding, and the crops themselves can sometimes be better for the environment (e.g. in terms of pesticide usage) than are their conventional equivalents.

However, this rationalistic argument reveals the power of worldviews thinking in that it may miss the point. Someone who holds that GMOs are 'unnatural' may well not be swayed by such reasoning. Indeed, my experience of some 25 years of public consultations and debates about GMOs is that scientific evidence one way or the other rarely seems to change the opinions of those who have what they regard as *principled* objections to GMOs.

Something similar is going on in respect of Roman Catholic objections to IVF (*in vitro fertilisation*). As is well known, IVF has allowed millions of babies to be born who would not otherwise have existed. At the same time, the procedure is expensive, often unsuccessful and not infrequently painful for the woman concerned. However, the Roman Catholic disagreement with IVF is somewhat different.

First of all, the Roman Catholic doctrine about conception is such that it is held that anything that interferes with the possibility that sexual intercourse can lead to conception is wrong. This, of course, is the reason why any form of contraception other than the rhythm method is considered unacceptable. As the Catechism puts it:

> Techniques involving only the married couple (homologous artificial insemination and fertilization) . . . remain morally unacceptable. They dissociate the sexual act from the procreative act. The act which brings the child into existence is no longer an act by which two persons give themselves to one another, but one that entrusts the life and identity of the embryo into the power of doctors and biologists and establishes the domination of technology over the origin and destiny of the human person. Such a relationship of domination is in itself contrary to the dignity and equality that must be common to parents and children. Under the moral aspect procreation is deprived of its proper perfection when it is not willed as the fruit of the conjugal act, that is to say, of the specific act of the spouses' union . . . Only respect for the link between the meanings of the conjugal act and respect for the unity of the human being make possible procreation in conformity with the dignity of the person.
>
> *(Vatican, 2017, para 2377)*

In addition, IVF almost inevitably results in the destruction of human embryos and, given the teaching of the Roman Catholic Church that human life begins at conception, this is held to be unacceptable.

It is important to note that questions about moral acceptability, human dignity and the point at which human life begins are not questions that can be answered by the use of science alone. In other words, no science education is going to be able to prove that the Roman Catholic position on IVF or embryo research is right or wrong. However, as I discuss below in the section on pedagogical implications, this does not mean that science education has no role to play.

Vaccination

Vaccination is frequently presented in school biology textbooks as an unproblematic success story. After an account of Edward Jenner's iconic 1796 experiment on 8-year-old Edward Phipps (sometimes with a nod to Lady Mary Wortley Montagu's earlier introduction of the Ottoman practice of 'variolation'—inoculation against smallpox—on her return from Turkey), graphs are presented showing dramatic decreases in the incidence of such diseases as smallpox and polio.

However, objections to vaccination began almost as soon as the practice was introduced. A famous 1802 cartoon by Gillray is sometimes reproduced in school textbooks (Figure 21.1). Nineteenth-century objections to vaccinations included arguments that they didn't work and were unsafe (Ernst & Jacobs, 2012) or that

FIGURE 21.1 Gillray's 1802 satirical anti-vaccination cartoon. Downloaded from Wikimedia Commons https://upload.wikimedia.org/wikipedia/commons/d/d6/The_cow_pock.jpg

their compulsory introduction (e.g. the 1853 Compulsory Vaccination Act in the UK) violated personal liberties (Durbach, 2000).

To this day, objections to vaccinations can exist for a number of reasons. In addition, still, to concerns about safety or the over-ruling of individual liberties, a distinctive issue about vaccinations is that the benefits from them can be obtained by individuals who have not been vaccinated providing enough other individuals have been. Parents who refuse to have their children vaccinated are therefore sometimes accused of selfishness.

It is perhaps questionable whether the concept of 'worldviews' applies to objections to vaccinations in quite the way that is may apply to the other issues considered above. In any event, it is important to realise that objections to vaccinations may not result from misconceptions but from deeply held views to do with individual autonomy and the limits of state authority.

Indigenous Knowledge

A final example that I will mention is so-called 'indigenous knowledge'. There is a growing literature on what is sometimes referred to as 'multicultural science education' (e.g. Reiss, 1993; Cobern & Loving, 2001; Aikenhead & Michell, 2011). Three main reasons are usually given as to why indigenous knowledge should be valued and taken seriously in school science lessons. First, it can motivate students from certain ethnic backgrounds in their study of science, who otherwise conclude that only Western science is worthwhile. Secondly, such knowledge gives a better historical account of how science develops. Thirdly, such knowledge retains a value even today, generally not so much in the claims it makes to be better science (understood narrowly and epistemologically) than Western science but in the claims that it makes to be better with respect, for example, to the way it may value non-human interests or cultural traditions.

However, such indigenous knowledge may conflict with much standard Western science. The approach of worldviews can therefore be useful in helping educators appreciate what it can be like for indigenous students when faced with this conflict.

Conclusions and Pedagogical Implications

So, what are the pedagogical implications of taking worldviews in biology education seriously? Perhaps the first thing to say is that whatever the precise subject matter or age range of a class, there is much to be said for a teacher bearing in mind that for some students, these issues are likely to be sensitive ones. It has been suggested to me that to describe an issue as 'sensitive' is to be somewhat disparaging. This, of course, is not my intention. Topics like death, sexuality and child abuse are examples of issues that are sensitive for many students, and many teachers are used to being respectful to students when dealing with such issues.

An advantage, if one thinks of evolution education, of shifting the discourse from controversy to sensitivity is that one shifts the focus from epistemology to pedagogy. One can be sensitive with someone in respect of an issue without implying that one shares the same perspective (or worldview) as the person to whom one is being respectful and considerate. Explicitly accepting the teaching of evolution as controversial, on the other hand, will be difficult for many science teachers as the distinction between this and accepting evolution as controversial is a fine one and many science teachers are likely to see it as selling out to creationists (cf. Hermann, 2008).

Of course, treating a topic as sensitive does not absolve teachers from the expectation that they teach the topic with rigour and with high expectations for student learning. Evolution is consistently presented in creationist books and articles as illogical (e.g. natural selection cannot, on account of the second law of thermodynamics, create order out of disorder; mutations are always deleterious and so cannot lead to improvements), contradicted by the scientific evidence (e.g. the fossil record shows human footprints alongside animals supposed by evolutionists to be long extinct; the fossil record does not provide evidence for transitional forms) and the product of non-scientific reasoning (e.g. the early history of life would require life to arise from inorganic matter—a form of spontaneous generation rejected by science in the nineteenth century; radioactive dating makes assumptions about the constancy of natural processes over aeons of time whereas we increasingly know of natural processes that affect the rate of radioactive decay). Similarly, students who object to the use of animals by humans may cite arguments to do with physiology (fundamental similarities between the nervous systems of different species, changes in blood cortisol levels in animals that are stressed), behaviour (the recoil that many animals show from noxious stimuli) and social structure (whether in pair-bonded swans or in groups of cetaceans). Responding to such objections makes high demands on teachers who may need to expand their own knowledge.

But what precisely should be the *aims* of such teaching? One possibility is to attempt to convince a student who does not accept the theory of evolution, the reality of anthropogenic climate change or the utility of vaccination (for example) that they are wrong and should change their mind. This approach makes considerable sense if we view students as possessing misconceptions. Indeed, many chapters in this volume contain helpful suggestions as to how students' biological misconceptions can be countered.

However, the argument I would like to advance is somewhat different. It is that when we teach about issues, such as evolution, climate change and new biotechnologies, our aim should not be to persuade students to *accept* the standard scientific position (Smith, 1994; Williams, 2015) but to *understand* it. There are two main reasons for this—one principled, the other pragmatic.

The principled reason is that a pedagogical model based on understanding rather than acceptance is more respectful of students. It entails a teacher doing what she can to ensure that a student fully appreciates a position (the scientific consensus) while recognising that the student is unlikely, certainly in the short term, to change their mind. The pragmatic reason flows from the observation that students often actively resist attempts to get them to abandon their own views—views they may share with close family members and neighbours—and accept the standard scientific view. In such circumstances, it may be sensible simply to strive for understanding.

However, students can and do change their mind as a result of the teaching they receive, even when the teaching is in such contentious areas (Reiss, 2014). One way of interpreting a move from a position where the scientific viewpoint is actively rejected to one where it is accepted is to see it as an example of Thomas Kuhn's well-known contention that there are times when existing understandings break down—such as the Ptolemaic model of the structure of the solar system (in which the Earth is at the centre) or the Newtonian understanding of motion and gravity—and a completely new *paradigm* takes over (Kuhn, 1970). At the time of such a crisis, a scientific revolution happens during which a new paradigm—such as the Copernican model of the structure of the solar system or Einstein's theory of relativity—begins to replace the previously accepted paradigm. Kuhn likens the switch from one paradigm to another to a *gestalt* switch (when we suddenly see something in a new way) or even a religious conversion. It alerts us to the fact that scientific understanding may change for reasons that cannot entirely be reduced to the rational acceptance of new information. How much more is this true for science education and for learning in general.

The science educator Lee Meadows is one who has made the journey from a position where one sees creationism as true to one where one sees the evolutionary understanding of life as valid:

> I know well this tension between the heart and the mind because I've lived it. I was raised in a Christian fundamentalist home and church, and I'm now a science teacher and educator. Working through this tension was a perspective I brought to the Hearts and Minds study. My own faith journey has led me away from fundamentalism, but I do still hold to the view that the Christian scriptures are the inspired words of God. I find truth in both worldviews. Science provides truth from the basis of evidence, but my faith also provides an intellectual, durable system of knowing the world.
>
> *(Meadows, 2007, p. 149)*

Recent research has shown that good quality teaching can help students from a fundamentalist position to learn about evolution and come to accept some of evolution's central tenets. David Long (2011) has undertaken anthropological research into the ways that university students interact with the concept of evolution during their introductory biology classes. He concluded that a few lessons on the topic are most unlikely to change the views of students of a fundamentalist religious persuasion.

However, other research suggests that careful and respectful teaching about evolution can make students who initially reject the theory of evolution considerably more likely to accept at least some aspects of it. Winslow et al. (2011) studied 15 biology-related majors or recent biology-related graduates at a mid-western Christian liberal arts university in the USA. The study found that most participants were raised by their families to believe in creationism but came to accept evolution through evaluating the evidence for it, negotiating the meanings of *Genesis*, recognising evolution as a non-salvation issue and observing the teachers as Christian role models who accept evolution. For example, one student reported:

> They can say they're Christian and be an evolutionist, but it would really help for Dr. [Origins professor] because she actually showed you. She'd talk about God and . . . then she also talked about evolution and so you kinda had to reconcile the two . . . My whole life it was just two things that were separate and they must stay separate, but with her they kinda came together and you had to reconcile them.
>
> *(Winslow et al., 2011, p. 1036)*

This story indicates the importance of teachers who do not separate out evolution and religion or see them in conflict. This is also clear from another project carried out by Lee Meadows involving 12 teachers in US Christian schools who explicitly attempted to teach evolutionary biology and also to nurture their students' religious faith. He found:

> Participating teachers refused to require their students to accept a single view of origins, just as adult Christians across America don't have a single view. Instead, they asked their students to understand scientific evidence, consider Biblical interpretation, and think for themselves. So often, evolution education seems to drive to a single point, whether in faith-based schools teaching a single creationist view or in a public college classroom where a professor advocates scientific materialism. Participating teachers refused to tie themselves to such agendas, and their trust in their students to think and make good decisions serves as a model for evolution educators in all school types, not just faith-based. Participating teachers appeared to have found a powerful key for diminishing controversy during the teaching of evolution by simply acknowledging what almost any teacher of teenagers already knows: teenagers make their own decisions. Students consider what their teachers say, but they come to their own conclusions, despite pressure from teachers or other adults with an agenda. These teachers lifted the reality of their students' decision making into the overt learning climate of their classrooms.

> Participating teachers were passionate about teaching evolution well to their students. Their passion makes clear the profound importance of caring, skilled teachers when students from faith backgrounds are learning evolution. These teachers' words remind us how learning evolution is a matter of the heart for many students, with profound and often eternal implications in students' minds. These teachers' practices give us a good image of how teaching evolution and caring for children's souls can be intertwined successfully.
>
> *(Meadows, 2014, pp. 239–240)*

Comparable findings are now beginning to emerge in other studies. For example, Barnes and Brownell (2017), in a study of the experiences and practices of 32 evolution instructors at Christian universities in the USA, found that such instructors were aware of their students' struggles with evolution, considered acceptance of evolution a goal of their instruction and said that their own experiences in reconciling their religious beliefs with evolution informed their instructional practices.

The concept of worldviews can be used to help understand why some students refuse to accept the standard scientific version of certain biology topics. Treating such worldviews as misconceptions that can straightforwardly be corrected can prove unsuccessful as a teaching strategy. Respecting students and seeking to help them understand, rather than accept, the scientific position may be more productive.

References

Aerts, D., Apostel, L., De Moor, B., Hellemans, S. Maex, E., Van Belle, H. & Van der Veken, J. (1994). *World Views: From Fragmentation to Integration*. Brussels: VUB Press.

Aikenhead, G. & Michell, H. (2011). *Bridging Cultures: Indigenous and Scientific Ways of Knowing Nature*. Ontario: Pearson.

Baker, S. (2010). Creationism in the classroom: a controversy with serious consequences. *Research in Education*, 83, 78–88.

Barnes, M. E. & Brownell, S. E. (2017). Experiences and practices of evolution instructors at Christian universities that can inform culturally competent evolution education. *Science Education*. doi:10.1002/sce.21317

Cobern, W. W. (1996). Worldview theory and conceptual change in science education. *Science Education*, 80, 579–610.

Cobern, W. W. & Loving, C. C. (2001). Defining "science" in a multicultural world: implications for science education. *Science Education*, 85, 50–67.

Durbach, N. (2000). They might as well brand us: working class resistance to compulsory vaccination in Victorian England. *The Society for the Social History of Medicine*, 13(1), 45–62.

Ecklund, E. H. (2012). *Science vs. Religion: What Scientists Really Think*. Oxford: Oxford University Press.

Ernst, K. & Jacobs, E. T. (2012). Implications of philosophical and personal belief exemptions on re-emergence of vaccine-preventable disease: the role of spatial clustering in under-vaccination. *Human Vaccines & Immunotherapeutics*, 8(6), 838–841.

Hermann, R. S. (2008). Evolution as a controversial issue: a review of instructional approaches. *Science & Education*, 17, 1011–1032.

Kuhn, T. S. (1970). *The Structure of Scientific Revolutions*, 2nd edn. Chicago, IL: University of Chicago Press.

LaHaye, T. & Jenkins, J. B. (1995). *Left Behind: A Novel of the Earth's Last Days*. Carol Stream, IL: Tyndale House.

Long, D. E. (2011). *Evolution and Religion in American Education: An Ethnography*. Dordrecht: Springer.

Matthews, M. R. (Ed.) (2009). *Science, Worldviews and Education. Reprinted from the Journal Science & Education*. Dordrecht: Springer.

Mayer, F., Adair, S. & Pfaff, A. (2013). *Americans Think the Climate is Changing and Support some Actions*. Durham, NC: Duke Nicholas Institute for Environmental Policy Solutions. Available at https://nicholasinstitute.duke.edu/sites/default/files/publications/ni_pb_13-01_0.pdf.

Meadows, L. (2007). Approaching the conflict between religion and evolution. In L. Jones & M. J. Reiss (Eds.), *Teaching about Scientific Origins: Taking Account of Creationism*. New York: Peter Lang, 145–157.

Meadows, L. (2014). Shepherding and strength: teaching evolution in American Christian schools. In J. Chapman, S. McNamara, M. J. Reiss & Y. Waghid (Eds.), *International Handbook of Learning, Teaching and Leading in Faith-based Schools*. Dordrecht: Springer, 223–240.

Miller, J., Scott, E. & Okamoto, S. (2006). Public acceptance of evolution. *Science*, 313, 765–766.

Reiss, M. J. (1993). *Science Education for a Pluralist Society*. Milton Keynes: Open University Press.

Reiss, M. J. (2009). Imagining the world: the significance of religious worldviews for science education. *Science & Education*, 18, 783–796.

Reiss, M. J. (2011). How should creationism and intelligent design be dealt with in the classroom? *Journal of Philosophy of Education*, 45, 399–415.

Reiss, M. J. (2014). Changing one's mind over evolution. In R. J. Berry (Ed.), *Christians and Evolution: Christian Scholars Change Their Mind*. Oxford: Monarch, 318–330.

Reiss, M. J. (2017). A framework within which to determine how we should use animals in science education. In M. P. Mueller, D. J. Tippins & A. J. Stewart (Eds.), *Animals and Science Education—Ethics, Curriculum and Pedagogy*. Dordrecht: Springer, 243–259.

Reiss, M. J. & Straughan, R. (1996). *Improving Nature? The Science and Ethics of Genetic Engineering*. Cambridge: Cambridge University Press.

Smith, M. U. (1994). Counterpoint: belief, understanding, and the teaching of evolution. *Journal of Research in Science Teaching*, 31, 591–597.

Stevenson, K. T., Peterson, M. L., Bondell, H. D., Moore, S. E. & Carrier, S. J. (2014). Overcoming skepticism with education: interacting influences of worldview and climate change knowledge on perceived climate change risk among adolescents. *Climate Change*, 126, 293–304.

Vatican. (2017) *Catechism of the Catholic Church: III. The Love of Husband and Wife*. Available at www.vatican.va/archive/ENG0015/_P86.HTM.

Williams, J. D. (2015). Evolution versus creationism: a matter of acceptance versus belief. *Journal of Biological Education*, 49, 322–333.

Winslow, M. W., Staver, J. R. & Scharmann, L. C. (2011). Evolution and personal religious belief: Christian university biology-related majors' search for reconciliation. *Journal of Research in Science Teaching*, 48, 1026–1049.

22

CONCLUDING REMARKS

Kostas Kampourakis and Michael J. Reiss

Introduction

We believe that the contributors to this book have managed to present in a concise and clear manner research in teaching and learning biology, with special attention to students' misconceptions about various biological phenomena and processes. We hope that biology teachers and biology educators will find the book useful, either while using it in teacher education courses or when reading it on their own. In this concluding chapter, based on our own reading of the chapters in the present book, we want to make two main suggestions about the directions that teaching biology in schools ought to take. Each of us is aware from our personal experience, both as biology teachers and as teacher educators, that biology curricula and textbooks are often heavily content-focused. In addition, information about biology content is relatively straightforward to transmit and assess; therefore, teachers often find it easy to 'teach' in this way, rather than help students to also develop skills and reflect about doing science. We therefore have two major suggestions to make. The first is that teachers ought to acquire the habit of beginning, and learn how to begin, each biology course with a documentation of their students' preconceptions. This should be done at the beginning of the school year, so that teachers have the opportunity to plan their lessons and activities in a manner that explicitly addresses the misconceptions that their students actually hold. The second suggestion is that a significant portion of each biology course ought to be devoted to reflections about how science is undertaken, the features of scientific knowledge and its implications for our lives. School students are the future adult citizens who will need a functional understanding of biology and not just rote knowledge.

Documenting Students' Preconceptions

As any good teacher knows, effective teaching is an interactive process that goes beyond the transmission of content knowledge to actively engage students in learning procedures. However, it is quite easy for teachers to design their teaching on a content-basis without addressing their students' deeper needs. As a result, students finish schools while holding erroneous conceptions about important biological phenomena (see, for example, Coley & Tanner, 2015). In our view, teachers should approach their students in a way similar to the way in which physicians approach their patients: with care and sensitivity about what they have, but in this case, not in their bodies but in their minds. Teachers should thus be able to diagnose how their students think and identify what conceptions they hold in order to plan their teaching in a manner that would address their conceptions, quite similarly to how a physician would prescribe a therapy to address a patient's symptoms after diagnosing this patient's problem. This being said, we certainly do not imply that students are in any way similar to patients, but only that teachers ought to care about and be sensitive to students' pedagogical needs. This is, of course, especially challenging and demanding, because physicians do this on an individual basis with each patient, whereas school teachers have to deal with 20–30 students at a time!

Therefore, teachers need to carefully diagnose their students' conceptions before their courses begin in order to be able to adjust or adapt their teaching accordingly. Content knowledge transmission is of course part of their job, as it is for them to help students understand the various biological phenomena under discussion. But this can only be done effectively, and this is a conclusion from biological education research, if students' misconceptions are explicitly challenged and appropriately restructured. This is why Chapter 2 of the present book is devoted to intuitive reasoning in biology and reviews research about cognitive biases that should be carefully considered before teaching. In Chapter 3, an important point is also made: we cannot start teaching biology in secondary schools by simply ignoring the conceptions, ideas and understandings that students developed during their childhoods. In contrast, students come to secondary schools with many intuitive ideas about heredity (the case study of Chapter 3), development, evolution and other biological phenomena and processes. To a certain extent, primary teachers can build the foundations for biology teaching in secondary schools or—unfortunately—strengthen students' misconceptions. Therefore, don't we want to know what we have to deal with before starting biology teaching at the secondary level?

Once this is done, teachers can look into research about various topics in books like this one, become aware of the most frequent students' misconceptions and obstacles that they face and then teach so as to address them. The contributors to this book have provided overviews of the research in teaching and learning molecular biology (Chapter 4), cell biology (Chapter 5), human biology (Chapter 6), health and disease (Chapter 7), reproduction and sex education (Chapter 8), development (Chapter 9), genetics (Chapter 10), plant biology (Chapter 11), energy, photosynthesis and respiration (Chapter 12), ecology (Chapter 13), evolution (Chapter 14), microbiology (Chapter 15) and biotechnology (Chapter 16). While this collection of topics is certainly not exhaustive, we believe that it covers most of the important aspects of teaching biology in schools, as well as some very important but often neglected ones such as energy and development. These chapters present the state of the art in biology education research. Even though research both in biology and in biology education advance at a fast pace, we are confident that the main conclusions of the authors of these chapters will not change significantly for at least a number of years. Of course, as research advances we will likely achieve an understanding of students' conceptions at a finer level of detail, and we will also likely figure out improved ways to address them while teaching. Nevertheless, the core of their conceptions and the concepts that they will need to learn at schools are unlikely to change in the near future.

Teachers ought to be aware of this research and acquire also the habit of a constructive, formative assessment. We have heard many teachers state with confidence that their students have understood all these phenomena. But when asking these teachers how they know this, their response is quite often not that they assessed their students' understanding; their response is frequently that their students know because they told them! Worse than that, many teachers often teach towards the test—that is, guide students to learn some material and perhaps be able to solve some problems—without having really arrived at sufficient conceptual understanding. Teachers should not simply be story-tellers; they should excite their students' minds and help them get the rewarding feeling of understanding. This, of course, requires a lot on behalf of teachers; they need to: master the relevant science content knowledge and pedagogical content knowledge; select teaching approaches aimed at appropriate conceptual change; learn to choose and use appropriate questioning before and after their courses in order to assess their students' understanding; take account of variation between their students; and more. There is a lot of work to do to teach effectively.

Reflecting About Science and Its Implications

A very important, but often implicit (if not neglected), aspect of biology, and science, teaching at schools is what students should be able to do with the conceptual understanding we want them to achieve and with the skills we want them to develop. Although only a minority of students will become scientists, all students will become citizens. Therefore, if we want schools to educate scientifically literate citizens, then the goal should not only be to acquire a particular level of conceptual understanding but also to be able to draw on this knowledge whenever necessary. However, this is neither simple not straightforward. Knowing a recipe and having understood how to implement it do not guarantee that one will be able to produce a delicious meal. This requires practice and reflection. This is why the present book includes chapters on topics that one does not typically find in books of this kind, about reading authentic scientific texts (Chapter 17), the nature of scientific knowledge and scientific inquiry (Chapter 18), explanation (Chapter 19), argumentation (Chapter 20) and worldviews (Chapter 21).

The first three of these chapters are about better understanding how science is done and what some of the main features of scientific knowledge are. Let us consider scientific models as an example. Many of the topics taught in biology education are about, or are related to, models. For instance, students learn about the double helical structure of the DNA molecule, the model of cell structure or the model of gene expression. However, how many teachers actually explain to students the properties of these models? It is very easy to implicitly convey the message that such models are true, whereas we should explain that these models are not 'true', in the sense of full and accurate depictions of reality; rather, they exhibit important similarities with the actual structures or processes that occur in nature and are of value in helping us to understand how nature behaves. In other words, teachers should not only teach about the specific features of each of these models, but also clearly explain to students that these models are a means of representation with particular strengths and limitations. Models exhibit similarities with aspects of real systems; they do not represent whole systems, nor do they show how they truly are; they are neither entirely precise, nor entirely accurate; and they are used by scientists for a particular purpose in a specific context (Giere, 2010).

Students will likely forget certain of the details of the models of DNA, cell structure or gene expression after finishing school. However, if they could end up understanding the general features of scientific models, this would help them to become scientifically literate citizens who would understand later in life why, for example, forecasts about ecology and diseases are not always confirmed. Understanding what to expect and what not to expect from a scientific model, details notwithstanding, is an important aspect of science literacy. It would be even better if students could understand the use of these models in science for explanations, and therefore appreciate why models have a central place in scientific theories. This in turn would help students understand certain aspects of scientific knowledge and scientific inquiry, as well as what one can and cannot expect from this knowledge and the inquiry processes that produce it. Uncertainty is an inherent feature of science; however, at the same time, science is the most robust kind of knowledge that we can have about the natural word. Students therefore need to be able to appreciate both the strengths of science and its limitations.

The next issue concerns the implications for our lives (Chapters 20 and 21). One kind of implication concerns decisions that we must make in our everyday lives. For instance, nowadays companies sell direct-to-consumer genetic tests, making promises that they often cannot really deliver. In order for students to be able to understand what is possible and what is (at least for the moment) not possible, they should be able to evaluate the arguments made by these companies, for instance the evidential basis for the claims they make. This requires more than content knowledge. It also requires the ability to analyse and understand an argument, and recognise that claims and evidence are distinct from each other: to use evidence to support or challenge claims or to test predictions and, if necessary, develop a counter-argument about taking or not taking a particular decision. To give one example, the statement that a person who carries a particular genetic variant has a probability of 20% of developing a particular disease also means that this very same person has a probability of 80% of not developing the disease in question. Students should be able to deconstruct the argument that such a person *should* take a genetic test, understand the actual risk (e.g., by taking into account the average risk in the population for people who do not carry the particular variant, and how rare or frequent the disease is) and thus make decisions.

Another important question concerns our perception and understanding of our existence. Biological knowledge certainly has implications for our views on where we come from and where we go. Many philosophical conclusions about religion and morality can, and indeed should, be informed by biological knowledge. Therefore, it is important for students to understand that each student has one or more particular worldviews, and this may need to be taken into account when teaching. Biology has implications for these worldviews, but which exactly these are depends on what exactly people think. For instance, some people are accommodationists and consider that biology and religion can co-exist. Others take more fundamentalist positions and reject either the one or the other. In our view, it is an important goal of biology education to explain to students the limits of science and clearly distinguish science from philosophy, religious fundamentalism and scientism.

Conclusions

Our intention has been to produce a book that can guide biology teachers and biology educators by presenting and discussing the relevant research. We hope that it can be used as a course-book in biology initial teacher education courses, and by individual teachers as an introduction to research in teaching and learning of biology. We are particularly grateful to the contributors to this book who are leading experts on their respective topics, and who are actively involved in research in teaching and learning biology. We believe that their chapters present up-to-date research in the teaching and learning of biology from an international perspective and discuss a wide range of topics that are important in the teaching and learning of biology, including ones that are all-too-often overlooked. We do hope that beginning and practising teachers of biology all over the world will find this book useful.

References

Coley, J. D. & Tanner, K. (2015). Relations between intuitive biological thinking and biological misconceptions in biology majors and nonmajors. *CBE-Life Sciences Education*, 14(1), p.ar8. doi:10.1187/cbe.14–06–0094.

Giere, R. N. (2010). *Scientific Perspectivism*. Chicago: The University of Chicago Press.

CONTRIBUTORS

Moriah Ariely is Head of the National Centre for High-School Biology and Environmental Sciences Teachers, a teachers' instructor in the Department of Science Teaching at the Weizmann Institute of Science, Israel, and a high school biology teacher. She holds an undergraduate degree in life sciences (Tel-Aviv University), and a MSc and a PhD in Science Education (Weizmann Institute of Science), and is currently finishing her PhD in Science Education. In her recent work she is focusing on the unique features of scientific texts in general and of Adapted Primary Literature (APL) in particular. She developed and implemented new teaching strategies for using APL, aimed at promoting students' disciplinary literacy.

Orit Ben Zvi Assaraf is Associate Professor in the Science Education Department at Ben Gurion University of the Negev, Israel. She holds a master's degree in agricultural sciences (The Hebrew University of Jerusalem) and a PhD in Science Education (Weizmann Institute of Science). Her work in science education focuses on issues including cognitive-based research into system thinking in the field of biology, ecology, and earth sciences. Understanding the complexity of natural systems is challenging; the goal of her research is therefore to identify and characterize the longitudinal learning processes that facilitate the development of system thinking. She is also currently involved in several major research projects that focus on the design of informal outdoor learning environments within science education.

Uwe Bertsch has been a biology and chemistry teacher at the Stiftung Louisenlund (Germany) since 2016. He trained as a biochemist at the University of Hannover and received his PhD in Molecular Biology from the University of Kiel. After a postdoctoral period at the Institute of Medicinal Biochemistry and Molecular Biology at the University Hospital in Hamburg Eppendorf, he changed to a position as Project Leader at the Institute of Neuropathology of the Ludwig-Maximilians-University in Munich. After six years he moved back to Kiel University. There he worked at the Department for Immunology, then at the Botanical Institute. His scientific expertise ranges from plant chaperonins through inositol phosphate signalling, neurodegenerative diseases, immune cell death receptors, and cancer biology to retrograde signalling by plant chloroplasts.

Dirk Jan Boerwinkel worked as a science teacher trainer before starting his research at the Freudenthal Institute for Science and Mathematics Education at Utrecht University. He has published on form-function thinking, molecular mechanistic reasoning, and genetic literacy, and has designed educational material that aims at preparing for personal and societal decision making on genetics related issues. His current research concerns teaching and learning metacognition with crosscutting concepts.

Jenny Byrne is Associate Professor in Education at the Southampton Education School, the University of Southampton, with expertise and research interests in science and health education. She graduated from Birmingham University with a BSc in Bacteriology. After completing a PGCE at Durham University she taught secondary science, eventually becoming Head of Science. She has taught in all phases of education, including a pupil referral unit. She has also worked as a health education officer and an adviser for health education/PSHE before moving to H.E. She joined Chichester University where she co-coordinated the undergraduate and PGCE primary science programmes. She moved to Southampton in 2000 and she has taught science on, and led, both primary and secondary science PGCE courses. Her research interests include children's affective and cognitive development and their impact on learning, socio-scientific issues, health education, and the connections between science and health education, including scientific and health literacy.

Katherine Carson is Research Officer in the School of Education at the University of Western Australia and has worked in science education research for ten years. In her previous career she worked as a small animal veterinarian and taught veterinary nurses. She has been Project Manager of Education Research conducted at Edith Cowan University, Curtin University and also the University of Western Australia. Her research interests include socio-scientific issues, scientific literacy and student engagement in science. She has a special interest in climate change education and is currently completing her master's thesis in youth climate change activism.

Vaille Dawson is Professor of Science Education in the School of Education at the University of Western Australia, where she conducts classroom-based research and teaches pre-service secondary science education. She has taught master's-level teaching and learning in science, supervised doctoral and master" research students and conducted science education research at the secondary and tertiary level. Vaille has co-edited three widely used pre-service teacher education textbooks: *The Art of Teaching Science* (2004), *The Art of Teaching Primary Science* (2007), and *The Art of Teaching Science for Middle and Secondary School* (2012) published by Allen and Unwin. Her research interests include scientific literacy, argumentation and decision-making, socio-scientific issues and supporting teachers in under-performing schools. She is a Fellow of the Royal Society of Biology.

Ravit Golan Duncan is an Associate Professor of Science Education with a joint appointment in the Graduate School of Education and the School of Environmental and Biological Sciences at Rutgers University. She received her PhD in Learning Sciences from Northwestern University. She currently has two main research strands: designing and studying inquiry-based learning environments in life sciences that engage students with modeling and argumentation, and studying learning progressions in science education, and specifically in genetics. Duncan is the recipient of several U.S. federal grants and has published in education research journals, practitioner journals and teacher education journals, including the *Journal of Research in Science Education, Science Education*, and *Journal of Science Teacher Education*.

Charbel N. El-Hani is Professor of History, Philosophy, and Biology Teaching at the Institute of Biology, Federal University of Bahia, Brazil, and Productivity in Research Grantee level 1-B from the Brazilian National Council for Scientific and Technological Development (CNPq). He is affiliated with the graduate studies programs in history, philosophy, and science teaching (Federal University of Bahia and State University of Feira de Santana), in ecology and biomonitoring (Federal University of Bahia), and in genetics and biodiversity (Federal University of Bahia). He coordinates the History, Philosophy, and Biology Teaching Lab at UFBA. His research interests are in science education research, philosophy of biology, biosemiotics, ecology, and animal behavior. He is a member of editorial boards of Brazilian and international journals in science education and philosophy of biology, and is Book Review Editor of *Science & Education*. Among his recent publications one finds: Mortimer, E. F. & El-Hani, C. N. (Eds.) (2014). *Conceptual Profiles: A Theory of Teaching and Learning Scientific Concepts*. Dordrecht: Springer.

Marida Ergazaki is currently Associate Professor of Biology Didactics at the Department of Educational Sciences and Early Childhood Education of the University of Patras in Greece. She is a member of the ERIDOB academic committee and a reviewer for a number of international journals. She is the supervisor of several master's and PhD students and has published several papers. Her research interests have to do with how young learners start building their own understanding about the biological world and how they might be supported with appropriately designed learning environments. She is also concerned with designing/testing learning environments for older learners and studying how they collaboratively build new biological knowledge through argumentative discourse. Reasoning skills and NOSK in the context of biology are also included in her research agenda.

Maria Evagorou is Associate Professor at the University of Nicosia, Cyprus. Her scholarly activity focuses on exploring and enhancing young students' and pre-service teachers' argumentation skills within socio-scientific issues. More specifically, the emphasis of her work is on students' and teachers' talk when they engage in the discussion of SSI; she aims to explore ways in which teachers can be supported in their effort to include SSI in their teaching. She has worked as a principal investigator and senior researcher on various EU and local projects, served as a member on the JRST editorial board and as a strand coordinator for ESERA, and has published widely.

Evelyn Margaret Evans is Research Scientist in the Center for Human Growth and Development at the University of Michigan. Her PhD in Developmental Psychology focused on the cultural and cognitive factors that influence the emergence of scientific and religious concepts in children and adults. Insights gained from these studies have informed research projects and exhibit development for four different exhibits on evolution and related topics, funded by the National Science Foundation and the National Institutes of Health. These studies have been published in cognitive science, developmental science, science education and museum, journals and edited books. She is a fellow of the Association of Psychological Science.

Niklas Gericke is Professor in Science Education at Karlstad University in Sweden and Professor II at NTNU in Trondheim, Norway. Niklas holds a master's degree in biology and a certification as a teacher. He has ten years of experience from teaching biology at secondary level and is currently engaged in teacher education. He earned a PhD in Biology Education in 2009. His research interests focus on biology education and sustainability education. Much of his work is centered around how disciplinary knowledge is transformed into school knowledge and what impact these transformations have on students' understanding. His research interests include history and philosophy of science, models and representations in science, genetic education, textbook research, scientific inquiry, and education for sustainable development. Niklas is Director of the SMEER (Science, Mathematics and Engineering Education Research) center at Karlstad University. He is also a member of the Academic Committee of ERIDOB (European Researcher in Didactics of Biology), elected Research Fellow of the Royal Society of Biology in London, UK and President of the Swedish Association for Research in Science Education (FND).

Marcus Grace is Professor of Science Education and Head of the Education School at the University of Southampton, UK. He is Chair of the Academic Committee of ERIDOB (European Research in Didactics of Biology) and Chair of the UK's Biology Education Research Group (BERG), a Royal Society of Biology special interest group. Marcus taught biology and science at secondary schools in London and was head Teacher at a Berlitz school in Tokyo. His main interests are in learning and teaching about socio-scientific issues, particularly health and environmental issues, and outdoor science education.

Marcus Hammann is Professor of Biology Education at Muenster University in Germany. He is a certified biology teacher and earned a PhD in Biology Education at the Leibniz Institute for Science and Mathematics Education (IPN) at Kiel University in 2003. Marcus was Chair of the Academic Committee of ERIDOB (European Researchers in Didactics of Biology) from 2004 to 2008, Member of the Science Expert Group (SEG) for the 2015 PISA Scientific Literacy Test, and Chair of the German Biology Education Research Group (FDdB im VBIO) from 2011 to 2017. His research focuses on cognitive and affective aspects of biology education, which includes research on students' and teachers' attitudes, students' interests, and their ability to think across the levels of biological organizations. He published his research in *Science Education*, *Science & Education*, and other peer-reviewed journals. He is a member of the International Advisory Board for Journal of Biological Education (JBE) and a member of the Editorial Board for International Journal of Science Education (IJSE). In 2014, he co-authored a book on student conceptions in biology education published in German.

Ute Harms is Director at the Leibniz Institute for Science and Mathematics Education (IPN), Full Professor for Biology Education at the University of Kiel (Germany) since 2007, and Fellow of the Royal Society of Biology (Great Britain). She has a PhD in Cell Biology and has worked as a high school teacher for several years. In 2000 she received her first Professorship for Biology Education at the Ludwig-Maximilians-University in Munich (Germany). From 2006 to 2007 she held a chair in Biology Education at the University of Bremen. Her main research interests are conceptual learning in biology and in science focusing on evolution and energy, biology teacher education, biology-related competitions, and transfer of contemporary topics in the life sciences to the public.

Dan Jenkins is Director of the Science and Plants for Schools project and Head of the Gatsby Plant Science Education Programme, based at the University of Cambridge. As a passionate botanist and science educator Dan has spent over ten years enthusing teachers to engage with plant science teaching resources. Science education research informs the Science and Plants for Schools (SAPS) approach to design and implementation of new teaching resources to support the UK 11–18 curriculum. Dan's work focuses on students' interest in plant biology topics and supporting the development of pedagogic content knowledge with pre-service teachers. Dan has developed teaching resources to consider, for example, alternative conceptions in plant science topics, developing animations and practical lab-based activities to address these concerns.

María Pilar Jiménez-Aleixandre is Science Education Ad Honorem Professor at the University of Santiago de Compostela, in Spain. She has published around 60 books or book chapters and about 70 papers in refereed journals in Spanish, English, French, Portuguese, and Galician, among them some seminal work on argumentation in science education. She has participated in several European Union funded projects about introducing argumentation and scientific practices in teacher education. She also has long-standing cooperation with Latin American research groups in Brazil, Chile, Mexico, and Argentina. Her work focuses on argumentation, epistemic practices, biology education, and socio-scientific issues. She currently serves on the boards of *Science Education* (as co-editor of the Issues and Trends section), *Science & Education*, and *Environmental Education Research*, among other journals.

Kostas Kampourakis is a researcher in science education and a lecturer at the University of Geneva, where he teaches courses at the University Teacher Training Institute and the Section of Biology. He is Editor-in-Chief of the journal *Science & Education* and the book series *Science: Philosophy, History and Education* (Springer). He is also the editor of *Newton's Apple and Other Myths about Science* (Harvard University Press, 2015) and *The Philosophy of Biology: A Companion for Educators* (Springer, 2013), as well as the author of *Turning Points: How Critical Events Have Driven Evolution, Life and Development* (Prometheus Books, 2018), *Making Sense of Genes* (Cambridge University Press, 2017), and *Understanding Evolution* (Cambridge University Press, 2014). His main research interests include the teaching, learning, and the public understanding of evolution, genetics, and nature of science.

Konstantinos Korfiatis is Associate Professor of Environmental Education at the Department of Education, University of Cyprus. He has been, since 2016, a member of the Academic Committee of ERIDOB (European Research in Didactics of Biology). His research interests extend to the areas of environmental education, ecology, and sustainability education. His current focus is on educational spaces that support participation, agency, and environmental citizenship. Concerning ecology education, he is particularly interested in the study of ecological misconceptions and how they can be destabilized through teaching.

Norman G. Lederman is Distinguished Professor and Chair of Mathematics and Science Education at Illinois Institute of Technology. He is also a former high school biology and chemistry teacher and university-level biology teacher. Dr Lederman has a PhD in Science Education, a master's degree in biology, a master's degree in secondary science teaching, and a bachelor's degree in biology. He is internationally known for his scholarship in the teaching and learning of nature of science and scientific inquiry. Dr Lederman has received many honors/awards, including the Illinois Biology Teacher of the Year, Association of Science Teacher Education Outstanding Mentor, National Science Teachers Association Distinguished Contribution to Science Education, and National Association for Research in Science Teaching Distinguished Contribution to Science Education through Research. He is also Fellow of the American Association for the Advancement of Science and the American Educational Research Association.

William F. McComas is the inaugural holder of the Parks Family Endowed Professorship in Science Education at the University of Arkansas following experiences as a secondary biology teacher and professor at the University of Southern California. He is interested in the intersection of philosophy of science on science instruction, the improvement of laboratory teaching, informal science learning, evolution education, and science for gifted learners. McComas edited *The Nature of Science in Science Education: Rationales and Strategies* and *The Language of Science Education* (with a new, extensively updated edition to be published in 2018), and recently co-authored a new pictorial account of *The Scopes Monkey Trial*. McComas was Fulbright Fellow at the Centre for the Advancement of Science and Mathematics Teaching and Learning at Dublin City University in Ireland and was recently elected Fellow of the American Association for the Advancement of Science (AAAS). He currently serves as Editor of the *American Biology Teacher*.

Ross H. Nehm is Associate Professor in the Department of Ecology and Evolution and the Institute for STEM Education, and Associate Director of the PhD Program in Science Education at Stony Brook University in New York. He received his PhD from the University of California at Berkeley. He is the author of more than 60 journal articles and book chapters. His work focuses on biology education with particular attention to evolution and natural selection. His major honors include a CAREER award from the U.S. National Science Foundation, a teaching award from UC Berkeley, and a mentoring award from CUNY. In 2013–2014 he was named an Education Mentor in the Life Sciences by the U.S. National Academies. Dr. Nehm currently serves as Associate Editor for the journals *Evolution Education and Outreach* and *Science & Education*, and as Monitoring Editor for *CBE-Life Sciences Education*. He has also served on the editorial boards of several journals, including the *Journal of Research in Science Teaching*, *Journal of Science Teacher Education*, and *Journal of Science Education and Technology*. He is a member of many advisory boards for federally funded science education projects and has served as Panel Chair for National Science Foundation programs.

Kai Niebert holds a PhD in Science Education from the Leibniz University Hannover. After spending his postdoc time at Curtin University of Technology and Penn State University, he is now Professor for Science Education at the University of Zurich. His research interests are the model of design-based-research approaches, educational reconstruction, embodied cognition, and its application on socio-scientific issues.

Michael J. Reiss is Professor of Science Education at UCL Institute of Education, University College London, Fellow of the Academy of Social Sciences, and Visiting Professor at the University of Kiel and York and the Royal Veterinary College. The former Director of Education at the Royal Society, he has written extensively about curricula, pedagogy, and assessment in science education and has directed a very large number of research, evaluation, and consultancy projects over the past 20 years funded by UK Research Councils, government departments, charities, and international agencies.

Karl S. Rosengren is Professor of Psychology at the University of Wisconsin-Madison. His PhD is in developmental psychology. He conducts research on cognitive and motor development across the lifespan, funded by the National Science Foundation and the National Institutes of Health. In the cognitive domain, his research focuses on children's reasoning about the physical world, especially biological change. He is the lead editor of two books: *Imagining the Impossible: Magical, Scientific and Religious Thinking in Children* and *Evolution Challenges: Integrating Research and Practice in Teaching and Learning about Evolution*. He has published over 100 journal articles and book chapters. He is a fellow of the Association of Psychological Science.

Dawn L. Sanders is Associate Professor in Biology Didactics at Göteborgs Universitet, Sweden. Dawn's doctoral research (Sussex University, 2004) examined the educational role of botanic gardens, a study inspired by her ten years as an educator in the Chelsea Physic Garden, London. She trained as an artist before completing her botanical education and draws on both in her academic work. Dawn is a scientific associate of the Natural History Museum, London, and a fellow of the Linnean Society, London. She currently leads a three-year research study, *Beyond Plant Blindness: Seeing the Importance of Plants for a Sustainable World*, funded by the Swedish Research Council. Dawn is on the editorial board of the journal *Plants People Planet*. Recent publication work includes an original contribution to the forthcoming book *Why Look at Plants?* by Giovanni Aloi, to be published by Brill in Spring 2018.

Zohar Snapir is a teachers' educator at Levinsky College of Education and a high school biology teacher. She holds a PhD in Life Science (Weizmann Institute of Science) and was a postdoctoral fellow in the field of science education (Ben Gurion University of the Negev). Her studies in life science involved protein degradation and cancer research, and her focus in science education is on systems understanding in biology, particularly the human body.

Panagiotis K. Stasinakis is a high school biology teacher with a PhD in Biology Didactics. Currently he is a supervisor in one of the Laboratory Centers for Natural Sciences (EKFE) in Greece. His research interests focus on science education and particularly on the understanding of biology (special topics of interest are evolution, phylogeny, human body, etc.), as well on the use of technology as a means of effective science teaching. He has participated in several training seminars for secondary education teachers and in many national working groups for the production of educational material related to biology curriculum in low-secondary education, learning objectives for biology teaching, and public promotion of biology through communication.

Grady Venville is Dean of Coursework Studies and Professor of Science Education at the University of Western Australia. She is responsible for the quality and structural integrity of the coursework curriculum across the university. Grady's research focuses on students' understanding of complex scientific phenomenon and conceptual change; the integration of the curriculum across traditional disciplines; and the teaching of thinking and its impact on cognition. Grady has published five books and more than 50 papers in international refereed journals. She is currently an associate editor of the *Journal of Research in Science Teaching*.

Anat Yarden is Professor of Science Education, Head of the Department, and Head of the Life Sciences Group at the Department of Science Teaching at the Weizmann Institute of Science, Israel. She holds an undergraduate degree in agricultural sciences (The Hebrew University of Jerusalem), a master's degree and a PhD in Molecular Biology (Weizmann Institute of Science), and has postdoctoral training in genetics (Stanford University). The primary theme in all of her academic activities has been the attempt to adapt practices employed by scientists to the processes by which students and teachers accumulate and advance their knowledge within the discipline of biology. Her group pioneered the adaption of primary scientific literature for the teaching and learning of biology in high schools. She has published in leading peer-reviewed journals, is a member of the editorial board of several leading science education journals, and serves as an associate editor for the *Journal of Research in Science Teaching* (*JRST*).

INDEX

Taylor & Francis Group
an informa business

Taylor & Francis eBooks

www.taylorfrancis.com

A single destination for eBooks from Taylor & Francis with increased functionality and an improved user experience to meet the needs of our customers.

90,000+ eBooks of award-winning academic content in Humanities, Social Science, Science, Technology, Engineering, and Medical written by a global network of editors and authors.

TAYLOR & FRANCIS EBOOKS OFFERS:

- A streamlined experience for our library customers
- A single point of discovery for all of our eBook content
- Improved search and discovery of content at both book and chapter level

REQUEST A FREE TRIAL

support@taylorfrancis.com

Routledge
Taylor & Francis Group

CRC Press
Taylor & Francis Group